Cell Lineage Choice During Haematopoiesis: A Commemorative Issue in Honor of Professor Antonius Rolink

Cell Lineage Choice During Haematopoiesis: A Commemorative Issue in Honor of Professor Antonius Rolink

Special Issue Editors

Geoffrey Brown
Rhodri Ceredig

MDPI • Basel • Beijing • Wuhan • Barcelona • Belgrade

MDPI   **DECIDE**

Special Issue Editors
Geoffrey Brown
University of Birmingham
UK

Rhodri Ceredig
National University of Ireland
Ireland

Editorial Office
MDPI
St. Alban-Anlage 66
Basel, Switzerland

This is a reprint of articles from the Special Issue published online in the open access journal *International Journal of Molecular Sciences* (ISSN 1422-0067) in 2018 (available at: https://www.mdpi.com/journal/ijms/special_issues/haematopoiesis#published)

For citation purposes, cite each article independently as indicated on the article page online and as indicated below:

LastName, A.A.; LastName, B.B.; LastName, C.C. Article Title. *Journal Name* **Year**, *Article Number*, Page Range.

ISBN 978-3-03897-274-7 (Pbk)
ISBN 978-3-03897-275-4 (PDF)

Cover image courtesy of Geoffrey Brown and Rhodri Ceredig.

Contents

About the Special Issue Editors

Geoffrey Brown is the Reader in Cellular Immunology at the College of Medical and Dental Sciences at the University of Birmingham, UK. He received a BSc in microbiology from Queen Elizabeth College, University of London and a Ph.D. in tumor biology from University College, London. Postdoctoral research was undertaken at the MRC Immunochemistry Unit, Oxford and the Nuffield Department of Clinical Medicine, Oxford. At Oxford, Geoffrey was also the IBM Fellow, University of Oxford and Research Lecturer of the House, Christ Church College. His early work described human B and T lymphocyte antigens leading to the discovery of the common acute lymphoblastic leukemia antigen. This led to designation of the childhood leukemia as common acute lymphoblastic leukemia. His research for many years has concerned the development of blood cells, including proposing the pairwise model for blood cell development. Geoffrey was the Director of the EU-funded Marie Curie FP7 Initial Training Network and consortium DECIDE (decision-making within cells and differentiation entity therapies).

Rhodri Ceredig, Prof Rhodri Ceredig studied medicine at the University of Birmingham and obtained a PhD from the Walter and Eliza Hall Institute, Melbourne University, Australia. Following postdoctoral studies at the Ludwig Institute for Cancer Research, Lausanne, Switzerland, he was Senior Research Fellow, John Curtin School of Medical Research, Australian National University, Canberra; Professor of Immunology at the Université Louis Pasteur, Strasbourg; Directeur de Recherches, INSERM, France and Stokes Professor of Immunology, National University of Ireland, Galway. Rhodri was the Director of Training of the EU-funded Marie Curie FP7 Initial Training Network and consortium DECIDE (Decision-making within cells and differentiation entity therapies).

International Journal of
Molecular Sciences

MDPI

Editorial

Cell Lineage Choice during Haematopoiesis: In Honour of Professor Antonius Rolink

Geoffrey Brown [1,*] and Rhodri Ceredig [2]

[1]   Institute of Clinical Sciences, Institute of Immunology and Immunotherapy, College of Medical and Dental Sciences, University of Birmingham, Edgbaston, Birmingham B15 2TT, UK
[2]   Discipline of Physiology, College of Medicine & Nursing Health Science, National University of Ireland, Galway H91 TK33, Ireland; rhodri.ceredig@nuigalway.ie
*   Correspondence: g.brown@bham.ac.uk; Tel: +44-0121-414-4082

Received: 5 September 2018; Accepted: 13 September 2018; Published: 17 September 2018

This volume of the *International Journal of Molecular Sciences* contains a collection of articles by colleagues of Antonius (Ton) Gerardus Rolink (19 April 1953–6 August 2017). Ton had participated in an FP7 Marie Curie Initial Training Network called DECIDE (decision-making within cells and differentiation entity therapies), and the DECIDE partners have submitted articles for this Special Issue. Articles have also been submitted by scientists outside the DECIDE network. We would like to thank all authors for their valuable contributions to this volume.

The DECIDE (decision-making within cells and differentiation entity therapies) network initially arose from the shared interests of Drs Ceredig, Brown, and Rolink in the subject of the process of blood cell formation, or haematopoiesis, and in particular, how progenitor cells differentiate to give rise to the different blood cell types. Haematopoiesis is an archetype cell-lineage system with which to study cell-lineage choice. Research in this area is driven by the need to understand this process at the basic scientific level, as well as abnormal haematopoiesis, in particular, the development of leukaemia. Based largely upon the sequential appearance of blood cells during ontogeny and the close and pairwise relationships between the blood cell types, Brown had initially proposed a radically different (developmental) view of haematopoiesis [1]. In addition, work with cell lines had demonstrated that the B-cell and myeloid lineages were more closely related than what was envisaged by the "classical" model of haematopoiesis. In this model, after an early bifurcation, lymphocytes derive from a common lymphocyte progenitor (CLP), whereas myeloid cells derive from a common myeloid progenitor (CMP). Clearly, cloned cell lines with both lymphoid and myeloid potential posed a problem for the classical model. Subsequently, the work by Amanda Fisher from Ceredig's laboratory [2] showed that cloned lymphoid tumours arising in an interleukin-7 (IL-7) transgenic mouse line possessed both lymphoid and myeloid potentials. Finally, the seminal papers from Ton's laboratory describing cell lines deficient in the transcription factor Pax-5 and with a multi-lineage (including lymphoid, in this case T-cell and myeloid) potential also contributed to the need to rethink models of haematopoiesis [3,4].

Thus, after some initial discussions, we three (Ceredig, Rolink, and Brown) decided to put some ideas into writing, and in 2009, we published an opinion article in Nature Reviews Immunology [5]. Based on ideas expressed in this article, we then decided to apply for funding in order to address these ideas. After several iterations and the inclusion of chemists and pharmacists interested in using vitamin D analogues for the control of myeloid cell differentiation and the treatment of leukaemia, funding for the Marie Curie DECIDE network was finally approved in 2013.

In all of the above processes, Ton showed immense enthusiasm for the science. In the meantime, his laboratory had identified an apparently homogeneous bone marrow-derived progenitor population with both lymphoid and myeloid potential, and termed it early progenitor with lymphoid and myeloid potential, or EPLM. With the help of two early stage researchers employed by the DECIDE network working in Ton's laboratory (Audrey Lilly von Münchow and Llucia Alberti Servera) and two in Brown's group (Ciaran Mooney and Alan Cunningham), it now transpires that the original EPLM population contains four phenotypically distinct subpopulations. At the genotypic level, the earliest EPLM subpopulation contains individual cells already committed to either lymphoid or myeloid lineage, with essentially no bipotent cells [6].

It must be mentioned that Ton made a massive contribution to the DECIDE network. His enthusiasm for the science has already been mentioned, and this continued throughout the lifetime of DECIDE, even when ill health prevented his full participation. Ton was always prepared to give advice to other members of the DECIDE consortium on projects that were not necessarily aligned with his own research interests. This typifies Ton's immense generosity and profound knowledge of areas of science outside his immediate domain. His participation in and organization of DECIDE meetings was much appreciated (Figure 1). All 12 Marie Curie Fellows were in receipt of his encouragement and advice regarding their projects. Unfortunately, from early 2017, health issues prevented his participation in the last DECIDE meeting held in Galway.

Figure 1. (L–R) Andrew Kutner, Rhodri Ceredig, Geoffrey Brown, Eva Marcinkowska, and Antonius (Ton) Gerardus Rolink at the Wroclaw DECIDE (decision-making within cells and differentiation entity therapies) consortium meeting in April 2014.

All of the DECIDE partners are eternally grateful for having had the privilege of knowing and working with Ton, and we dedicate this Special Issue to his memory. In this issue, five DECIDE partners—Brown, Ceredig, Kutner, Sánchez-García, and Marcinkowska—describe some of their findings from the DECIDE work. In 2009, Brown, Ceredig, and Rolink first published their pairwise model for haematopoiesis, and this formed the major scientific basis for establishing the DECIDE consortium [6]. The pairwise model replaces the "classical", bifurcating lineage tree models with a continuum-like view of the spectrum of fate options open to each hematopoietic stem cell (HSC). In tree models, the progeny of HSCs progress through a series of intermediate hematopoietic progenitors, progressively closing down the lineage options. In the pairwise model, each HSC either self-renews or chooses directly from all of the end-cell options, and then "merely" differentiates. HSCs are also versatile; even soon after their progeny has selected a lineage, they can still step "sideways" to adopt alternative, closely related fates. In their review article in this issue, Brown and Ceredig examine the importance of the developmental ancestry and environmental nurture of HSCs, and argue that stem and progenitor cells are sensitive to lineage guidance by environmental cues. Thus, a cell's environmental history is important to the specification of lineage [7].

A proposition by Sánchez-García, developed from studies using in vivo models of leukaemogenesis and of carcinomas, relates to the role(s) of oncogenes in the leukaemia and cancer cells-of-origin. He argues that some oncogenes prime the epigenome of leukaemia-initiating cells, but they need not be active thereafter for tumour progression and maintenance. This oncogene "hit" programs the HSC epigenome towards a defined leukemic cell fate. All of the progeny of the resulting clonal leukaemia stem cells (LSCs) then progress towards one leukemic lineage. Vicente-Dueñas, Sánchez-García, and colleagues review the notion of epigenetic stem cell rewiring as a driver of cancer, emphasising that this mechanism represents a common mechanism at work in epithelial tumours [8]. In the case of leukaemia, rewiring fixes the identity of leukaemia cells at the level of HSCs. As HSCs are versatile, an interesting proposition is that the acquisition of a stable oncogene-initiated block to such lineage versatility is a key initiating step to the generation of LSCs and cancer stem cells. This may be a cardinal feature of cancer.

$1\alpha,25$-dihydroxyvitamin D_3 (1,25D3)—a physiologically active metabolite of vitamin D—is a potent differentiating agent for both normal and malignant cells, and vitamin D prevents malignant transformation and reduces tumour progression in experimental models, and may be important in human disease. 1,25D3 is also a central regulator of calcium homeostasis, with high doses leading to hypercalcaemia. During DECIDE, Kutner designed and synthesised a panel of novel analogues of vitamin D2, which separate the differentiating and calcaemic actions of 1,25D3, and are substantially more potent differentiating agents. At the DECIDE meetings, Ton often asked the question of how could this be, with just the one receptor for 1,25D3, namely the vitamin D receptor (VDR)? Presently, we do not know the answer to this important question. Kutner and Brown outline some of the rules for eliminating calcaemic action while retaining potency for cell differentiation. A-ring chair β-conformation and (24E) side geometry are important for differentiating activity, an aromatic modification of the CD-ring reduces the calcaemic action, and a rigid and straight (24E) sidechain confers resistance to catabolism [9].

1,25D3 and all-trans-retinoic acid (ATRA), the active metabolite of vitamin A, are both inducers of myeloid differentiation. Marcinkowska and colleagues have shown that the retinoic acid receptor α (RAR α, for ATRA) regulates VDR expression, and that the outcome of this interaction depends on the developmental status of the cells. For KG-1 (stem-like) and NB-4 (pro-myeloid) cells, activated RAR α upregulates the VDR expression, rendering the KG-1 cells sensitive to 1,25D3-driven differentiation towards monocytes. HL60 and U937 cells typify later myeloid development and, by contrast, an activated RAR α down-regulates VDR [10]. The CCAAT/enhancer-binding protein (C/EBP) family of transcription factors is important for the expression of myeloid-associated genes. Marchwicka and Marcinkowska report that ATRA induces *CEBPB* and *CEBPE* expression; a high level of RAR α results in a strong and sustained *CEBPB* expression. A high VDR expression is required for the strong and sustained upregulation of the *CEBPB* gene, whereas a moderate level of active VDR is sufficient for the expression of *CEBPD*. *CEBPB* is, therefore, the major VDR- and RAR α-regulated gene among the C/EBP family [11]. The transcription factor GATA-1 is important for the erythroid differentiation, which also requires an adequate supply of iron for haemoglobin production. The ferritin heavy subunit maintains iron in a non-toxic form. The article by Zolea and colleagues reveals that this protein does not merely act as a mere iron-metabolism-related factor. Instead, in response to the inducer haemin and via the miR-150 up-regulation and repression of GATA-1, the silencing of the ferritin heavy subunit in the K562 erythroid/myeloid cells blocks the commitment of these cells to erythroid differentiation [12].

For many years, Ton worked on B-cell development, providing important information with which to unravel this process. The review by Sigvardsson brings to attention the developmental trajectories to B-cell development, the complex regulatory networks, and the targeting of the networks in human B-lineage malignancies [13]. Nature versus nurture considerations highlight the roles of bone marrow niches in the development of B-cells from HSCs. Thus, the article by Aurrand-Lions and Mancini emphasises the importance of the marrow environment in maintaining stem cells, as well as their differentiation into mature cells. Cross talk between B-cells and the niches for early pro-B, pre-B, immature B, recirculating B, and plasma cells, either via direct contact and/or secreted specific factors, all contribute to a dynamic process, which is important for the commitment and differentiation of hematopoietic stem and progenitor cells towards a particular pathway [14]. Interleukin-7 (IL-7) is essential for B- and T-lymphocyte development, although there appears to be a species difference in the dependence of B lymphopoiesis. The article by Kasai and colleagues identifies a cytoplasmic region of the mouse IL-7 receptor α subunit (IL-7R α) that is essential for B-cell development, as revealed by a series of deletion mutants of IL-7R α. Amino acids 414–441 in the IL-7Rα chain form a critical subdomain [15]. Studies of antibody-secreting plasma cells have been continuously hampered by the lack of surface molecules with which to identify them. The article by Trezise and colleagues reports mining of the transcriptome of plasma cells to identify novel and cell surface proteins. Three surface proteins, Plpp5, Clptm1l, and Itm2c, represent potential targets for novel treatments for multiple myeloma, a tumour of antibody-secreting cells. In this regard, and as revealed by the analysis of mouse strains with a loss-of-function mutation for each protein, these proteins are dispensable for normal B-cell development and antibody production by antibody secreting cells [16]. Lastly, the review article by Urbanczyk addresses an area of pro- to pre-B-lymphocyte development that developmental biologists frequently neglect, namely the regulation of energy metabolism, specifically glycolysis and oxidative phosphorylation. The hypoxia-inducible transcription factor HIF1α plays an important role in early B-cell development by promoting glycolysis in B-cell progenitors. By contrast, Urbanczyk and colleagues have shown that the cell surface expression of the pre-B cell receptor down-regulates EFhd1, a Ca2$^+$-binding protein that localises on the inner mitochondrial membrane and that limits glycolysis in pro-B cells. They therefore speculate on the importance of Ca2$^+$ fluctuation-mediated mitochondrial flashes (mitoflashes) for the pro- to pre-B-cell transition [17].

Concluding Remarks

Many articles in this Special Issue concern the research interests of Ton. We take this opportunity to thank all of the DECIDE and non-DECIDE authors for their efforts in preparing these articles. Ton often commented on whether there was really a need to make a clear distinction between HSCs and their immediate progeny. HSCs can make an immediate choice of cell lineage, and as we might view HSCs as, at the very least, lineage affiliated, the distinction between HSCs and their immediate progeny becomes blurred. Ton and others showed that bone marrow stromal cells play a vital role in B-lymphocyte development. Appropriately, this issue highlights the increased interest and importance of bone marrow niches and hematopoietic cytokines in instructing lineage affiliation and cell differentiation, rather than HSCs and progenitors following a wired/prescribed developmental pathways in a stochastic manner. Some hematopoietic cytokines, including the ligand for FMS-like tyrosine kinase 3 (FL, myeloid versus lymphoid, as shown by Ton and colleagues), erythropoietin (erythroid), macrophage colony stimulating factor (CSF; monocyte), and granulocyte-colony stimulating factor (granulocyte) can instruct the HSC fate. There has been a long debate about whether IL-7 is instructive for B lymphocyte development. One of Ton's last papers addressed this matter by showing that IL-7 and FL enhance the production of already "pre-decided" (committed) B-cell progenitors. IL-7 promoted their survival, whereas FL made these progenitors proliferate [18]. Opinions about IL-7 and B-cell niches and metabolism are also very important for our understanding of normal immunity, and possibly autoimmunity. For Ton, science was always meant to be fun; he did this by inspiring and encouraging his colleagues and students. This is perhaps best exemplified by a group photograph of Ton's lab, taken in 2014 (Figure 2).

Figure 2. Ton stands to one side and lets the students take centre stage.

Finally, with new research tools becoming available, in particular single cell transcriptomic and proteomic analysis, it does seem that we are moving towards an entirely new architecture for haematopoiesis. This will guarantee that there will be lots of fun in the future trying to dissect this fascinating cell lineage. As emphasized in the review by Sánchez-García, a new view on normal haematopoiesis also has important implications for the way we view the initiation of cancer, perhaps as a fixed lineage-choice in a LSC/cancer stem cell that an oncogenic event wires. When considered in the context of this view of cancer initiation, understanding how to "normalise" the behaviour of LSCs and other cancer stem cells might give rise to potentially valuable therapeutic leads.

Acknowledgments: This project received funding from the European Union's Seventh Framework Programme for research, technological development, and demonstration, under grant agreement no. 315902. G.B. and Rh.C. were partners within the Marie Curie Initial Training Network DECIDE (decision-making within cells and differentiation entity therapies).

Conflicts of Interest: The authors declare no conflicts of interest.

References

1. Brown, G.; Bunce, C.M.; Guy, G.R. Sequential determination of lineage potentials during haematopoiesis. *Br. J. Cancer* **1985**, *52*, 681–686. [CrossRef] [PubMed]
2. Fisher, A.G.; Burdet, C.; Bunce, C.; Merkenschlager, M.; Ceredig, R. Lymphoproliferative disorders in IL-7 transgenic mice: Expansion of immature B-cells which retain macrophage potential. *Int. Immunol.* **1995**, *7*, 415–423. [CrossRef] [PubMed]
3. Nutt, S.L.; Heavey, B.; Rolink, A.G.; Busslinger, M. Commitment to the B-lymphoid lineage depends on the transcription factor Pax5. *Nature* **1999**, *401*, 556–562. [CrossRef] [PubMed]
4. Rolink, A.G.; Nutt, S.L.; Melchers, F.; Busslinger, M. Long-term in vivo reconstitution of T-cell development by Pax5-deficient B-cell progenitors. *Nature* **1999**, *401*, 603–606. [CrossRef] [PubMed]

5. Ceredig, R.; Rolink, A.G.; Brown, G. Models of hematopoiesis: Seeing the wood for the trees. *Nat. Rev. Immunol.* **2009**, *9*, 293–300. [CrossRef] [PubMed]
6. Alberti-Servera, L.; von Muenchow, L.; Tsapogas, P.; Capoferri, G.; Eschbach, K.; Beisel, C.; Ceredig, R.; Ivanek, R.; Rolink, A. Single-cell RNA sequencing reveals developmental heterogeneity among early lymphoid progenitors. *EMBO J.* **2017**, *36*, 3619–3633. [CrossRef] [PubMed]
7. Brown, G.; Ceredig, R. The making of hematopoiesis: Developmental ancestry and environmental nurture. *Int. J. Mol. Sci.* **2018**, *19*, 2122. [CrossRef] [PubMed]
8. González-Herrero, I.; Rodríguez-Hernández, G.; Luengas-Martínez, A.; Isidro-Hernández, M.; Jiménez, R.; García-Cenador, M.B.; García-Criado, F.J.; Sánchez-García, I.; Vicente-Dueñas, C. The making of leukemia. *Int. J. Mol. Sci.* **2018**, *19*, 1494. [CrossRef] [PubMed]
9. Kutner, A.; Brown, G. Vitamins D: Relationship between structure and biological activity. *Int. J. Mol. Sci.* **2018**, *19*, 2119. [CrossRef] [PubMed]
10. Marchwicka, A.; Malgorzata, M.; Lasziewica, A.; Sniezewska, L.; Brown, G.; Marcinkowska, E. Regulation of vitamin D receptor expression by retinoic acid receptor alpha in acute myeloid leukemia cells. *J. Steroid Biochem. Mol. Biol.* **2016**, *159*, 121–130. [CrossRef] [PubMed]
11. Marchwicka, A.; Marcinkowska, E. Regulation of expression of *CEBP* genes by variably expressed vitamin D receptor and retinoic acid receptor α in human acute myeloid leukemia cell lines. *Int. J. Mol. Sci.* **2018**, *19*, 1918. [CrossRef] [PubMed]
12. Zolea, F.; Battaglia, A.M.; Chiarella, E.; Malanga, D.; De Marco, C.; Bond, H.M.; Morrone, G.; Costanzo, F.; Biamonte, F. Ferritin heavy subunit silencing blocks the erythroid commitment of K562 cells via miR-150 up-regulation and GATA-1 repression. *Int. J. Mol. Sci.* **2018**, *18*, 2167. [CrossRef] [PubMed]
13. Sigvardsson, M. Molecular regulation of differentiation in early B-lymphocyte development. *Int. J. Mol. Sci.* **2018**, *19*, 1928. [CrossRef] [PubMed]
14. Aurrand-Lions, M.; Mancini, S.J. Murine bone marrow niches from hematopoietic stem cells to B cells. *Int. J. Mol. Sci.* **2018**, *19*, 2353. [CrossRef] [PubMed]
15. Kasai, H.; Kuwabara, T.; Matsui, Y.; Nakajima, K.; Kondo, M. Identification of an essential cytoplasmic region of interleukin-7 receptor α subunit in B-cell development. *Int. J. Mol. Sci.* **2018**, *19*, 2522. [CrossRef] [PubMed]
16. Trezise, S.; Karnowski, A.; Fedele, P.L.; Mithraprabhu, S.; Liao, Y.; D'Costa, K.; Kueh, A.J.; Hardy, M.P.; Owczarek, C.O.; Herold, M.J.; et al. Mining the Plasma Cell Transcriptome for Novel Cell Surface Proteins. *Int. J. Mol. Sci.* **2018**, *19*, 2161. [CrossRef] [PubMed]
17. Urbanczyk, S.; Stein, M.; Schuh, W.; Jäck, H.-M.; Mougiakakos, D.; Millenz, D. Regulation of energy metabolism during early B lymphocyte development. *Int. J. Mol. Sci.* **2018**, *19*, 2192. [CrossRef] [PubMed]
18. Von Muenchow, L.; Alberti-Servera, L.; Klein, F.; Capoferri, G.; Finke, D.; Ceredig, R.; Rolink, A.; Tsapogas, P. Permissive roles of cytokines interleukin-7 and Flt3 ligand in mouse B-cell lineage commitment. *Proc. Natl. Acad. Sci. USA* **2016**, *113*, E8122–E8130. [CrossRef] [PubMed]

International Journal of
Molecular Sciences

MDPI

Review

The Making of Hematopoiesis: Developmental Ancestry and Environmental Nurture

Geoffrey Brown [1,*], Rhodri Ceredig [2] and Panagiotis Tsapogas [3]

[1] Institute of Clinical Sciences, Institute of Immunology and Immunotherapy, College of Medical and Dental Sciences, University of Birmingham, Edgbaston, Birmingham B15 2TT, UK
[2] Discipline of Physiology, College of Medicine & Nursing Health Science, National University of Ireland, Galway, Ireland; rhodri.ceredig@nuigalway.ie
[3] Developmental and Molecular Immunology, Department of Biomedicine, University of Basel, 4058 Basel, Switzerland; panagiotis.tsapogas@unibas.ch
* Correspondence: g.brown@bham.ac.uk; Tel.: +44-0121-414-4082

Received: 10 July 2018; Accepted: 18 July 2018; Published: 21 July 2018

Abstract: Evidence from studies of the behaviour of stem and progenitor cells and of the influence of cytokines on their fate determination, has recently led to a revised view of the process by which hematopoietic stem cells and their progeny give rise to the many different types of blood and immune cells. The new scenario abandons the classical view of a rigidly demarcated lineage tree and replaces it with a much more continuum-like view of the spectrum of fate options open to hematopoietic stem cells and their progeny. This is in contrast to previous lineage diagrams, which envisaged stem cells progressing stepwise through a series of fairly-precisely described intermediate progenitors in order to close down alternative developmental options. Instead, stem and progenitor cells retain some capacity to step sideways and adopt alternative, closely related, fates, even after they have "made a lineage choice." The stem and progenitor cells are more inherently versatile than previously thought and perhaps sensitive to lineage guidance by environmental cues. Here we examine the evidence that supports these views and reconsider the meaning of cell lineages in the context of a continuum model of stem cell fate determination and environmental modulation.

Keywords: blood and immune cells; cell lineage; cytokines; fate determination; hematopoiesis; stem cells

1. Introduction

In 1983, Sulston and colleagues produced a tree map—a series of bifurcations—that delineated the origins of the 671 cells of the newly hatched roundworm *Caenorhabditis elegans* [1]. In this organism, cell lineages and the fates of cells are largely invariant and ancestry therefore determines the end fate of a cell. The apparent rigidity of a tree lineage map ensures tissues develop reliably and consistently. An autonomous lineage programme is also likely to generate the cell types required in a manner that is both efficient and economical. To add to lessons learned from *Caenorhabditis elegans*, there is "unpredictability" in the developmental nature of cells within this organism. Numerous sub-lineages give rise to various populations of motor neurons, rather than these cells descending from the same progenitor in the map. Moreover, when one of the progeny of a terminal division is a motor neuron its "sibling" is unlikely to be a cell of the same type [1].

For over 30 years, tree maps have also been used to describe blood and immune cell development [2] and the development of other tissues (e.g., neuronal-crest-derived cells) [3]. Accordingly, investigations of decision-making by developing cells have tended to examine how they make either-or choices. In the case of hematopoiesis, attention focused on how the progeny of hematopoietic stem cells (HSC) first become common lymphoid progenitors (CLP) and hence generate B cells, T cells and natural killer (NK)/innate lymphoid (ILC) cells, or common myeloid progenitors (CMP), which gives rise to all the other blood cell types [2]. It is then envisaged that later in development progenitors choose, for example, between the neutrophil and monocyte pathways [4] and between the megakaryocyte and erythroid pathways [5]. For years, we have defined hematopoietic cell lineages in terms of just two initial types of oligo-potential cells, CLP and CMP, giving rise to all of the various end-cell types.

In recent years, descriptions of the architecture of haematopoiesis have been moving away from the idea of HSC development progressing via: a) a series of intermediate oligo-potent progenitors; and b) these cells making a series of binary choices that stepwise restrict lineage options, ultimately to one final fate. In a new scenario, the suggestion is that HSC can make an immediate lineage choice from a continuum that encompasses all of the end-cell options. Our pairwise model, proposed in 2009, (Figure 1A) [6], does not prescribe an invariant route to each end-cell type via definitive intermediate progenitors. It presumes, quite simply, that all of the options are open to HSCs. This is in keeping with decision-making, regarding lineage affiliation, at the level of the HSC and very different from HSC making an immediate and binary choice, for example, between the myeloid and lymphoid fates. In support of this new viewpoint is the recent finding that HSC in mice can undergo restriction to the myeloid, megakaryocyte/erythroid and megakaryocyte pathways without dividing or entering S phase of the cells cycle [7]. Previously, we showed that the human promyeloid cell line HL60 makes a choice between the neutrophil and monocyte fates without dividing and when held in the G_1 and S phases of the cell cycle [8,9]. Cell fate decisions are therefore uncoupled from cell division and the mechanisms that control fate determination/differentiation and proliferation are at least partly independent. An inverse relation between cell cycle and differentiation can actually occur, as has been shown in the case of the transcription factor PU.1 in myeloid differentiation. Increased PU.1 levels can lead to prolongation of the cell cycle resulting in further accumulation of PU.1 and thus commitment of progenitors to the macrophage lineage [10].

The concept of a continuum of developmental options is important and applies to the description of various changes to the status of cells. A longstanding viewpoint is that embryonic stem cells make a binary choice between pluripotency and specification to a germ cell layer. However, the stem cell compartment of human embryonic stem cell cultures is significantly heterogeneous and Hough and colleagues have argued that a continuum of states spans pluripotency and lineage commitment of these cells [11]. Moreover, the existence of self-renewal as a canonical state is debatable: cells progressively decrease their likelihood of self-renewal as the expression of stem cell- and pluripotency-related genes decreases and that of genes encoding differentiation attributes increases [11]. Epiblast cells of the early mammalian embryo are able to make mesoderm, endoderm or ectoderm fate decisions and as development progresses they mature through a continuum of lineage potency states [12]. Additionally, a continuum best describes the progression of cell differentiation. For example, early studies revealed gradual changes in the expression of a large number of surface antigens during the different stages of neutrophil development [13]. Similarly, maturation changes occur along a continuum during erythropoiesis [14]. To return to *Caenorhabditis elegans*, a continuum framework has been used to model epithelial morphogenesis and elongation of the embryo to a worm [15].

The pairwise model also emphasizes that there are particular nearest neighbour relationships between the various cell lineages. The known sets of fates of cells termed lymphoid-primed multipotent progenitors (LMPP), early progenitors with lymphoid and myeloid potential (EPLM) and common myeloid progenitors (CMP) and of other downstream progenitors (see Figure 1A,B), guided the construction of the relationships of progenitors in the pairwise model. However, recent genotypic studies have revealed that LMPP, EPLM and CMP although phenotypically relatively homogenous are not homogeneous populations and instead are mixtures of cells with different lineage affiliations [16–20]. We have examined the affiliations of EPLM, which were originally described as a population of cells that could generate B and T lymphocytes, NK cells, dendritic cells (DC) and macrophages but that lacked megakaryocyte and erythroid potentials [21]. Thus, by using the surface markers Ly6D, SiglecH and CD11c and by RNA sequencing single cells, we separated EPLM into four subpopulations each with lineage biases. We divided the most primitive EPLM, lacking these markers, into cells that already have myeloid, DC or lymphoid signatures [16]. Likewise, Naik and colleagues have similarly divided LMPP into lymphoid-, myeloid- and DC lineage-biased cell populations [18]. Hoppe and colleagues used RNA expression data to classify CMP into granulocyte/monocyte and megakaryocyte/erythroid progenitors [17] and Paul and colleagues have assigned bone marrow progenitors into seven groups with transcriptional characteristics of neutrophils, basophils, eosinophils, monocytes, DC, erythrocytes and megakaryocytes [20]. Figure 1B shows the sub-populations of LMPP, EPLM and CMP as cells added to their arcs of potentials. The particular close relationships between cell lineages remain the same as lineages adjacent to one another share the usage of transcription factors and cell responsiveness to promiscuous cytokines [6,22]. This fits with the notion of visualising cell specification as a continuum and infers that adjacent elements are less different from each other whereas the extremes are quite distinct.

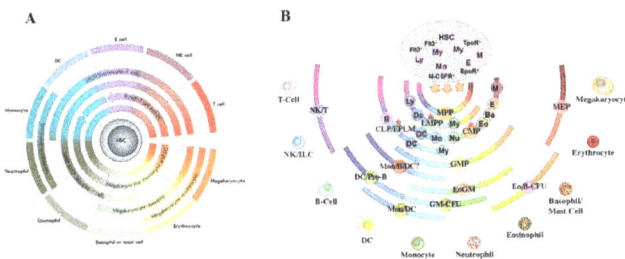

Figure 1. A continuum or pairwise model for hematopoiesis. (**A**) The model envisages a continuum of fates is available to hematopoietic stem cells (HSC), with pairwise relationships between the various cell fates. The pairwise model, replaces a rigid and bifurcating lineage tree with a spectrum of fate options open to HSC. In the "classic" model, HSCs progress stepwise through a series of intermediate progenitors, to close down developmental options in a binary manner. In the new scenario, HSC make an immediate lineage choice from all of the end-cell options. The figure is, with permission, from [6] ©Macmillan Magazines Ltd. HSC, hematopoietic stem cell; DC, dendritic cell; NK cell, natural killer cell. (**B**) The partial arcs represent the close relationships between cell lineages that we inferred from the different groups of fates that are available to known intermediary progenitor cells (marked with an asterisk). Investigators initially described Lymphoid-primed multipotent progenitors (LMPP), early progenitors with lymphoid and myeloid potential (EPLM) and common myeloid progenitors (CMP) as homogeneous population of cells. However, they are a mixture of cells with the lineage affiliations shown as cells added to the arcs for LMPP, EPLM and CMP. Affiliations include B lymphocyte (B), DC, monocyte (Mo), eosinophil (Eo), basophil (Ba), erythroid (E) and megakaryocyte (M) [16–20]. HSC are shown to include the following: lymphoid biased cells (Ly) [23,24] that express the fms-like tyrosine kinase 3 receptor (Flt3$^+$) [25]; myeloid biased or committed cells (My) [26,27] that express Flt3$^+$ [25] and/or the receptor for thrombopoietin (TpoR$^+$) [28]; cells committed to the erythroid pathway [19] and affiliated as to expression of the receptor for erythropoietin (EpoR$^+$) [25,29]; and cells that express the receptor for macrophage colony stimulating factor (M-CSFR$^+$) and monocyte-affiliated (Mo) [25,30]. CLP, common lymphoid progenitor; CMP, common myeloid progenitor; DC/Pro-B, dendritic cell and B lymphocyte progenitor; Eo/B-CFU, eosinophil and basophil progenitor; EPLM, early progenitor with lymphoid and myeloid potential; GMP, granulocyte and macrophage progenitor; LMPP, lymphoid-primed multipotent progenitor; MEP, megakaryocyte and erythrocyte progenitor; Mon/B/DC?, monocyte, B lymphocyte and dendritic cell? progenitor; Mon/DC, monocyte and dendritic cell progenitor, NK/T, natural killer cell and T lymphocyte progenitor and NK/ILC, natural killer cell and innate lymphoid cell progenitor.

There is good evidence to support the view that the lineage affiliations seen in recent investigations of LMPP, EPLM and CMP become apparent earlier during development, even as early as in HSC. In 2010, Ichi and colleagues reported that the transcriptional profiles of individual HSCs show considerable variation and some HSCs are poor precursors of lymphocytes whilst others generate a balanced output of cell types [31]. Subsequently, investigators examined the progeny of single cells transferred into irradiated mice and described myeloid- and lymphoid-biased mouse HSC [24,26,32,33]. Notta and colleagues have shown that the progenitor cell compartment of human bone marrow is a mixture of cells with uni-potent myeloid or erythroid potential alongside multipotent cells [19]. The Jacobsen group described a sub-set of murine HSC with a bias towards platelets and myeloid cells, requiring thrombopoietin for their maintenance [28]. Similarly, sub-sets of murine HSC express the receptor for macrophage colony-stimulating factor (M-CSF), mRNA for the erythropoietin (Epo) receptor [25,29,30] and the fms-like tyrosine kinase 3 (Flt3) receptor, for Flt3 ligand (Flt3L) [25]. The selective expression of these receptors is suggestive of a predisposition and/or affiliation of some HSC to a lineage because the corresponding cytokines (M-CSF, Epo and Flt3L) can instruct end-cell fate ([30,34] and see later).

2. The Mature End-Cell Populations Are Not Homogeneous

The variety of end-cell types is more complex than is usually shown in models. For example, the DC family includes Langerhans cells (LC), two types of interstitial DC (iDC) [35,36], CD8$^+$ DC [37] and CD8$^-$ conventional DC (cDC) [37] (Figure 2A). The delineation of these DC sub-populations depends on their precise functions. They include antigen transport, immune regulation and the control of infections e.g., yeast. In addition, there are type 1 interferon-producing plasmacytoid pDC [38] and monocytes that are a form of pre-DC [39] and both are classified within mononuclear phagocytes (see later). The local environments in which DC precursors reside may be important for their fate specification (Figure 2). Satpathy and colleagues have highlighted the importance of various cytokines to DC development, including whether their actions within distinct stromal niches influence fate choice [40].

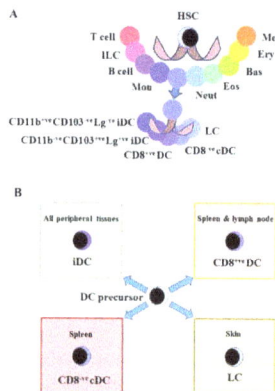

Figure 2. Heterogeneity of dendritic cells. (**A**) Surface markers are used to describe the various end-cell DCs and their nature is a matter of gradations. (**B**) Where each population resides is important to their function within the immune system and the local environment may determine the fate of DC precursors [41]. HSC, hematopoietic stem cell, ILC, innate lymphoid cell, Mon, monocyte; Neut, neutrophil; Eos, eosinophil, Bas, basophil; Ery, erythrocyte; Meg, megakaryocyte, DC, dendritic cells, iDC, interdigitating dendritic cell; LC, Langerhans cells; Lg, Langerin.

There is heterogeneity among cells of the mononuclear phagocyte system [42]. Human monocyte sub-populations are discriminated on the basis of CD14 and CD16 expression and those in mice by the use of markers such as F4/80, Ly6C (Gr-1), CD11b, CD43 and the chemokine receptor CX3CR1 [43]. Following trans-endothelial migration, monocytes enter the tissues where they give rise to macrophages. There is debate about how best to distinguish macrophages from DCs: is there any real difference between the capacity of some macrophages and DCs to present antigens [44]? Live imaging studies have not provided clear evidence for a distinction [45]. Additionally, a population of cutaneous Langerin[+ve] DC develops from blood monocytes, rather than from Langerhans cells [35,46].

The markers used to define macrophage sub-sets include the macrophage scavenger receptors (CD36, CD14, signal-regulatory protein family (SIRP)), Toll-like receptors, various integrins, epidermal growth factor-seven transmembrane (EGF-TM7) proteins, other immunoglobulin superfamily receptors (Siglecs) and multiple C-type lectins. Our perception of the assortment of these markers is that it is apparently random [47]. The use of quantitative flow cytometry (FC) profiles reveals a spectrum of phenotypic variability, rather than precisely defined sub-sets of cells [42]. Some of this phenotype noise may be due to stochastic fluctuations in the levels of expression of transcription factors. Even so, the different macrophage sub-sets, as defined by different combinations of markers, provide distinct signals to T cell sub-sets and are therefore important physiologically.

Macrophages encounter multiple signals in the various tissues that are important for shaping their phenotype [48]. For example, macrophages that reside in the lung and in the peritoneum have distinct gene expression profiles and chromatin landscapes. Lavin and colleagues transferred macrophages that had been isolated from the mouse peritoneum into the alveolar cavity, whereupon they upregulated the macrophage-specific genes that typify lung macrophages and downregulated the peritoneal macrophage-specific genes. Expression profiling confirmed this transition: the transferred cells resembled lung macrophages more than peritoneal macrophages. So, the originally peritoneal macrophages had adapted to their new environment [49]. Further evidence of macrophage adaptability is that CD8[+ve] and CD8[−ve] antigen-presenting cells in the spleen that arise from the same progenitor are interconvertible [50]. These findings both highlight the adaptability of differentiated cells and show that the microenvironment can influence cell specialization. Signals from the tissue environment may modulate the chromatin landscape that, in turn, prescribes a tissue-related gene expression pattern within macrophages.

3. Some End-Cells Are Interconvertible

CD4[+ve] effector cells include T helper 1 cells (Th1), T helper 2 cells (Th2), interleukin (IL) 17-producing T helper cells (Th17), follicular T helper cells (Tfh) and regulatory T cells (Treg). These sub-types have different functions. They include the control of bacterial, helminth and fungal infections and B cell maturation in germinal centres and the suppression of immune responses. CD4[+ve] effector cells produce many different cytokines, including IL-2, IL-3, IL-4, IL-13, IL-17, IL-21 and granulocyte-macrophage colony-stimulating factor (GM-CSF) and to a lesser extent interferon g (IFNγ) [51]. The nature of the cytokines produced is indicative of the type of CD4[+ve] helper cell. CD4[+ve] cell types arise in multiple environments and the prevailing cytokine environment is important for their differentiation. For example, IL-12 is required for Th1 cell differentiation and IFNγ amplifies this process, whereas IL-4 drives the differentiation of Th2 cells [52].

The characteristics of some of the mature CD4[+] effector cells are interconvertible (Figure 3). IL-21 is produced by Th2 cells [53] and TFh cells [54] and Th2 cells can give rise to Tfh cells [55]. Induced regulatory T cells (iTreg) can convert to pro-inflammatory Th17 [56] and when transferred into T cell-deficient CD3[−/−] or Rag2[−/−] mice respectively give rise to Tfh cells in Peyer's patches [57] and Th2 cells in the spleen [58]. Memory Th2 cells convert in vitro to iTreg, in response to treatment with transforming growth factor β (TGF-β) and blockade of IFNγ and IL-4 signalling [59]. In all of these examples, the switching between the different types of CD4[+] cells requires an external set of signals, suggesting that environmental signals may underlie the adaptability of CD4[+] cells to the type of the immune response required. Eizenberg-Magar and colleagues have constructed mathematical models that predict how different cytokine inputs to CD4[+] T cells determine their differentiation state [60].

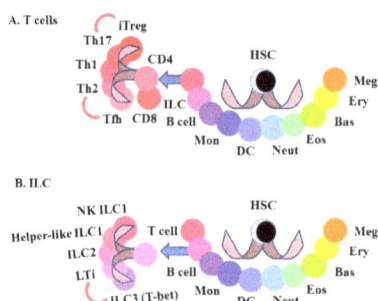

Figure 3. The heterogeneity and interconversion of CD4^{+ve} helper T lymphocytes and innate lymphoid cells. CD4^{+ve} effector cells and innate lymphoid (ILC) cells include various sub-populations. For both of these types of cells, the characteristics of some of the mature effector cells are interconvertible (**A**) The arrows connecting the sub-types of CD4 T cells show interconversions. (**B**) The arrows connecting the sub-types of ILC show interconversions. HSC, hematopoietic stem cell; iTreg, induced-regulatory T cells; Th17, Interleukin 17 producing T helper cells, Th1, T helper 1 cells; Th2, T helper 2 cells, Tfh, follicular helper T cells; NK, natural killer cell, Mon, monocyte; Neut, neutrophil; Eos, eosinophil, Bas, basophil; Ery, erythrocyte; Meg, megakaryocyte, DC, dendritic cells.

The sub-types of ILC have natural killer and helper-like functions that are important as a first line defence against pathogens, the genesis of lymphoid organs and tissue modelling [61]. There are three main groups of ILC. Group 1 includes natural killer cells and helper-like ILC and groups 2 and 3 are helper-like. The the expression of transcription factors defines two populations within groups 1 and 2. The developmental relationship between the various ILC populations is presently unclear and their characteristics are interconvertible (Figure 3). Expression of master regulators of transcription, surface receptors and the ILs produced define group 3 sub-types. There are cells that either: (1) express the transcription factor retinoid-related orphan receptor γt (RORγt), CD4 and the cytokine receptor CCR6 at their surface and produce IL-17 and IL-22 (lymphoid tissue inducer (LTi)-like ILC3) or (2) express the transcription factor T-bet, the NK receptors NKp46 and NK1.1 at their surface and produce IFNγ and TNFα. However, LTi-like ILC3 (RORγt^{+}, CCR6^{+}) cells express T-bet when exposed in vitro to a Notch stimulus [62]. There may be a true bi-directional plasticity of ILC3 as NKp46^{+} ILC3 may downregulate NKp46 in vivo giving rise to NKp46^{-} ILC3 [63]. The relative levels of expression of RORγt and T-bet may determine the effector functions of group 3 ILC3 and underlie the plasticity of the phenotype of these cells. However, there is controversy on this matter as the subsets of group 3 ILC3 may be separate lineages that develop from different progenitors [64]. To add to the plasticity of ILC, mouse studies have shown that ILC2 that reside in the lung become ILC1 in response to infection by influenza virus and *Staphylococcus aureus* and by cigarette smoke [65].

4. How Might We Classify the Types of Cells?

One purpose of classifying blood cells is to aid the understanding of their development: we have no hope of understanding cell diversification without categorizing a cell's identity. The conventional use of the term cell lineage and cell type, refers to the developmental history of a cell. For example, a progenitor cell that is committed to the B lymphocyte developmental pathway gives rise to cells we denote as a B lymphocyte. However, ancestry does not always resolve cell identity where there is inconsistency between the attribution of cells to a lineage and classification with regard to a phenotype. For cells viewed collectively as ILCs, there are two separate origins; a progenitor that gives rise to the NK precursor and NK cells and another for all the helper-like ILC [61]. Similarly, it is not clear to what extent there are separate progenitors for the different DCs. They appear to arise from two separate -lymphoid and monocytic- origins but the surface phenotypes and gene transcription profiles of DCs derived in vitro from purified CLPs or purified CMPs are indistinguishable [66]. A Common Dendritic cell Progenitor (CDP) with the ability to give rise to both cDC and pDC has been identified [67,68]. Several other phenotypically distinct cells have been proposed as progenitors of different DC sub-populations [67,69–71]. However, it appears that multiple developmental pathways are at play in generating the different DCs, and, in some cases, they converge into phenotypically homogeneous but transcriptionally and functionally distinct mature DC [72,73]. The delineation of cell type with regard to ancestry is also confounded if we accept that HSCs predispose to a lineage by expressing, for example, the receptor for M-CSF but might step sideways and adopt a different trajectory.

In the case of the mature immune cells, an answer to the problem of their classification, their attributes or conversely the absence of a characteristic(s), is the unique function of each type of cell. In other words, members of a cell type serve a function that is different from members of another cell type. However, immune cell types can share functional attributes that confounds ascribing cell identity on this basis and blurs the boundaries between cell lineages. A cytotoxic capacity brings together some T cells and some ILC, whereas macrophages, DC and B cells can phagocytose, pinocytose, process and present foreign antigens. Additionally, cells of the immune system cooperated to perform their role and it is therefore not too surprising that different types of cells share, for example, the chemokine receptors that dictate the location of cells to a particular environment and the cytokine receptors for survival.

5. What Are the Differences between Types of Cells?

So, what are features that allow us to specify a population of immune cells? Distinguishing one cell type from another is in essence a matter of how many phenotypic markers we use to define a cell type. The use of two surface markers can clearly differentiate one type of cell from another. However, and as mentioned above, use of additional surface markers reveals substantial heterogeneity regarding mononuclear phagocytes and investigators must rely on their judgment as to how best to classify cell populations. Traditionally, the basis for the identification of early progenitors, in particular, is the use of a limited number of cell surface markers, which appear in many cases to have a graded, rather than discontinuous expression pattern. Examples of graded expression are the markers used for FC-based identification of Lineage-negative, CD117$^+$ (kit), Sca-1$^+$ cells (LSK), LMPP and CLP populations.

CD117, Sca-1 and Flt3 show a graded, continuous expression, thus making the identification and isolation of cells with "high" or "intermediate" expression of markers almost arbitrary and dependent on the staining prowess and judgement of the investigator (Figure 4). This problem becomes even larger for genetically modified mice, where it might prove difficult to distinguish between true changes in the proportions of gated cells (representing progenitor stages) and any potential alterations, caused by the genetic modification to the mice, in the expression level of markers that are used to "ring-fence" these cells. Modern and comprehensive analyses, particularly of single cells, have revised the approach to categorize cell types. Tools currently available for the examination of cellular heterogeneity include global gene expression analysis and mass cytometry where the number of individual surface markers analysed is extremely large. This leads to an even more detailed consideration of how finely to draw lines between different cell types and perhaps this becomes arbitrary for some of the different types of immune cells.

Regarding the use of global gene expression analyses, the conceptual problem is how fine should the criteria be that we use to distinguish one type of cell from another? A caveat is that it is impossible to be certain that two cells that, using a set of phenotypic attributes are ring-fenced as identical, are identical genotypically [77]. A simple reason is that cells of any given end-cell type are not all interacting with just one environment: some will be inside a particular environment and some left outside. Even within an environment, a graded input of signals might well lead to a graded output regarding the level of acquisition/stabilization of a phenotype. Moreover, a population of cells, even if clonal, is heterogeneous. Chang and colleagues have shown that cells of cloned populations of mouse hematopoietic progenitors express different levels of Sca-1. This relates to a state that fluctuates but is meaningful because cells that express extremely high or low levels of Sca-1 have distinct transcriptomes and different leanings towards the myeloid or erythroid lineage [78]. Heterogeneity of cloned cells is also the case for neuronal stem cells whereby in vitro subsets have distinct developmental commitments [79]. The heterogeneity of cloned stem cells presumably reflects the inherent versatility of multipotent cells. Of particular interest to lineage-predisposition of these cells are the events that regulate the expression of transcription factors. It would thus seem that none of the classically accepted features alone is reliable enough to unequivocally define a particular lineage or cell type. Rather, we should consider a combination of several attributes, including morphology, phenotype, function (for mature cell types), genetic signature and developmental ancestry.

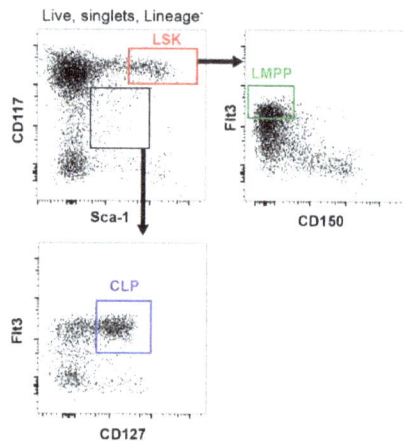

Figure 4. Markers used to identify hematopoietic progenitors display graded, rather than discontinuous, expression. CD117, Sca-1 and Flt3 show a continuous expression making the identification of cells with "high" or "intermediate" expression of markers almost arbitrary. Figure shows a typical gating strategy for the identification of LSK (red), LMPP (green) and CLP (blue) progenitor populations based on the expression of CD117, Sca-1, Flt3 and CD127 (IL-7Rα). LSK are identified as Lineage⁻CD117^{high}Sca-1^{high} [2] and LMPP as the LSK with the highest expression of Flt3 [74]. CLP are gated as Lineage⁻CD117^{int}Sca-1^{int}Flt3⁺CD127⁺ [75,76]. LSK: Lineage-negative, Sca-1-positive, kit-positive cells; LMPP: Lymphoid-primed Multi-Potent Progenitors; CLP: Common Lymphoid Progenitors.

For years, the issue of delineation has applied for how to best classify organisms into species, such as Darwin's many finches. One solution is that each species of finch breeds true. In a similar vein, whether cells can or cannot readily, even when "pushed" in vivo or in vitro, interconvert provides a way of resolving the delineation of cell types. The matter of definition of cell type and lineage, if we view this a synonymous, is simple if both refer to a cell that has progressed along a pathway to a stage of development and a phenotype that is irreversible. For example, and to the best of our knowledge, a cell, including its phenotypically identifiable progeny in the bone marrow, destined to generate mature B cells does not normally give rise to a mature T cell. Interestingly, the importance of the extent of developmental progression to irreversible lineage affiliation is exemplified by the finding that the Double Negative 2 T cell progenitor population that have already progressed some way along the T cell pathway can still give rise to NK and myeloid cells when "pushed" using the right conditions [21,80]. Nevertheless, and by applying the rule of irreversibility of phenotype, we would view the different ILC and CD4⁺ T cells that can interchange/reverse mature fates under the right conditions as sub-types of cells belonging to the ILC and CD4⁺ T cell lineages respectively, rather than as separate lineages. As to their taxonomy, we might view their differentiation and end phenotype as a much more flexible and continuous process that circumvents the need to draw overly strict boundaries. In this case, there is a continuum of intermediate states between cell sub-types.

6. Does the Environment That Cells Reside in Instruct Cell Fate?

As mentioned above, cells and their genes, do interact with and respond to their environment. The environmental history of a cell is therefore important to the specification of phenotype. Whilst for cell identity we have attached prime importance to developmental history, namely who begat whom, this might merely serve to keep offspring within a particular nurturing environment. In some instances, we might have mistakenly read environmental nurture as developmental history.

Some hematopoietic cytokines are important for shaping hematopoiesis. Generally, we accept that they are essential for the survival and proliferation of hematopoietic cells at all stages of their development. However, in some cases they are also important for the differentiation of hematopoietic lineages. Thus, their function as merely survival and/or proliferation agents (permissive role) or inducers of differentiation (instructive role) has been a matter of debate for long time [81–83]. Several investigations have provided evidence in support of both modes of cytokine action, with the data in some cases being contradictory (reviewed in: [84–86]). Genetic deletion of cytokines or their receptors has resulted in a reduced production of immune cells whose generation is subject to regulation by the cytokine in question but even though the disruption in developmental output can be quite severe, it has never led to the complete absence of the lineage. This indicates either a permissive role of these cytokines or some level of redundancy, with other cytokines compensating for the absent instructive signal. Furthermore, over-expression of anti-apoptotic signals, such as B-cell lymphoma 2 (Bcl2), in order to provide a strong survival signal to cells, has been on some occasions sufficient to rescue the affected lineage, suggesting a permissive cytokine function. Thus, whilst the absence of IL-7 signalling severely disrupts T cell development, Bcl2 over-expression can significantly rescue T cell output in these mice [87–89], suggesting a survival role of IL-7 in early T cell development. Similarly, the B cell defect observed in mice lacking Stat5 (the crucial signalling mediator for IL-7) can be fully recovered by Bcl2 over-expression [90], while sustained and increased Flt3L levels can rescue B cell development in *Il7*$^{-/-}$ mice [91], thus demonstrating the permissive role of IL-7 in both T and B cell development. Moreover, early in vivo studies with chimeric receptors provided further evidence for a permissive role of hematopoietic cytokines. The replacement of the intra-cellular part of the TPO receptor (mpl) with that of G-CSF [92] or the signalling domain of G-CSF with that of Epo-R [93] did not result in any lineage skewing, as would be expected regarding an instructive role of the corresponding cytokines.

Other transgenic approaches, mainly by ectopic expression of cytokine receptors, have nevertheless suggested instruction of lineage fate by cytokines. Thus, expression of M-CSF receptor in multi-potent hematopoietic cell lines resulted in skewing of their developmental output [94,95], while ectopic GM-CSF receptor expression on CLP or pro-T cells increased their myeloid differentiation potential [96–98]. Similarly, Flt3 expression in MEP up-regulated myeloid-specific transcription factors and promoted their differentiation towards granulocyte/macrophage lineages [99]. More recently, investigators have provided in vitro and in vivo evidence for an instructive role of cytokines in lineage fate. Our studies support an instructive role of Flt3L at an early stage of hematopoiesis [34]. Transgenic mice over-expressing Flt3L ubiquitously are anemic and thrombocytopenic. Investigation of the progenitor cell populations in these mice revealed that Flt3L acts to skew hematopoiesis towards the myeloid and lymphoid lineages, as there is a lack of generation of erythroid and megakaryocyte progenitors. The cells diverted lack expression of cell lineage markers (Lin⁻) but express the Sca-1 antigen and the receptor for the stem cell factor CD117 or c-kit and are termed LSK. Likewise, increasing the level of Epo in vivo, by transgenesis, has revealed an instructive role for this cytokine in erythropoiesis [29]. In 1982, Metcalf and Burgess provided evidence supporting the view that G- and GM-CSF instruct bi-potent granulocyte-macrophage colony forming cells in their choice between granulocyte and macrophage pathways respectively [4]. Rieger and colleagues added to these finding by showing that G- and M-CSF instruct granulocyte and macrophage progenitors to follow each of the pathways respectively [100].

Overall, there is evidence to support the view that some of the hematopoietic cytokines can instruct cell fate and therefore guide the generation of a particular type(s) of cell as required. The effect of cytokines seems to be cell-context dependent, highlighting the interplay between extra-cellular signals from the environment and intra-cellular fate-determining factors, such as transcription factors and the epigenetic landscape [86]. Additionally, some level of promiscuity in the use of cytokines by hematopoietic cells makes the identification of their specific mode of action challenging. In that context, studies at the single cell are necessary. As evident from the data discussed above, most of the experimental evidence for an instructive role of hematopoietic cytokines comes either from in vivo over-expression of cytokines and/or their receptors or from in vitro culture systems. It is conceivable that, depending on the strength of the signal, a particular cytokine may act to provide a signal for survival, to proliferate or to instruct lineage choice. The response elicited within a cell, as to which of the three outcomes is undertaken, would accord with an increasing level of signal intensity. Thus, in order to demonstrate their instructive role, it is necessary to use a high level of cytokine in in vitro experiments. A graded response of cells to cytokines may be important for steady state versus hematopoiesis under stress conditions, such as infection, whereby the different requirements are survival/expansion versus survival/expansion coupled to a rapid and emergency, diversion of cells towards a required end cell type [101]. Recent evidence to support this viewpoint is that Singh and colleagues have observed that during chronic erythroid stress, in Epo transgenic mice, HSC exhibit a vastly committed erythroid progenitor profile together with enhanced cell division [102].

This makes physiological sense when one considers that the hematopoietic system is extremely flexible, responding rapidly to situations of stress, infection and blood loss. Whilst there is debate as to how this operates, it is important to bear in mind that cells we identify as HSC that can reconstitute the entire hematopoietic/immune system long term (LT-HSC) express the receptors for the proposed instructive cytokines. These include the receptors for Epo, M-CSF and Flt3. Around 19% of LT-HSC express the receptor for M-CSF on their cell surface and ~5% express cell surface Flt3, with ~12% expressing *Flt3* mRNA. Around 13% express mRNA for the Epo receptor and the level of surface protein expression is unknown due to the lack of a suitable antibody reagent. These are essentially sub-populations of LT-HSC [25]. We might assume that engagement of the cytokine with the receptor provides a signal for survival and proliferation. We do not know whether the presence of the appropriate cytokine instructs LT-HSC to adopt either a macrophage, myeloid or erythroid fate.

7. The Events That Shape Cell Identity

It is reasonable to conclude that the nature of immune cell diversity is not merely a matter of affiliation to an ancestral pathway but that environmental nurture is important. A mixture of (i) predisposition towards/affiliation of HSC to a lineage and (ii) nurture at a later stage of development, by environmental cues, appear to govern the end-fate of a cell (Figure 5). This perhaps explains the many sub-types of a number of different immune cells. Nurture is also important for HSC. It is likely that the localization of HSC within diverse and supportive bone marrow niches and their exposure to different development growth factors and morphogen gradients plays an important role in driving their heterogeneity [103].

Waddington's epigenetic landscape has provided a longstanding model for the general nature of the events that shape cell identity [104]. Cells roll down bifurcating valleys and once a cell has chosen a fate there are ridges that help maintain this fate. Changes to the epigenome dictate the hills and valleys that exclude other fates and which the microenvironment might impose. Ferrell has proposed an alternative landscape whereby at the start there are numerous valleys and ridges, commitment to a fate relates to the disappearance of some of the valleys and ridges and new valleys and ridges arise from cell-cell competition [105]. Sieweke examines a landscape that he likens to a small group of Pacific islands—Captain Cook's Islands—in the middle of an ocean. These can be reached from many directions such that the routes to a fate are flexible [106]. Waddington's landscape and the alternative landscape both involve stepwise bifurcations but there does not seem to be the need for HSC to undergo this progressive restriction. These models and a "small group of islands" model mean that there are close relations between particular cell lineages, as envisaged in the pairwise model. A "small groups of islands" model is more in keeping with the pairwise model except that each island can be approached separately rather than by Captain Cook's route from island to island and from east to west.

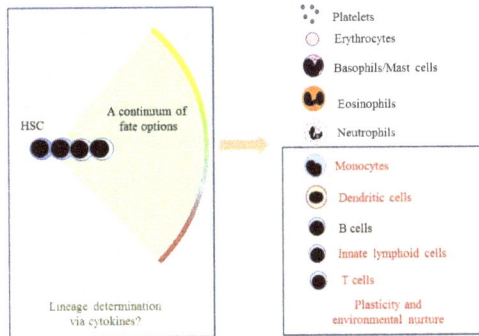

Figure 5. The developmental versus environmental history of cell types. A continuum of options is open to HSC and they are able to choose a fate without having to progress stepwise through a series of intermediate progenitors in order to close down developmental options. There is good evidence to support the viewpoint that environmental nurture is important to the generation of diversity of some of the mature cell types.

8. Implications for Leukemia

Normal HSCs and their immediate progeny are versatile in their choice of developmental pathway. If we accept that HSC can commit themselves directly to one pathway, we might presume that each pathway is equally available. Versatility is also the case for the progeny of HSC. The colonies observed for the progeny of HSC, particularly multipotent HPC, grown in semi-solid agar or methylcellulose medium contain many different cell types. We have argued that these findings are misleading as to how HSC and HPC behave in vivo, because single cells dispersed in agar are not in their normal social environment and lack the influence of niches, the extracellular matrix and the appropriate cytokines [101]. Nevertheless, the findings from colony assays support the notion of inherent versatility of HPC and perhaps indicate that environmental influences in vivo are important for guiding and/or narrowing the trajectories of HSC. HSC are versatile in the pairwise model, because they can step sideways to adopt alternative, closely related, fates, even after they have "made a lineage choice." The making of the architecture of leukemia and lymphoma, is very different because the lineage output of the transformed counterparts is restricted to a particular pathway [107–109]. In contrast to the flexibility of HSC, the transformed counterpart essentially dumps cells down a particular pathway. The rationale for this and perhaps the cardinal aspect of cell transformation in cancer is that a genetic insult fixates the epigenome to a lineage. Sanchez-Garcia has proposed that the contribution of oncogenes to leukemia is via epigenetic priming of the cells that initiate leukemia and the initial oncogenic insult(s) is/are then dispensable for tumour progression and maintenance [109]. In other words, there appears to be an insult-wired block to versatility in the leukemia stem cell.

9. Concluding Remarks

In the pairwise model, HSC can immediately predispose towards or affiliate to a single lineage, yet retain the ability to divert to an adjacent lineage. We might view the adjacent options as easily accessible or latent in some way. In this case, predisposed or affiliated HSC are essentially progressively differentiating towards an outcome, including narrowing their trajectory from the continuum of options. In the original diagram for the pairwise model, we showed DC, monocytes and NK cells as a single population. Each of these cell types is clearly a heterogeneous population and, therefore, it seems that some cells have a lifelong capacity to differentiate and diversify. Questions that are perhaps more important than lineage affiliation are (i) what channels a cell towards a particular phenotype and (ii) how does a cell stabilize a pattern of attributes, some of which are shared by different cell types. Perhaps the signals they receive from the environment in which they reside, that "mark" the epigenome, are as important as developmental history. Our view of the architecture of haematopoiesis has important implications to our understanding of the initiation of leukaemia. Whereas HSC are versatile, an insult(s) appears to wire the epigenome of HSC regarding lineage choice to a fixed leukaemia cell fate [109].

Funding: This project received funding from the European Union's Seventh Framework Programme for research, technological development and demonstration under grant agreement no 315902.

Acknowledgments: Geoffrey Brown and Rhodri Ceredig were Partners within the Marie Curie Initial Training Network DECIDE (Decision-making within cells and differentiation entity therapies). We dedicate this article to our friend, colleague and mentor, Antonius G Rolink whose incisive comments greatly influenced our way of viewing hematopoiesis. We thank Bob Michell for his comments on the manuscript.

Conflicts of Interest: The authors declare no conflict of interest.

References

1. Sulston, J.E.; Schierenberg, E.; White, J.G.; Thomson, J.N. The embryonic cell lineage of the nematode caenorhabditis elegans. *Dev. Biol.* **1983**, *100*, 64–119. [CrossRef]
2. Weissman, I.L.; Anderson, D.J.; Gage, F. Stem and progenitor cells: Origins, phenotypes, lineage commitments and transdifferentiations. *Annu. Rev. Cell. Dev. Biol.* **2001**, *17*, 387–403. [CrossRef] [PubMed]
3. Westin, J.A. Motile and social behaviour of neural crest cells. In *Cell Behavior*; Bellairs, R., Curtis, A., Dunn, G., Eds.; Cambridge University Press: London, UK, 1981; pp. 429–470.
4. Metcalf, D.; Burgess, A.W. Clonal analysis of progenitor cell commitment of granulocyte or macrophage production. *J. Cell. Physiol.* **1982**, *111*, 275–283. [CrossRef] [PubMed]
5. Pang, L.; Weiss, M.J.; Poncz, M. Megakaryocyte biology and related disorders. *J. Clin. Invest.* **2005**, *115*, 3332–3338. [CrossRef] [PubMed]
6. Ceredig, R.; Rolink, A.G.; Brown, G. Models of haematopoiesis: Seeing the wood for the trees. *Nat. Rev. Immunol.* **2009**, *9*, 293–300. [CrossRef] [PubMed]
7. Grinenko, T.; Eugster, A.; Thielecke, L.; Ramasz, B.; Kruger, A.; Dietz, S.; Glauche, I.; Gerbaulet, A.; von Bonin, M.; Basak, O.; et al. Hematopoietic stem cells can differentiate into restricted myeloid progenitors before cell division in mice. *Nat. Commun.* **2018**, *9*, 1898. [CrossRef] [PubMed]
8. Brown, G.; Hughes, P.J.; Michell, R.H. Cell differentiation and proliferation—Simultaneous but independent? *Exp. Cell. Res.* **2003**, *291*, 282–288. [CrossRef]
9. Drayson, M.T.; Michell, R.H.; Durham, J.; Brown, G. Cell proliferation and cd11b expression are controlled independently during hl60 cell differentiation initiated by 1,25α-dihydroxyvitamin D3 or all-trans-retinoic acid. *Exp. Cell. Res.* **2001**, *266*, 126–134. [CrossRef] [PubMed]
10. Kueh, H.Y.; Champhekar, A.; Nutt, S.L.; Elowitz, M.B.; Rothenberg, E.V. Positive feedback between pu.1 and the cell cycle controls myeloid differentiation. *Science* **2013**, *341*, 670–673. [CrossRef] [PubMed]
11. Hough, S.R.; Laslett, A.L.; Grimmond, S.B.; Kolle, G.; Pera, M.F. A continuum of cell states spans pluripotency and lineage commitment in human embryonic stem cells. *PLoS ONE* **2009**, *4*, e7708. [CrossRef] [PubMed]
12. Morgani, S.; Nichols, J.; Hadjantonakis, A.K. The many faces of pluripotency: In vitro adaptations of a continuum of in vivo states. *BMC Dev. Biol.* **2017**, *17*, 7. [CrossRef] [PubMed]
13. Elghetany, M.T. Surface antigen changes during normal neutrophilic development: A critical review. *Blood Cells Mol. Dis.* **2002**, *28*, 260–274. [CrossRef] [PubMed]
14. McGrath, K.E.; Catherman, S.C.; Palis, J. Delineating stages of erythropoiesis using imaging flow cytometry. *Methods* **2017**, *112*, 68–74. [CrossRef] [PubMed]

15. Ciarletta, P.; Ben Amar, M.; Labouesse, M. Continuum model of epithelial morphogenesis during caenorhabditis elegans embryonic elongation. *Philos. T. Roy. Soc. A.* **2009**, *367*, 3379–3400. [CrossRef] [PubMed]

16. Alberti-Servera, L.; von Muenchow, L.; Tsapogas, P.; Capoferri, G.; Eschbach, K.; Beisel, C.; Ceredig, R.; Ivanek, R.; Rolink, A. Single-cell RNA sequencing reveals developmental heterogeneity among early lymphoid progenitors. *EMBO J.* **2017**, *36*, 3619–3633. [CrossRef] [PubMed]

17. Hoppe, P.S.; Schwarzfischer, M.; Loeffler, D.; Kokkaliaris, K.D.; Hilsenbeck, O.; Moritz, N.; Endele, M.; Filipczyk, A.; Gambardella, A.; Ahmed, N.; et al. Early myeloid lineage choice is not initiated by random pu.1 to gata1 protein ratios. *Nature* **2016**, *535*, 299–302. [CrossRef] [PubMed]

18. Naik, S.H.; Perie, L.; Swart, E.; Gerlach, C.; van Rooij, N.; de Boer, R.J.; Schumacher, T.N. Diverse and heritable lineage imprinting of early haematopoietic progenitors. *Nature* **2013**, *496*, 229–232. [CrossRef] [PubMed]

19. Notta, F.; Zandi, S.; Takayama, N.; Dobson, S.; Gan, O.I.; Wilson, G.; Kaufmann, K.B.; McLeod, J.; Laurenti, E.; Dunant, C.F.; et al. Distinct routes of lineage development reshape the human blood hierarchy across ontogeny. *Science* **2016**, *351*, aab2116. [CrossRef] [PubMed]

20. Paul, F.; Arkin, Y.; Giladi, A.; Jaitin, D.A.; Kenigsberg, E.; Keren-Shaul, H.; Winter, D.; Lara-Astiaso, D.; Gury, M.; Weiner, A.; et al. Transcriptional heterogeneity and lineage commitment in myeloid progenitors. *Cell* **2015**, *163*, 1663–1677. [CrossRef] [PubMed]

21. Balciunaite, G.; Ceredig, R.; Massa, S.; Rolink, A.G. A b220+ cd117+ cd19- hematopoietic progenitor with potent lymphoid and myeloid developmental potential. *Eur. J. Immunol.* **2005**, *35*, 2019–2030. [CrossRef] [PubMed]

22. Brown, G.; Hughes, P.J.; Michell, R.H.; Ceredig, R. The versatility of haematopoietic stem cells: Implications for leukaemia. *Crit. Rev. Clin. Lab. Sci.* **2010**, *47*, 171–180. [CrossRef] [PubMed]

23. Ema, H.; Morita, Y.; Suda, T. Heterogeneity and hierarchy of hematopoietic stem cells. *Exp. Hematol.* **2014**, *42*, 74–82.e2. [CrossRef] [PubMed]

24. Shimazu, T.; Iida, R.; Zhang, Q.; Welner, R.S.; Medina, K.L.; Alberola-Lla, J.; Kincade, P.W. Cd86 is expressed on murine hematopoietic stem cells and denotes lymphopoietic potential. *Blood* **2012**, *119*, 4889–4897. [CrossRef] [PubMed]

25. Mooney, C.J.; Cunningham, A.; Tsapogas, P.; Toellner, K.M.; Brown, G. Selective expression of flt3 within the mouse hematopoietic stem cell compartment. *Int. J. Mol. Sci.* **2017**, *18*, 1037. [CrossRef] [PubMed]

26. Gekas, C.; Graf, T. Cd41 expression marks myeloid-biased adult hematopoietic stem cells and increases with age. *Blood* **2013**, *121*, 4463–4472. [CrossRef] [PubMed]

27. Oguro, H.; Ding, L.; Morrison, S.J. Slam family markers resolve functionally distinct subpopulations of hematopoietic stem cells and multipotent progenitors. *Cell Stem Cell* **2013**, *13*, 102–116. [CrossRef] [PubMed]

28. Sanjuan-Pla, A.; Macaulay, I.C.; Jensen, C.T.; Woll, P.S.; Luis, T.C.; Mead, A.; Moore, S.; Carella, C.; Matsuoka, S.; Bouriez Jones, T.; et al. Platelet-biased stem cells reside at the apex of the haematopoietic stem-cell hierarchy. *Nature* **2013**, *502*, 232–236. [CrossRef] [PubMed]

29. Grover, A.; Mancini, E.; Moore, S.; Mead, A.J.; Atkinson, D.; Rasmussen, K.D.; O'Carroll, D.; Jacobsen, S.E.; Nerlov, C. Erythropoietin guides multipotent hematopoietic progenitor cells toward an erythroid fate. *J. Exp. Med.* **2014**, *211*, 181–188. [CrossRef] [PubMed]

30. Mossadegh-Keller, N.; Sarrazin, S.; Kandalla, P.K.; Espinosa, L.; Stanley, E.R.; Nutt, S.L.; Moore, J.; Sieweke, M.H. M-csf instructs myeloid lineage fate in single haematopoietic stem cells. *Nature* **2013**, *497*, 239–243. [CrossRef] [PubMed]

31. Ichii, M.; Shimazu, T.; Welner, R.S.; Garrett, K.P.; Zhang, Q.; Esplin, B.L.; Kincade, P.W. Functional diversity of stem and progenitor cells with b-lymphopoietic potential. *Immunol. Rev.* **2010**, *237*, 10–21. [CrossRef] [PubMed]

32. Beerman, I.; Bhattacharya, D.; Zandi, S.; Sigvardsson, M.; Weissman, I.L.; Bryder, D.; Rossi, D.J. Functionally distinct hematopoietic stem cells modulate hematopoietic lineage potential during aging by a mechanism of clonal expansion. *Proc. Natl. Acad. Sci. USA* **2010**, *107*, 5465–5470. [CrossRef] [PubMed]

33. Challen, G.A.; Boles, N.C.; Chambers, S.M.; Goodell, M.A. Distinct hematopoietic stem cell subtypes are differentially regulated by tgf-beta1. *Cell Stem Cell* **2010**, *6*, 265–278. [CrossRef] [PubMed]

34. Tsapogas, P.; Swee, L.K.; Nusser, A.; Nuber, N.; Kreuzaler, M.; Capoferri, G.; Rolink, H.; Ceredig, R.; Rolink, A. In vivo evidence for an instructive role of fms-like tyrosine kinase-3 (flt3) ligand in hematopoietic development. *Haematologica* **2014**, *99*, 638–646. [CrossRef] [PubMed]
35. Ginhoux, F.; Collin, M.P.; Bogunovic, M.; Abel, M.; Leboeuf, M.; Helft, J.; Ochando, J.; Kissenpfennig, A.; Malissen, B.; Grisotto, M.; et al. Blood-derived dermal langerin+ dendritic cells survey the skin in the steady state. *J. Exp. Med.* **2007**, *204*, 3133–3146. [CrossRef] [PubMed]
36. Nestle, F.O.; Zheng, X.G.; Thompson, C.B.; Turka, L.A.; Nickoloff, B.J. Characterization of dermal dendritic cells obtained from normal human skin reveals phenotypic and functionally distinctive subsets. *J. Immunol.* **1993**, *151*, 6535–6545. [PubMed]
37. De Smedt, T.; Pajak, B.; Muraille, E.; Lespagnard, L.; Heinen, E.; De Baetselier, P.; Urbain, J.; Leo, O.; Moser, M. Regulation of dendritic cell numbers and maturation by lipopolysaccharide in vivo. *J. Exp. Med.* **1996**, *184*, 1413–1424. [CrossRef] [PubMed]
38. Liu, Y.J. Ipc: Professional type 1 interferon-producing cells and plasmacytoid dendritic cell precursors. *Annu. Rev. Immunol.* **2005**, *23*, 275–306. [CrossRef] [PubMed]
39. Taylor, P.R.; Gordon, S. Monocyte heterogeneity and innate immunity. *Immunity* **2003**, *19*, 2–4. [CrossRef]
40. Satpathy, A.T.; Murphy, K.M.; Kc, W. Transcription factor networks in dendritic cell development. *Semin. Immunol.* **2011**, *23*, 388–397. [CrossRef] [PubMed]
41. Naik, S.H. Demystifying the development of dendritic cell subtypes, a little. *Immunol. Cell Biol.* **2008**, *86*, 439–452. [CrossRef] [PubMed]
42. Hume, D.A. Differentiation and heterogeneity in the mononuclear phagocyte system. *Mucosal Immunol.* **2008**, *1*, 432–441. [CrossRef] [PubMed]
43. Geissmann, F.; Jung, S.; Littman, D.R. Blood monocytes consist of two principal subsets with distinct migratory properties. *Immunity* **2003**, *19*, 71–82. [CrossRef]
44. Hume, D.A. The mononuclear phagocyte system. *Curr. Opin. Immunol.* **2006**, *18*, 49–53. [CrossRef] [PubMed]
45. Lindquist, R.L.; Shakhar, G.; Dudziak, D.; Wardemann, H.; Eisenreich, T.; Dustin, M.L.; Nussenzweig, M.C. Visualizing dendritic cell networks in vivo. *Nat. Immunol.* **2004**, *5*, 1243–1250. [CrossRef] [PubMed]
46. Bursch, L.S.; Wang, L.; Igyarto, B.; Kissenpfennig, A.; Malissen, B.; Kaplan, D.H.; Hogquist, K.A. Identification of a novel population of langerin+ dendritic cells. *J. Exp. Med.* **2007**, *204*, 3147–3156. [CrossRef] [PubMed]
47. Ravasi, T.; Wells, C.; Forest, A.; Underhill, D.M.; Wainwright, B.J.; Aderem, A.; Grimmond, S.; Hume, D.A. Generation of diversity in the innate immune system: Macrophage heterogeneity arises from gene-autonomous transcriptional probability of individual inducible genes. *J. Immunol.* **2002**, *168*, 44–50. [CrossRef] [PubMed]
48. Mantovani, A.; Sica, A.; Sozzani, S.; Allavena, P.; Vecchi, A.; Locati, M. The chemokine system in diverse forms of macrophage activation and polarization. *Trends Immunol.* **2004**, *25*, 677–686. [CrossRef] [PubMed]
49. Lavin, Y.; Winter, D.; Blecher-Gonen, R.; David, E.; Keren-Shaul, H.; Merad, M.; Jung, S.; Amit, I. Tissue-resident macrophage enhancer landscapes are shaped by the local microenvironment. *Cell* **2014**, *159*, 1312–1326. [CrossRef] [PubMed]
50. Fogg, D.K.; Sibon, C.; Miled, C.; Jung, S.; Aucouturier, P.; Littman, D.R.; Cumano, A.; Geissmann, F. A clonogenic bone marrow progenitor specific for macrophages and dendritic cells. *Science* **2006**, *311*, 83–87. [CrossRef] [PubMed]
51. Yamane, H.; Paul, W.E. Cytokines of the γ_C family control CD4$^+$ t cell differentiation and function. *Nat. Immunol.* **2012**, *13*, 1037–1044. [CrossRef] [PubMed]
52. Abbas, A.K.; Murphy, K.M.; Sher, A. Functional diversity of helper t lymphocytes. *Nature* **1996**, *383*, 787–793. [CrossRef] [PubMed]
53. Wurster, A.L.; Rodgers, V.L.; Satoskar, A.R.; Whitters, M.J.; Young, D.A.; Collins, M.; Grusby, M.J. Interleukin 21 is a t helper (th) cell 2 cytokine that specifically inhibits the differentiation of naive th cells into interferon gamma-producing th1 cells. *J. Exp. Med.* **2002**, *196*, 969–977. [CrossRef] [PubMed]
54. Nurieva, R.I.; Chung, Y.; Hwang, D.; Yang, X.O.; Kang, H.S.; Ma, L.; Wang, Y.H.; Watowich, S.S.; Jetten, A.M.; Tian, Q.; et al. Generation of t follicular helper cells is mediated by interleukin-21 but independent of t helper 1, 2, or 17 cell lineages. *Immunity* **2008**, *29*, 138–149. [CrossRef] [PubMed]
55. Glatman Zaretsky, A.; Taylor, J.J.; King, I.L.; Marshall, F.A.; Mohrs, M.; Pearce, E.J. T follicular helper cells differentiate from th2 cells in response to helminth antigens. *J. Exp. Med.* **2009**, *206*, 991–999. [CrossRef] [PubMed]

56. Singh, K.; Gatzka, M.; Peters, T.; Borkner, L.; Hainzl, A.; Wang, H.; Sindrilaru, A.; Scharffetter-Kochanek, K. Reduced cd18 levels drive regulatory t cell conversion into th17 cells in the cd18hypo pl/j mouse model of psoriasis. *J. Immunol.* **2013**, *190*, 2544–2553. [CrossRef] [PubMed]

57. Tsuji, M.; Komatsu, N.; Kawamoto, S.; Suzuki, K.; Kanagawa, O.; Honjo, T.; Hori, S.; Fagarasan, S. Preferential generation of follicular b helper t cells from foxp3+ T cells in gut peyer's patches. *Science* **2009**, *323*, 1488–1492. [CrossRef] [PubMed]

58. Wang, Y.; Souabni, A.; Flavell, R.A.; Wan, Y.Y. An intrinsic mechanism predisposes foxp3-expressing regulatory t cells to th2 conversion in vivo. *J. Immunol.* **2010**, *185*, 5983–5992. [CrossRef] [PubMed]

59. Kim, B.S.; Kim, I.K.; Park, Y.J.; Kim, Y.S.; Kim, Y.J.; Chang, W.S.; Lee, Y.S.; Kweon, M.N.; Chung, Y.; Kang, C.Y. Conversion of th2 memory cells into foxp3+ regulatory t cells suppressing th2-mediated allergic asthma. *Proc. Natl. Acad. Sci. USA* **2010**, *107*, 8742–8747. [CrossRef] [PubMed]

60. Eizenberg-Magar, I.; Rimer, J.; Zaretsky, I.; Lara-Astiaso, D.; Reich-Zeliger, S.; Friedman, N. Diverse continuum of CD4+ T-cell states is determined by hierarchical additive integration of cytokine signals. *Proc. Natl. Acad. Sci. USA* **2017**, *114*, E6447–E6456. [CrossRef] [PubMed]

61. Gronke, K.; Kofoed-Nielsen, M.; Diefenbach, A. Innate lymphoid cells, precursors and plasticity. *Immunol. Lett.* **2016**, *179*, 9–18. [CrossRef] [PubMed]

62. Rankin, L.C.; Groom, J.R.; Chopin, M.; Herold, M.J.; Walker, J.A.; Mielke, L.A.; McKenzie, A.N.; Carotta, S.; Nutt, S.L.; Belz, G.T. The transcription factor t-bet is essential for the development of nkp46+ innate lymphocytes via the notch pathway. *Nat. Immunol.* **2013**, *14*, 389–395. [CrossRef] [PubMed]

63. Rankin, L.C.; Girard-Madoux, M.J.; Seillet, C.; Mielke, L.A.; Kerdiles, Y.; Fenis, A.; Wieduwild, E.; Putoczki, T.; Mondot, S.; Lantz, O.; et al. Complementarity and redundancy of il-22-producing innate lymphoid cells. *Nat. Immunol.* **2016**, *17*, 179–186. [CrossRef] [PubMed]

64. Sawa, S.; Cherrier, M.; Lochner, M.; Satoh-Takayama, N.; Fehling, H.J.; Langa, F.; Di Santo, J.P.; Eberl, G. Lineage relationship analysis of rorgammat+ innate lymphoid cells. *Science* **2010**, *330*, 665–669. [CrossRef] [PubMed]

65. Silver, J.S.; Kearley, J.; Copenhaver, A.M.; Sanden, C.; Mori, M.; Yu, L.; Pritchard, G.H.; Berlin, A.A.; Hunter, C.A.; Bowler, R.; et al. Inflammatory triggers associated with exacerbations of copd orchestrate plasticity of group 2 innate lymphoid cells in the lungs. *Nature immunology* **2016**, *17*, 626–635. [CrossRef] [PubMed]

66. Ishikawa, F.; Niiro, H.; Iino, T.; Yoshida, S.; Saito, N.; Onohara, S.; Miyamoto, T.; Minagawa, H.; Fujii, S.; Shultz, L.D.; et al. The developmental program of human dendritic cells is operated independently of conventional myeloid and lymphoid pathways. *Blood* **2007**, *110*, 3591–3660. [CrossRef] [PubMed]

67. Liu, K.; Victora, G.D.; Schwickert, T.A.; Guermonprez, P.; Meredith, M.M.; Yao, K.; Chu, F.F.; Randolph, G.J.; Rudensky, A.Y.; Nussenzweig, M. In vivo analysis of dendritic cell development and homeostasis. *Science* **2009**, *324*, 392–397. [CrossRef] [PubMed]

68. Naik, S.H.; Sathe, P.; Park, H.Y.; Metcalf, D.; Proietto, A.I.; Dakic, A.; Carotta, S.; O'Keeffe, M.; Bahlo, M.; Papenfuss, A.; et al. Development of plasmacytoid and conventional dendritic cell subtypes from single precursor cells derived in vitro and in vivo. *Nat. Immunol.* **2007**, *8*, 1217–1226. [CrossRef] [PubMed]

69. Onai, N.; Kurabayashi, K.; Hosoi-Amaike, M.; Toyama-Sorimachi, N.; Matsushima, K.; Inaba, K.; Ohteki, T. A clonogenic progenitor with prominent plasmacytoid dendritic cell developmental potential. *Immunity* **2013**, *38*, 943–957. [CrossRef] [PubMed]

70. Rodrigues, P.F.; Alberti-Servera, L.; Eremin, A.; Grajales-Reyes, G.E.; Ivanek, R.; Tussiwand, R. Distinct progenitor lineages contribute to the heterogeneity of plasmacytoid dendritic cells. *Nat. Immunol.* **2018**, *19*, 711–722. [CrossRef] [PubMed]

71. Schlitzer, A.; Loschko, J.; Mair, K.; Vogelmann, R.; Henkel, L.; Einwachter, H.; Schiemann, M.; Niess, J.H.; Reindl, W.; Krug, A. Identification of CCR9- murine plasmacytoid DC precursors with plasticity to differentiate into conventional DCs. *Blood* **2011**, *117*, 6562–6570. [CrossRef] [PubMed]

72. Sathe, P.; Vremec, D.; Wu, L.; Corcoran, L.; Shortman, K. Convergent differentiation: Myeloid and lymphoid pathways to murine plasmacytoid dendritic cells. *Blood* **2013**, *121*, 11–19. [CrossRef] [PubMed]

73. Shigematsu, H.; Reizis, B.; Iwasaki, H.; Mizuno, S.; Hu, D.; Traver, D.; Leder, P.; Sakaguchi, N.; Akashi, K. Plasmacytoid dendritic cells activate lymphoid-specific genetic programs irrespective of their cellular origin. *Immunity* **2004**, *21*, 43–53. [CrossRef] [PubMed]

74. Adolfsson, J.; Mansson, R.; Buza-Vidas, N.; Hultquist, A.; Liuba, K.; Jensen, C.T.; Bryder, D.; Yang, L.; Borge, O.J.; Thoren, L.A.; et al. Identification of flt3+ lympho-myeloid stem cells lacking erythro-megakaryocytic potential a revised road map for adult blood lineage commitment. *Cell* **2005**, *121*, 295–306. [CrossRef] [PubMed]

75. Karsunky, H.; Inlay, M.A.; Serwold, T.; Bhattacharya, D.; Weissman, I.L. Flk2+ common lymphoid progenitors possess equivalent differentiation potential for the b and t lineages. *Blood* **2008**, *111*, 5562–5570. [CrossRef] [PubMed]

76. Kondo, M.; Weissman, I.L.; Akashi, K. Identification of clonogenic common lymphoid progenitors in mouse bone marrow. *Cell* **1997**, *91*, 661–672. [CrossRef]

77. Elsasser, W.M. Outline of a theory of cellular heterogeneity. *Proc. Natl. Acad. Sci. USA* **1984**, *81*, 5126–5129. [CrossRef] [PubMed]

78. Chang, H.H.; Hemberg, M.; Barahona, M.; Ingber, D.E.; Huang, S. Transcriptome-wide noise controls lineage choice in mammalian progenitor cells. *Nature* **2008**, *453*, 544–547. [CrossRef] [PubMed]

79. Suslov, O.N.; Kukekov, V.G.; Ignatova, T.N.; Steindler, D.A. Neural stem cell heterogeneity demonstrated by molecular phenotyping of clonal neurospheres. *Proc. Natl. Acad. Sci. USA* **2002**, *99*, 14506–14511. [CrossRef] [PubMed]

80. Porritt, H.E.; Rumfelt, L.L.; Tabrizifard, S.; Schmitt, T.M.; Zuniga-Pflucker, J.C.; Petrie, H.T. Heterogeneity among dn1 prothymocytes reveals multiple progenitors with different capacities to generate t cell and non-t cell lineages. *Immunity* **2004**, *20*, 735–745. [CrossRef] [PubMed]

81. Enver, T.; Heyworth, C.M.; Dexter, T.M. Do stem cells play dice? *Blood* **1998**, *92*, 348–351. [PubMed]

82. Enver, T.; Jacobsen, S.E. Developmental biology: Instructions writ in blood. *Nature* **2009**, *461*, 183–184. [CrossRef] [PubMed]

83. Metcalf, D. Lineage commitment and maturation in hematopoietic cells: The case for extrinsic regulation. *Blood* **1998**, *92*, 345–347. [PubMed]

84. Endele, M.; Etzrodt, M.; Schroeder, T. Instruction of hematopoietic lineage choice by cytokine signaling. *Exp. Cell Res.* **2014**, *329*, 207–213. [CrossRef] [PubMed]

85. Metcalf, D. Hematopoietic cytokines. *Blood* **2008**, *111*, 485–491. [CrossRef] [PubMed]

86. Sarrazin, S.; Sieweke, M. Integration of cytokine and transcription factor signals in hematopoietic stem cell commitment. *Semin. Immunol.* **2011**, *23*, 326–334. [CrossRef] [PubMed]

87. Akashi, K.; Kondo, M.; von Freeden-Jeffry, U.; Murray, R.; Weissman, I.L. Bcl-2 rescues t lymphopoiesis in interleukin-7 receptor-deficient mice. *Cell* **1997**, *89*, 1033–1041. [CrossRef]

88. Kondo, M.; Akashi, K.; Domen, J.; Sugamura, K.; Weissman, I.L. Bcl-2 rescues t lymphopoiesis but not b or nk cell development, in common gamma chain-deficient mice. *Immunity* **1997**, *7*, 155–162. [CrossRef]

89. Maraskovsky, E.; O'Reilly, L.A.; Teepe, M.; Corcoran, L.M.; Peschon, J.J.; Strasser, A. Bcl-2 can rescue t lymphocyte development in interleukin-7 receptor-deficient mice but not in mutant rag-1-/- mice. *Cell* **1997**, *89*, 1011–1019. [CrossRef]

90. Malin, S.; McManus, S.; Cobaleda, C.; Novatchkova, M.; Delogu, A.; Bouillet, P.; Strasser, A.; Busslinger, M. Role of stat5 in controlling cell survival and immunoglobulin gene recombination during pro-b cell development. *Nat. Immunol.* **2010**, *11*, 171–179. [CrossRef] [PubMed]

91. von Muenchow, L.; Alberti-Servera, L.; Klein, F.; Capoferri, G.; Finke, D.; Ceredig, R.; Rolink, A.; Tsapogas, P. Permissive roles of cytokines interleukin-7 and flt3 ligand in mouse B-cell lineage commitment. *Proc. Natl. Acad. Sci. USA* **2016**, *113*, E8122–E8130. [CrossRef] [PubMed]

92. Stoffel, R.; Ziegler, S.; Ghilardi, N.; Ledermann, B.; de Sauvage, F.J.; Skoda, R.C. Permissive role of thrombopoietin and granulocyte colony-stimulating factor receptors in hematopoietic cell fate decisions in vivo. *Proc. Natl. Acad. Sci. USA* **1999**, *96*, 698–702. [CrossRef] [PubMed]

93. Semerad, C.L.; Poursine-Laurent, J.; Liu, F.; Link, D.C. A role for g-csf receptor signaling in the regulation of hematopoietic cell function but not lineage commitment or differentiation. *Immunity* **1999**, *11*, 153–161. [CrossRef]

94. Borzillo, G.V.; Ashmun, R.A.; Sherr, C.J. Macrophage lineage switching of murine early pre-b lymphoid cells expressing transduced fms genes. *Mol. Cell. Biol.* **1990**, *10*, 2703–2714. [CrossRef] [PubMed]

95. Pawlak, G.; Grasset, M.F.; Arnaud, S.; Blanchet, J.P.; Mouchiroud, G. Receptor for macrophage colony-stimulating factor transduces a signal decreasing erythroid potential in the multipotent hematopoietic eml cell line. *Exp. Hematol.* **2000**, *28*, 1164–1173. [CrossRef]

96. Iwasaki-Arai, J.; Iwasaki, H.; Miyamoto, T.; Watanabe, S.; Akashi, K. Enforced granulocyte/macrophage colony-stimulating factor signals do not support lymphopoiesis but instruct lymphoid to myelomonocytic lineage conversion. *J. Exp. Med.* **2003**, *197*, 1311–1322. [CrossRef] [PubMed]

97. King, A.G.; Kondo, M.; Scherer, D.C.; Weissman, I.L. Lineage infidelity in myeloid cells with tcr gene rearrangement: A latent developmental potential of prot cells revealed by ectopic cytokine receptor signaling. *Proc. Natl. Acad. Sci. USA* **2002**, *99*, 4508–4513. [CrossRef] [PubMed]

98. Kondo, M.; Scherer, D.C.; Miyamoto, T.; King, A.G.; Akashi, K.; Sugamura, K.; Weissman, I.L. Cell-fate conversion of lymphoid-committed progenitors by instructive actions of cytokines. *Nature* **2000**, *407*, 383–386. [CrossRef] [PubMed]

99. Onai, N.; Obata-Onai, A.; Tussiwand, R.; Lanzavecchia, A.; Manz, M.G. Activation of the flt3 signal transduction cascade rescues and enhances type i interferon-producing and dendritic cell development. *J. Exp. Med.* **2006**, *203*, 227–238. [CrossRef] [PubMed]

100. Rieger, M.A.; Hoppe, P.S.; Smejkal, B.M.; Eitelhuber, A.C.; Schroeder, T. Hematopoietic cytokines can instruct lineage choice. *Science* **2009**, *325*, 217–218. [CrossRef] [PubMed]

101. Brown, G.; Tsapogas, P.; Ceredig, R. The changing face of hematopoiesis: A spectrum of options is available to stem cells. *Immunol. Cell Biol.* **2018**. [CrossRef] [PubMed]

102. Singh, R.P.; Grinenko, T.; Ramasz, B.; Franke, K.; Lesche, M.; Dahl, A.; Gassmann, M.; Chavakis, T.; Henry, I.; Wielockx, B. Hematopoietic stem cells but not multipotent progenitors drive erythropoiesis during chronic erythroid stress in epo transgenic mice. *Stem Cell Rep.* **2018**, *10*, 1908–1919. [CrossRef] [PubMed]

103. Crisan, M.; Dzierzak, E. The many faces of hematopoietic stem cell heterogeneity. *Development* **2016**, *143*, 4571–4581. [CrossRef] [PubMed]

104. Waddington, C.H. *The Strategy of the Genes; A Discussion of Some Aspects of Theoretical Biology*; Allen & Unwin: London, UK, 1957.

105. Ferrell, J.E., Jr. Bistability, bifurcations and waddington's epigenetic landscape. *Curr. Biol.* **2012**, *22*, R458–R466. [CrossRef] [PubMed]

106. Sieweke, M.H. Waddington's valleys and captain cook's islands. *Cell Stem Cell* **2015**, *16*, 7–8. [CrossRef] [PubMed]

107. Brown, G.; Sanchez-Garcia, I. Is lineage decision-making restricted during tumoral reprograming of haematopoietic stem cells? *Oncotarget* **2015**, *6*, 43326–43341. [CrossRef] [PubMed]

108. Brown, G.; Sánchez-García, I. *Diversity, Versatility and Leukaemia*; Nova Biomedical: New York, NY, USA, 2016.

109. Gonzalez-Herrero, I.; Rodriguez-Hernandez, G.; Luengas-Martinez, A.; Isidro-Hernandez, M.; Jimenez, R.; Garcia-Cenador, M.B.; Garcia-Criado, F.J.; Sanchez-Garcia, I.; Vicente-Duenas, C. The making of leukemia. *Int. J. Mol. Sci.* **2018**, *19*, 1494. [CrossRef] [PubMed]

International Journal of
Molecular Sciences

MDPI

Review

The Making of Leukemia

Inés González-Herrero [1,2,†], Guillermo Rodríguez-Hernández [1,2,†], Andrea Luengas-Martínez [1,2],
Marta Isidro-Hernández [1,2], Rafael Jiménez [2,3], Maria Begoña García-Cenador [2,4],
Francisco Javier García-Criado [2,4,*,†], Isidro Sánchez-García [1,2,*,†] and
Carolina Vicente-Dueñas [2,*,†]

[1] Experimental Therapeutics and Translational Oncology Program, Instituto de Biología Molecular y Celular
     del Cáncer, CSIC/Universidad de Salamanca, Campus M. de Unamuno s/n, 37007 Salamanca, Spain;
     ighe@usal.es (I.G.-H.); guillermorh@usal.es (G.R.-H.); andrealuengas@usal.es (A.L.-M.);
     martaisidroh@gmail.com (M.I.-H.)
[2] Institute of Biomedical Research of Salamanca (IBSAL), 37007 Salamanca, Spain; rajim@usal.es (R.J.);
     mbgc@usal.es (M.B.G.-C.)
[3] Departamento de Fisiología y Farmacología, Universidad de Salamanca, Edificio Departamental, Campus M.
     de Unamuno s/n, 37007 Salamanca, Spain
[4] Departamento de Cirugía, Universidad de Salamanca, 37007 Salamanca, Spain
*   Correspondence: fjgc@usal.es (F.J.G.-C.); isg@usal.es (I.S.-G.); cvd@usal.es (C.V.-D.);
    Tel.: +34-923-294-813 (F.J.G.-C., I.S.-G. & C.V.-D.)
†   These authors contributed equally to this work.

Received: 23 April 2018; Accepted: 14 May 2018; Published: 17 May 2018

Abstract: Due to the clonal nature of human leukemia evolution, all leukemic cells carry the same
leukemia-initiating genetic lesions, independently of the intrinsic tumoral cellular heterogeneity.
However, the latest findings have shown that the mode of action of oncogenes is not homogeneous
throughout the developmental history of leukemia. Studies on different types of hematopoietic
tumors have shown that the contribution of oncogenes to leukemia is mainly mediated through the
epigenetic reprogramming of the leukemia-initiating target cell. This driving of cancer by a malignant
epigenetic stem cell rewiring is, however, not exclusive of the hematopoietic system, but rather
represents a common tumoral mechanism that is also at work in epithelial tumors. Tumoral epigenetic
reprogramming is therefore a new type of interaction between genes and their target cells, in which
the action of the oncogene modifies the epigenome to prime leukemia development by establishing a
new pathological tumoral cellular identity. This reprogramming may remain latent until it is triggered
by either endogenous or environmental stimuli. This new view on the making of leukemia not only
reveals a novel function for oncogenes, but also provides evidence for a previously unconsidered model
of leukemogenesis, in which the programming of the leukemia cellular identity has already occurred
at the level of stem cells, therefore showing a role for oncogenes in the timing of leukemia initiation.

Keywords: leukemia; oncogenes; reprogramming; stem cells; cancer therapy; leukemia stem cell;
mouse model

1. The Making of Leukemia: The Concept of Epigenetic Reprogramming

In spite of the enormous amount of data that we have gathered in the last four decades about the
biology of tumor cells, our capacity to control the development of the disease is still very limited [1,2].
We still do not know how to prevent the conversion of a precancerous cell into a tumor, mainly due to
the fact that the early events triggering the tumoral fate and the commitment to a new cancerous lineage
remain basically unknown [3]. This lack of knowledge is sadly illustrated by the cases of women
carrying *BRCA1* or *BRCA2* mutations, whose only chance to reduce their probability of developing
breast cancer is to undergo prophylactic tissue amputation [4]. It is clear that the most crucial point

in the biological history of a cancer is the transition from a normal target cell to a cancerous one. However, the developmental mechanisms controlling the establishing of a tumoral cellular identity, which are essential for the cancer to arise in the first place, have received little attention. The main focus of both basic and translational research has been the altered controls of cellular proliferation in malignant cells. This has been reflected in the therapeutic approaches used to treat the patients: in general, most of the anti-cancerous treatments are directed against the mechanisms behind the abnormal proliferation of cancerous cells. However, these therapies are unspecific, with many side effects caused by their high toxicity and, in the end, unable of eradicating the disease in a large percentage of the cases. Therefore, an unmet need in cancer research is to understand how to neutralize the mechanism(s) that convert a normal cell into a cancerous one in the first place.

Douglas Hanahan and Robert Weinberg condensed the complex biology of cancer cells into nine hallmarks, "nine essential alterations in cell physiology that collectively dictate malignant growth" [5]. Cancer cells are the basis of the tumoral disease: they give rise to the tumors and drive the progression of the disease, carrying the oncogenic and tumor suppressor mutations that define cancer as a genetic disease [5]. In spite of their importance, we still do not fully understand the mechanisms that lead to their appearance, or at least we do not know enough so as to have a significant impact on cancer mortality [6]. As a consequence, our advances in cancer treatment are incremental and mainly empirical, with successive clinical trials leading to slightly better therapeutic options that, although they might provide some benefit, do not bring an end to the disease [7].

Therefore, an in-depth understanding of cancer requires a more detailed knowledge of the mechanisms triggering malignant growth, and it is essential if we want to identify the molecular culprits of cancer maintenance [3]. Despite this, all the aspects related to the deregulation of the normal developmental mechanisms that take place in tumorigenesis have received little attention when trying to define the main features of cancer. However, this is a key aspect since, if cellular fate could not be changed, cancer would be impossible, since only normal, non-pathological cell types would exist. Therefore, the mechanisms establishing and regulating cellular identity play an essential role in allowing the appearance of aberrant cancerous cell types; hopefully, the understanding of these mechanisms might be the key to the total elimination of cancer cells in the patients. In this review, we discuss the importance that oncogenes have in establishing the identity of the tumor cells, and how reprogrammed cells participate in the disease evolution. A deeper knowledge of this, so far largely neglected, mode of action of the oncogenes should help us to develop new ways to attack cancer.

2. Oncogene Addition versus Reprogramming Leukemia Predisposition

Many years of research have shown that oncogene expression is necessary not only at the earliest stages of cancer development, but also for the posterior maintenance of the disease. Therefore, since the discovery that human cancers carry mutated oncogenes, these have been regarded as primary potential therapeutic targets. Indeed, in mouse models in which the expression of the oncogenes is driven by tissue-specific promoters, tumors arise frequently, but they regress when the oncogenic stimulus is switched off [8–10], suggesting that cancer cells are oncogene-addicted [11]. These findings seem to point to a homogeneous mode of action of oncogenes throughout tumor life, since the removal of the cancer-inducing oncogene will lead to tumor regression in these models (Figure 1). This model is fitting with the fact that, in human cancers, due to the clonal nature of the disease, all cancer cells carry the same initiating oncogenic lesions.

However, human cancers also present a very high degree of cellular heterogeneity [12], an indication that, in the oncogenic process, the nature and identity of the target cells suffering the action of the oncogene can be of great importance, especially since therapies based on the aforementioned current working model of cancer are incapable of eradicating cancer in humans [7]. On the contrary, the observations suggesting that the mode of action of oncogenes is not homogeneous throughout all the different cancer cell types would explain why anti-oncogene targeted therapies present different efficiencies against the different cellular stages of cancer evolution [7].

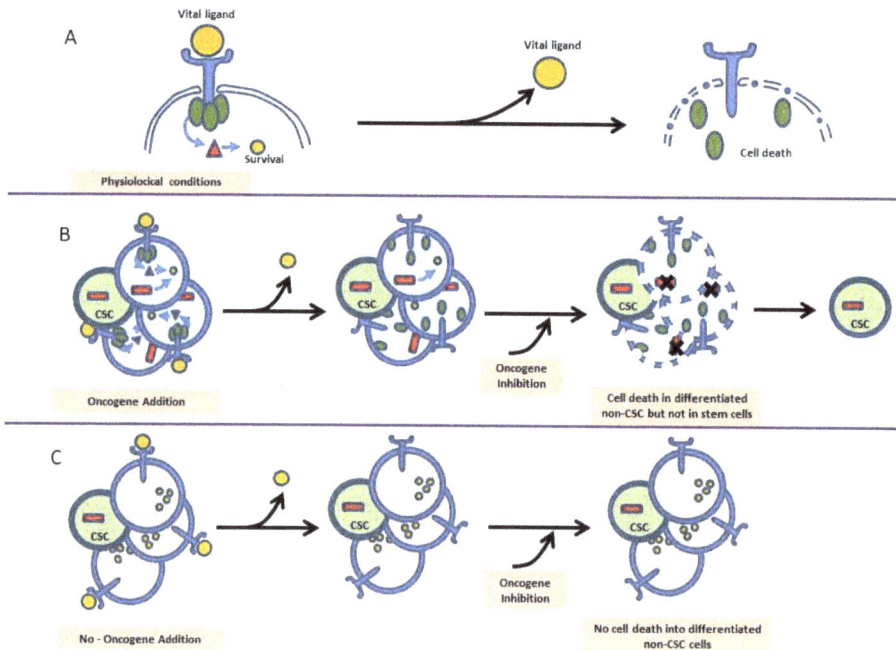

Figure 1. Oncogene addition versus reprogramming cancer predisposition. (**A**) In physiological conditions, the survival of a cell is determined by a specific survival pathway that needs to be activated by a vital ligand recognized by a specific receptor. If this vital ligand is removed from the niche/environment, the cell will die. (**B**) The oncogene addition hypothesis implies that a tumoral cell is able to grow independently of the survival signals because the oncogene activates survival pathways that maintain the tumor cell alive without vital ligands in the niche. The oncogene addition model implies that the inhibition of the oncogene in the tumor cells leads to the tumor cell death, as the survival pathways are no longer active. But under this scenario the evidence from the clinic indicates that only non-CSCs die. (**C**) The tumor reprogramming model of cancer initiation suggests that tumor stem cells are not oncogene addicted because there is a different function for oncogenes within CSCs. In this model, the target cancer cell of origin is not addicted neither to the oncogenes nor to the environmental signals.

This concept fits well with a model of cancer in which the tumor is generated and maintained in a hierarchical manner similar to that of the normal stem cell-driven tissues, like the hematopoietic system (Figure 2). In such a tissue, genetic programming of the stem cells is all what is required to give rise to all the differentiated cells forming the tissue, and the genetic information responsible for the stem cell programming does not need to be anymore present within those mature cells that form the tissue. When extrapolated to cancer formation, this concept would imply a potentially different role for the oncogenes at the level of the cancer stem cells (CSCs) [3,13]. Indeed, if cancer is generated by a malignant stem cell reprogramming process, the oncogenes initiating tumor formation might not be required for tumor progression [14,15].

This model also explains how, in the evolution of several human tumors, a pre-cancerous lesion can be stably maintained as the only proto-oncogenic alteration in an abnormal cell population that will only give rise to a full-blown tumor in response to secondary hits [16,17]. This initiating lesion is the driving force in the reprogramming process, essential for the acquisition of a tumoral phenotype. However, once its function in oncogenic reprogramming has been performed, this initiating hit would become just a passenger mutation within the CSC, having either no function or even performing a different one, unrelated to the initial reprogramming one, for example in tumor proliferation. This reprogramming model would explain why targeted therapies focused on the oncogenes can fail in eliminating all cancer cells, in spite of their initial efficacy against the cells composing the main tumor mass; a good example of this apparent paradox is imatinib, that fails to kill BCR-ABL$^+$ CSCs because somehow it cannot block the reprogramming capacity of the fusion oncogene in the stem cell compartment [15,18,19].

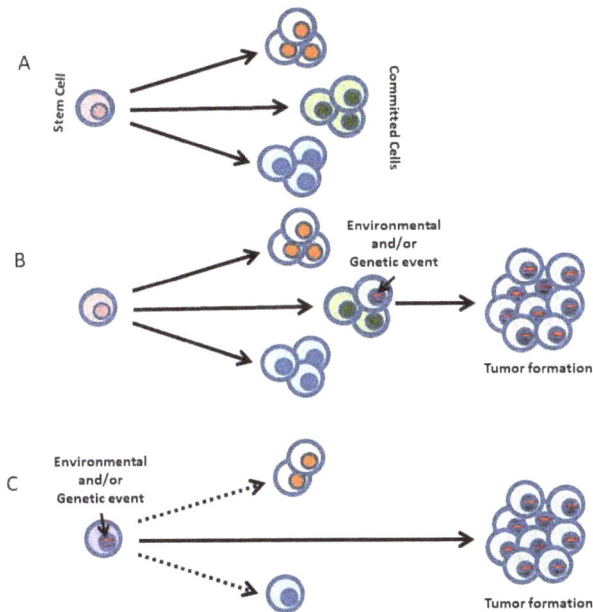

Figure 2. Proposed model for the role of human cancer gene defects in the making of a leukemia. (**A**) In the development of every normal tissue, a small pool of multipotent stem cells maintains multiple cell lineages. (**B**) Traditionally, human cancer genetic defects have been thought to act on cells already committed to a differentiation program. Hence, the cancer phenotype closely resembles that of the initial differentiated target cell. (**C**) Normal uncommitted stem and progenitor cells are the targets for transformation in some human cancers, and the human cancer-gene defects into these cells are the instigators of lineage choice decisions, which are therefore dictated by the oncogene and not by the cell-of-origin phenotype. Consistent with this is the finding that forced expression of these genes in stem cells can select or impose a specific cancer-lineage outcome. This explains why specific gene defects are usually found only in one type of cancer (see text for details). Dotted lines depict cell lineage choices that were not imposed by the oncogene.

In tumoral cells from human patients it is impossible to separately study the potential different roles of the oncogenes at the different stages of tumor development, due to the advanced stage of the tumors at diagnosis and the accumulation and superposition of numerous driver and passenger mutations. Indeed, in order to reveal the existence of a lack of homogeneity in the action of oncogenes throughout the biological history of the tumor, we need a system that allows us to isolate the function that the oncogene is playing in the first steps of cancer, in the cancer cell-of-origin. Ideally, one would need a system in which the oncogene expression was restricted to the stem/progenitor compartment in order to demonstrate that the posterior expression of the oncogene (as happens in human tumors) is in fact not necessary for tumor progression once the tumoral reprogramming has already taken place, and is therefore dispensable for cancer progression or maintenance (Figure 2).

This conceptual framework has recently given rise to experimental settings in which different oncogenic lesions, each linked to a specific type of hematopoietic cancers, have been targeted to the hematopoietic stem/progenitor cellular compartment of genetically engineered mice by using the locus control region of the *Sca1* ("Stem cell antigen-1") gene. It is a mouse glycosyl phosphatidylinositol-anchored cell surface protein (GPI-AP) of the *LY6* gene family. It is the common biological marker used to identify hematopoietic stem cell (HSC) along with other markers and plays a role in hematopoietic progenitor/stem cell lineage fate. In this setting, it has been shown that the different lesions can epigenetically re-program the targeted stem cells and create a differentiation state from which tumor cells with different properties emerge heterogeneously [15,20–24] (Figure 3). Overall, this new view on oncogenesis not only reveals a novel function for oncogenes in cancer, but also provides evidence for a previously unconsidered model of tumorigenesis, in which the programming of the cancerous cellular identity has already occurred at the level of stem/uncommitted cells, therefore showing a role for oncogenes in the timing of cancer initiation.

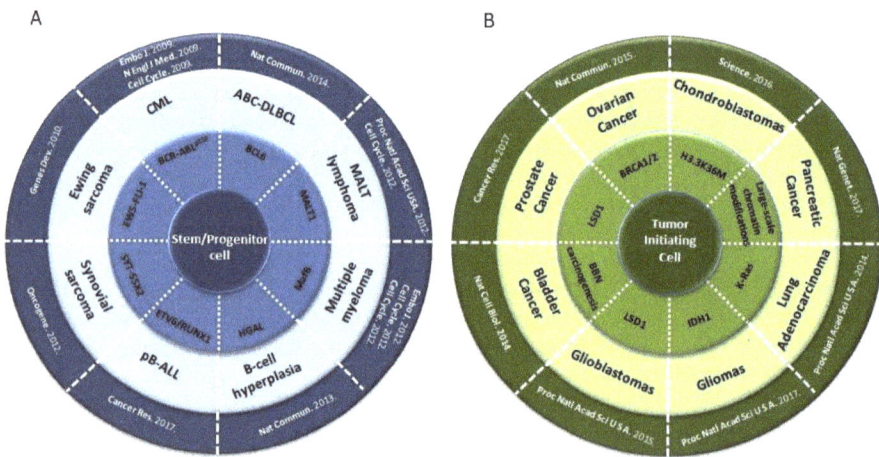

Figure 3. Driving of cancer by a malignant epigenetic stem cell rewiring. Different types of mesenchymal (**A**) and epithelial (**B**) malignancies have shown that the contribution of oncogenes to cancer development is mainly mediated through the epigenetic reprogramming of the cancer-initiating target cell. As illustrated, specific genotype alterations associated to human cancer (medium circle) give rise to specific phenotypes (outer circle) when targeted to the stem cell/progenitor compartment (see text for details). Chronic Myeloid Leukemia (CML), Activated B-Cell Diffuse Large B-Cell Lymphoma (ABC-DLBCL), precursor B-cell Acute Lymphoblastic Leukemia (pB-ALL), N-butyl-N-4-hydroxybutyl nitrosamine (BBN).

3. Restriction of Lineage Options during the Making of Leukemia

A conceptually clarifying example of the power of the aforementioned experimental setting to recapitulate the characteristics of human cancers is chronic myelogenous leukemia (CML), a widely accepted stem cell disorder characterized by the presence of the chimeric *BCR-ABLp210* oncogene. When the expression of BCR-ABL is restricted to the Sca1+ cells in mice, the animals develop CML [15]. This model is designed so that the oncogene expression is switched off in the differentiated cells that form the main mass of the tumor, although leukemia initiation has taken place within the stem cell/progenitor population. The fact that CML arises in mice under these circumstances indicates that the absence of BCR-ABL expression is not required for the generation of differentiated tumor cells (Figure 2). These results connect tumorigenesis with the reprogramming of early progenitors and strongly support the existence of a reprogramming-like mechanism in cancer development.

CML is a paradigmatic stem-cell-driven cancer in humans, but we reasoned that, assuming the tumoral reprogramming theory as we have previously mentioned, a similar experimental approach as the one used to model *BCR-ABLp210*⁺ CML could also be used to reproduce in the mouse the genotype-phenotype correlation (specific oncogene/specific tumor) found in other human cancers. A challenging system to test this hypothesis would be a tumor whose main cell type is a mature differentiated cell, like in the case of multiple myeloma (MM) or mature B-cell lymphoma. In fact, it has been shown that both MM (induced by the *MafB* oncogene) and B-cell lymphoma (induced either by the *MALT1* or the *BCL6* oncogenes) phenotypes and biology can be accurately mimicked in mice with the same *Sca1*-mediated stem cell targeting system described before [20–23] (Figure 3). These results implicated for the first time the stem cells in the pathogenesis of MM and B-cell lymphoma. Also, the fact that both hematopoietic tumors can be generated in mice by limiting oncogene expression to hematopoietic stem/progenitor cells (HS/PCs) implies that eliminating oncogene function beyond the stem cell stage does not interfere with the generation of later tumoral developmental cell types, and suggest that the oncogene is programming in the stem cells an epigenetic program that in some way persists during differentiation and will finally lead to a mature tumoral phenotype of MM or B-cell lymphoma [21–23,25,26]. Therefore, we postulate that cancer-initiating oncogenes epigenetically modify target genes that remain in this "poised" state in the mature tumor even when the oncogene is not present anymore (Figure 2).

4. Epigenetic Reprogramming in Non-Hematopoietic Tumors

Other examples of tumoral stem cell reprogramming, in which the induction of a new tumoral fate by the oncogene takes place at the stem cell level (as opposed to reprogramming to pluripotency, which is initiated from a differentiated cell) have been described for other types of non-hematopoietic tumors. For example, the *EWS-FLI-1* fusion gene, associated with most Ewing sarcoma tumors, triggers the expression of the embryonic stem cell genes *OCT4*, *SOX2*, and *NANOG* when present in human pediatric mesenchymal stem cells but not in adult ones, and it reprograms them to give rise to Ewing sarcoma cancer stem cells [27]. Similarly, the synovial sarcoma-associated oncogene SYT-SSX2 can reprogram mesenchymal stem cells by promoting their differentiation towards a pro-neural lineage, in what most likely constitutes the primary tumorigenic event in this type of cancer [28]. A similar scenario has recently been described in the genesis of chondroblastomas [29].

This driving of cancer by a malignant epigenetic stem cell rewiring is, however, not exclusive to mesenchymal-derived cancers, but rather represents a common tumoral mechanism that is also at work in epithelial tumors like lung carcinomas [30], bladder cancer [31], skin carcinomas [32], ovarian carcinomas [33], pancreatic carcinomas [34], brain tumors [35,36], and prostate carcinomas [37] (Figure 3).

These results prove that, when oncogenic proteins are expressed in stem or progenitor cells, they can have a highly selective impact in differentiation. This, in turn, helps explaining the strikingly consistent associations between each given chromosomal translocation, its resulting chimeric oncogene and the final phenotype of the cancer it triggers. Altogether, the evidence supports a new vision of cancer mainly as a disease of cellular differentiation, much more than just a proliferative disorder, and asks for a reconsideration of the function of oncogenes. We should also insist on the fact that this 'hit-and-run' reprogramming model for oncogene activity is not something happening only in pathological conditions. Indeed, during normal hematopoietic development, for example, molecular cues such as IL7 and erythropoietin are required to trigger specific differentiation programs but are not required once the programs have been established.

This new conceptual framework, supported by the experimental findings and by the frequent therapeutic failures in cancer human patients, also has important implications in the clinical management of cancer. Certainly, if cancers develop through a reprogramming mechanism, then the oncogenes, although necessary to initiate the tumor, might be dispensable for posterior tumor survival and/or progression. Oncogenes would then have a driving role in the reprogramming process, but be only passenger mutations afterwards, or have a secondary, unrelated role in more evolved tumor cell clones. For example, in human CLL, the susceptibility to generate malignant B cells is already present at the hematopoietic stem cell (HSC) stage, long before the cells become B cells [38]; consequently, patient-derived HSCs show an abnormal expression of lymphoid-related genes, reflecting their cell-autonomous aberrant priming into the B-cell lineage. These findings, therefore, have important implications for the therapeutic targeting of tumoral cells.

For example, in the case of CML, with one of the most commercially successful oncogene-targeted therapies so far, the data from *Sca1-BCR-ABLp210* animal models showed that the survival of CML stem cells was independent on the kinase activity of BCR-ABL, therefore indicating that curative approaches in CML should most probably focus on kinase-independent mechanisms of resistance [15]. These results on the failure of imatinib to eradicate CML in *Sca1-BCR-ABLp210* mice have been later confirmed in human patients [39–43], and this is a clear example of how a good preclinical model can anticipate the human CSC-therapeutic response. These results on the role of *BCR-ABLp210* in CML development show that leukemia stem cells might not be oncogene-addicted (Figure 1), and are most likely relevant to many other cancers (multiple myeloma, MALT lymphoma, CLL, etc.). Furthermore, these findings challenge the current accepted/working model of the role of oncogenes and support the hypothesis that mature hematological malignancies may be initiated by an inappropriate lineage-decision making process at the HSC level.

5. Hematopoiesis and Leukemia Are Both Lineage Decision-Making Processes

The most important functional characteristics of HSCs are their capacity for self-renewal and their multilineage differentiation potential. Traditionally, the generation of differentiated cells from HSCs was thought to occur through a series of dichotomic branching steps diverting into mutually exclusive stable progenitor states. However, recent work has shown that hematopoiesis occurs through a mechanism of continuous lineage priming [44] and therefore the architecture of the system is much less compartmentalized than previously considered, and also more versatile in terms of lineage plasticity, since developing progenitors can use different unusual pathways and/or have hidden potentials, so that it could very well be that "defined" progenitor populations are in fact mixtures of cells with several differentiation capabilities. This vision also changes our idea of cellular commitment, if developing hematopoietic cells are in fact being gradually biased towards a certain fate without sudden black-or-white developmental steps. A very important element of this vision is that these progressive developmental biases are physiological rules that can be bent or even broken by both intrinsic and extracellular factors in pathological conditions, a fact that has clear implications to the understanding of the origins and progression of malignant transformation in a setting in which, as we have stated, cancer would mainly be a disease of cellular differentiation (Figure 2).

This new point of view deepens our understanding of the biology of leukemia and its origins. In several types of leukemia it has been shown that the pre-leukemic stem cell (pre-LSC) possesses multilineage potential; this is the case in CML, where pre-leukemic *BCR-ABLp210*$^+$ stem cells can give rise to all different blood cell types (Figure 4). This is also the case for CMLs associated with a mutant *RAS* allele, in which this mutant *RAS* can be found in all mature lineages [45]. In myelodysplastic syndromes, a multipotent malignant stem cell is behind the development of refractory anemia (RA), RA with ringed sideroblasts or RA with excess blasts [46]. In all these cases, the existence of a pre-LSC with multilineage differentiation potential suggests that initiating mutations arise in a normal HSC and that afterwards, through the acquisition of additional mutations triggered by secondary events, the initiated clone will evolve to produce a sub-clone of lineage-restricted malignant blasts (Figure 4). This two-hit model of leukemogenesis relies on the stepwise acquisition and collaboration between two main groups of mutations: (i) those affecting genes of transcriptional or epigenetic regulators that can modify or restrict lineage options (e.g., generation of chimeric oncogenes such as *RUNX1-RUNX1T1* or *PML-RARA, BCR-ABLp190, ETV6-RUNX1* or mutations in *CEBPA, PAX5* or *NPM1*) and, (ii) those activating signal-transduction pathways that confer survival or proliferative advantages (e.g., mutations in *FLT3*, *RAS* or *KIT*). Therefore the pre-leukemic oncogenic lesion is stably maintained as a single alteration in an abnormal cell population, but will only progress to an open leukemia when secondary hits occur [16,17]. Therefore, although the cells that suffer the initiating leukemic hit possess multi-lineage potential, LSCs are reprogrammed by this oncogenic hit and their lineage decision-making becomes restricted or biased.

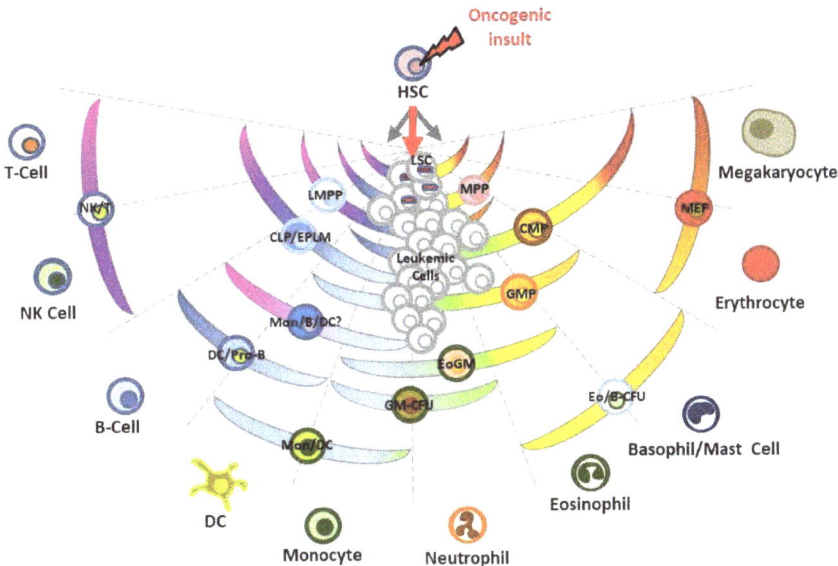

Figure 4. Schematic representation of the emergence of LSCs in the making of a leukemia. A mutation occurs in HSCs leading to the emergence of aberrant pre-leukemic HSCs. These aberrant pre-leukemic HSCs self-renew and expand within the HSC compartment. Pre-leukemic HSCs give rise to a high number of lineage-committed progenitors harboring this identical mutation. This leads to an increased chance of acquiring the additional oncogenic/environmental events, which finally transform the aberrant progenitor cells from pre-leukemic HSCs into the leukemic stem cells (LSCs). Loss of differentiation potentials is essential for the emergence of LSCs. LSCs are reprogrammed by an oncogenic insult to an invariant cell lineage. Hematopoietic stem cell (HSC), Multipotent progenitor cell (MPP), Lymphoid-primed MPP compartment (LMPP), Common myeloid progenitor (CMP), Common lymphoid progenitor (CLP), Early progenitor with lymphoid and myeloid potential (EPLM), Granulocyte-macrophage progenitor (GMP), Monocytes (Mon), Eosinophil-granulocyte-macrophage (EoGM), Granulocyte-macrophage progenitor-colony forming unit (GM-CFU), Dendritic cell (DC), Megakaryocyte-erythroid progenitor (MEP), Natural Killer cell (NK), Eosinophil/Basophil-colony forming unit (Eo/B-CFU).

It seems therefore that tumoral reprogramming and aberrant lineage-programming are crucial characteristics at the root of cancers including leukemia. The neural stem cells from malignant glioblastoma can be reprogrammed to induced-pluripotent stem cells (iPSC). These iPSC can then be differentiated into mesodermal lineages, and along this developmental pathway they lose their malignant nature, but they maintain it when they are differentiated into neural cells [47]. Similarly, primary human Philadelphia chromosome-positive B cell acute lymphoblastic leukemia (B-ALL) cells can be reprogrammed into non-leukemic macrophages, overriding the malignant differentiation block into the B cell lineage [48]. These findings underscore the fact that the cancerous condition is in some way linked to the fact that the cells have been programmed to adopt a specific given lineage. Then the key questions are: Which is the normal developmental stage that is being programmed? and at what stage does this occur within the leukemic development itself?

The leukemic conversion is only possible if the normal cell that gives rise to the leukemia, the leukemia cell-of-origin (LOC), has the necessary developmental plasticity to tolerate the reprogramming and react to it by changing its fate. On the other hand, the oncogenic event(s) triggering LOC malignant conversion cancer must also have a reprogramming capacity to be able to promote such a change in cellular identity [49].

It is generally accepted that tumoral progression is a multi-hit process and, also from a tumoral reprogramming perspective, the different aspects of normal cellular biology must be progressively altered to finally give rise to a full-blown tumor [5]. Under normal conditions, HS/PCs are slowly moving towards lineage biases, diversifying and differentiating towards their final cellular identities. The requirement of multiple hits for full tumor development is in relationship to the fact that the changes required to revert or deviate cells from their normal non-pathogenic fate are inherently disfavored developmentally, and biological barriers are in place to ensure that cells do not easily change their identity in order to minimize the risk of malignant transformation.

This biological reluctance of the cells to being reprogrammed by an oncogene to a tumor phenotype is illustrated by recent studies on stem-cell-based animal models of human cancer. The loss of the p53 tumor suppressor is a frequent occurrence in malignancy, and it has a clear function in facilitating pathological reprogramming to a malignant phenotype. In a stem-cell-based transgenic model of multiple myeloma, the loss of p53 accelerates the appearance of the disease by allowing the *MafB* oncogene to drive a much more efficient malignant transformation [22,25]. Something similar happens in the case of mucose-associated lymphoid tissue (MALT) lymphoma driven by the *MALT1* oncogene [21]. In a stem-cell-based model of CML [50], the restoration of p53 activity in already established cancers slowed the progression of the disease and prolonged the survival of diseased animals by causing the apoptotic death of leukemic progenitors, one more demonstrating the importance of reprogramming in lineage decision-making towards a tumoral fate.

6. The Importance of Environmental Signals in the Making of Leukemia

We have seen that precancerous lesions can exist as stable single alterations maintained in an abnormal, but not cancerous, cell population that will only progress to full-blown cancer as a result of secondary hits [16,17]. One of the best examples of this is childhood B-ALL, in which the first oncogenic hit, through a stem cell tumoral epigenetic reprogramming mechanism, gives rise to a preleukemic clone that remains harmless until its carrier is exposed to common infections [24,51,52]. This infection exposure would never cause leukemic development in healthy individuals (i.e., persons not carrying a preleukemic clone). Also, human epidemiological studies show a positive association between body weight at birth and the risk of developing childhood leukemia. This implies that, although some epigenetic reprogramming can be observed immediately after exposure to exogenous or endogenous agents, both aberrant epigenetic programming and altered disease susceptibility may manifest only later in life, long after the exposure took place. However, there are not known differences in epigenetic reprogramming between childhood and adults with leukemia. Together, these data lead us to propose that leukemia as a result of epigenetic reprogramming is a type of gene–environment interaction that can cooperate with a genetic predisposition, not by inducing mutations, but by reprogramming the epigenome to modulate gene expression in order to promote leukemia development. By dissecting how epigenetic reprogramming increases leukemia risk, we may not only be able to better identify who has an increased risk of developing leukemia from early life environmental exposures, but may also be able of developing interventions that can reverse the epigenetic effects of the tumor epigenetic reprogramming to decrease leukemia risk associated with this type of gene–environment interaction.

7. Therapeutic Intervention in Leukemia and the Prospect of Modifying the Making of Leukemia

The understanding of cancer as a LSC-dependent aberrant tissue has deep implications for cancer treatment. Obviously, under the LSC conceptual framework, the LSCs should the primary targets of anti-cancerous therapy. However, since LSCs share most of their basic biological properties with normal, non-pathologic stem cells, therapies directed against LSC pathways might also unintentionally eliminate normal resident stem cells.

We have seen how the main contribution of oncogenes to tumor development is not their proliferation-inducing capability, but rather their capacity for reprogramming the LSC epigenome. This capacity of making leukemia in such a way that the maintenance of oncogene expression is not required for the posterior generation of differentiated tumoral cells seems to be a common mechanism of determination of cancerous identity and, as such, it should change our understanding of how the "hallmarks of cancer" are acquired during tumor development. In this sense, it has recently been shown that epigenetic reprogramming can be the driving force behind intra-tumoral heterogeneity [53], and can also be the mechanism used by tumors to evade CD19 CAR immune therapy [54,55] (Figure 5).

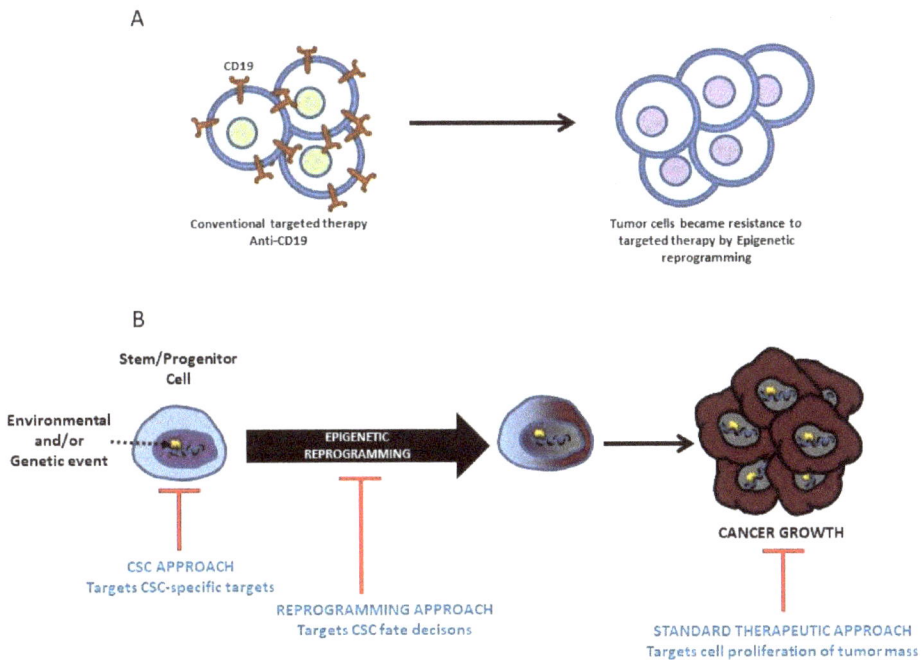

Figure 5. Epigenetic reprogramming and cancer therapy. (**A**) Epigenetic reprogramming can be the mechanism used by tumors to evade CD19 CAR immune therapy. (**B**) Epigenetic reprogramming can be exploited in therapy to kill leukemia/cancer stem cells. Recent findings indicate that rewiring the epigenetic programming of tumor cells is a viable prospect (see text for details).

With the new animal models generated within this stem cell reprogramming paradigm, we can now study how different cancerous stages develop from the very beginning, and we could unlock the potential to provide great advances in human cancer medicine. Since assessing the effects of therapies on the rare LSCs that are the responsible for relapse is almost impossible in the patients, the development of these treatments will have to heavily rely on the use of accurate preclinical models and preclinical assays.

The key factor in this new view of cancer specification is the setting of a new regulatory circuitry by epigenetic reprogramming. This opens a new door for therapeutic opportunities since in these last tears we are learning more and more about how to genetically or pharmacologically manipulate the epigenetic status of cells. In fact, epigenetic therapeutic protocols have already been incorporated in some cases to standard chemotherapy regimens as a potential improvement in the treatment of, for example, relapsed pediatric acute lymphoblastic leukemia [56].

Also, in more experimental settings, cancer cells have been reprogrammed to non-tumoral fates, losing their malignancy. For example, it is possible to produce even mouse embryos from brain-tumor-derived cells [57] and to reprogram embryonal carcinomas [58] or melanoma cells by using nuclear transplantation [59]. Similarly, B-ALL cells have been reprogrammed to an alternative lineage cell fate without a malignant phenotype [56]. Also, it has recently been shown that epigenetic reprogramming can be exploited in therapy to kill leukemia stem cells [60] and can also be used to treat pediatric brain cancer [61,62]. These findings support the underlying theory and indicate that rewiring the epigenetic programming of tumor cells is a viable prospect. Also, it is not unreasonable to expect that LSCs from different cancer types will share many similarities, so that similar LSC-based therapeutic approaches could be successfully employed against different cancer types (Figure 5). In any case, like for any other type of therapy, a detailed understanding of the epigenetic rewiring is a prerequisite for any potential intervention.

8. Future Opportunities and Challenges

The new perspective of leukemia that is arising from the most recent results from advanced animal models is leading us to a better understanding of the biology of the disease and, at the same time, is forcing us to question long-standing beliefs about the role of oncogenes in leukemia generation.

We have seen that the exposure of plastic stem cells to the epigenetic reprogramming capacity of some oncogenes works as a new type of gene–target cell interaction in which oncogene exposure poises the epigenome to induce leukemia development [20–24]. In this model of action, oncogene-activating mutations would have a driving role in the reprogramming process at the leukemia cell-of-origin, but may become passenger alterations (or have a different, secondary role) at later stages. To increase the complexity of the problem, the phenotypic consequences of the epigenetic reprogramming can remain silent until triggered by later exposures (genetic and/or environmental) [24,51,63–65]. Of great importance is the fact that the setting of the epigenetic circuits that lead to tumor cell development is unidirectional. This implies that even brief exposure to an environmental agent can disrupt the normal epigenetic developmental programs and alter the epigenome for life [66]. Now we have the ability to model tumor stem cell generation in vivo for different types of cancer, with their respective inducing oncogenes. This opens up new possibilities for studying how the different cancer stages develop from the start. If we can understand how the oncogene–target cell interaction is regulated, then we might learn how to manipulate tumoral cellular identities and stages experimentally, a knowledge that could lead to tremendous advances in human cancer medicine.

Looking into the future, we are faced with the paradox that we still do not understand how the balance is regulated between cell intrinsic and environmental agents in developmental processes. Cellular pluripotency is a pre-requisite for the versatility of the organisms, giving them the capacity of evolving different types of specialized cells to face different challenges. However, pluripotency is a force that needs to be tamed in order to give rise to an adult organism composed of highly differentiated (and not therefore pluripotent) cells. In the last decade, we have learnt that the architecture of adult tissues is much more versatile and plastic than previously thought, and this is particularly important, for example, when responding to the demand triggered by infectious agents as to the specialized cells required to fight them.

New findings open new questions. For example, is the decision to initiate leukemia made at one single time point during the tumoral differentiation process, or is composed by a series of consecutive decisions required to switch to a leukemia cell fate? What is the precise nature of the epigenetic

Int. J. Mol. Sci. **2018**, *19*, 1494

pathway triggered by the leukemia-initiating gene defect(s)? Last but not least, these findings on the mechanisms of cellular commitment to a tumoral fate are relevant not only to the understanding of the stem cell properties of leukemia and the development of new therapies, but also to regenerative medicine, since in this context it will be essential to have full control over the potential malignancy of reprogrammed cells.

Acknowledgments: We are indebted to all members of our groups for useful discussions and for their critical reading of the manuscript. Research in C.V.-D. group is partially supported by FEDER, "Miguel Servet" Grant (CP14/00082-AES 2013–2016) from the Instituto de Salud Carlos III (Ministerio de Economía y Competitividad), "Fondo de Investigaciones Sanitarias/Instituto de Salud Carlos III" (PI17/00167) and by the Lady Tata International Award for Research in Leukaemia 2016–2017. Research in I.S.G. group is partially supported by FEDER and by MINECO (SAF2012-32810, SAF2015-64420-R and Red de Excelencia Consolider OncoBIO SAF2014-57791-REDC), and by Junta de Castilla y León (UIC-017, and CSI001U16). I.S.G. lab is a member of the EuroSyStem and the DECIDE Network funded by the European Union under the FP7 program. I.S.G. has been supported by the German Carreras Foundation (DJCLS R13/26).

Conflicts of Interest: The authors declare no conflict of interest.

References

1. Chabner, B.A.; Roberts, T.G., Jr. Timeline: Chemotherapy and the war on cancer. *Nat. Rev. Cancer* **2005**, *5*, 65–72. [CrossRef] [PubMed]
2. Etzioni, R.; Urban, N.; Ramsey, S.; McIntosh, M.; Schwartz, S.; Reid, B.; Radich, J.; Anderson, G.; Hartwell, L. The case for early detection. *Nat. Rev. Cancer* **2003**, *3*, 243–252. [CrossRef] [PubMed]
3. Vicente-Duenas, C.; Romero-Camarero, I.; Cobaleda, C.; Sanchez-Garcia, I. Function of oncogenes in cancer development: A changing paradigm. *EMBO J.* **2013**, *32*, 1502–1513. [CrossRef] [PubMed]
4. Rebbeck, T.R.; Friebel, T.; Lynch, H.T.; Neuhausen, S.L.; van't Veer, L.; Garber, J.E.; Evans, G.R.; Narod, S.A.; Isaacs, C.; Matloff, E.; et al. Bilateral prophylactic mastectomy reduces breast cancer risk in brca1 and brca2 mutation carriers: The prose study group. *J. Clin. Oncol.* **2004**, *22*, 1055–1062. [CrossRef] [PubMed]
5. Hanahan, D.; Weinberg, R.A. Hallmarks of cancer: The next generation. *Cell* **2011**, *144*, 646–674. [CrossRef] [PubMed]
6. Jemal, A.; Siegel, R.; Ward, E.; Hao, Y.; Xu, J.; Thun, M.J. Cancer statistics, 2009. *CA Cancer J. Clin.* **2009**, *59*, 225–249. [CrossRef] [PubMed]
7. Booth, C.M.; Del Paggio, J.C. Approvals in 2016: Questioning the clinical benefit of anticancer therapies. *Nat. Rev. Clin. Oncol.* **2017**, *14*, 135–136. [CrossRef] [PubMed]
8. Boxer, R.B.; Jang, J.W.; Sintasath, L.; Chodosh, L.A. Lack of sustained regression of c-myc-induced mammary adenocarcinomas following brief or prolonged myc inactivation. *Cancer Cell* **2004**, *6*, 577–586. [CrossRef] [PubMed]
9. Chin, L.; Tam, A.; Pomerantz, J.; Wong, M.; Holash, J.; Bardeesy, N.; Shen, Q.; O'Hagan, R.; Pantginis, J.; Zhou, H.; et al. Essential role for oncogenic ras in tumour maintenance. *Nature* **1999**, *400*, 468–472. [CrossRef] [PubMed]
10. Huettner, C.S.; Zhang, P.; Van Etten, R.A.; Tenen, D.G. Reversibility of acute b-cell leukaemia induced by bcr-abl1. *Nat. Genet.* **2000**, *24*, 57–60. [CrossRef] [PubMed]
11. Weinstein, I.B. Cancer. Addiction to oncogenes—The achilles heel of cancer. *Science* **2002**, *297*, 63–64. [CrossRef] [PubMed]
12. Hamburger, A.; Salmon, S.E. Primary bioassay of human myeloma stem cells. *J. Clin. Investig.* **1977**, *60*, 846–854. [CrossRef] [PubMed]
13. Vicente-Duenas, C.; Hauer, J.; Ruiz-Roca, L.; Ingenhag, D.; Rodriguez-Meira, A.; Auer, F.; Borkhardt, A.; Sanchez-Garcia, I. Tumoral stem cell reprogramming as a driver of cancer: Theory, biological models, implications in cancer therapy. *Semin. Cancer Biol.* **2015**, *32*, 3–9. [CrossRef] [PubMed]
14. Krizhanovsky, V.; Lowe, S.W. Stem cells: The promises and perils of p53. *Nature* **2009**, *460*, 1085–1086. [CrossRef] [PubMed]
15. Perez-Caro, M.; Cobaleda, C.; Gonzalez-Herrero, I.; Vicente-Duenas, C.; Bermejo-Rodriguez, C.; Sanchez-Beato, M.; Orfao, A.; Pintado, B.; Flores, T.; Sanchez-Martin, M.; et al. Cancer induction by restriction of oncogene expression to the stem cell compartment. *EMBO J.* **2009**, *28*, 8–20. [CrossRef] [PubMed]

16. Hong, D.; Gupta, R.; Ancliff, P.; Atzberger, A.; Brown, J.; Soneji, S.; Green, J.; Colman, S.; Piacibello, W.; Buckle, V.; et al. Initiating and cancer-propagating cells in tel-aml1-associated childhood leukemia. *Science* **2008**, *319*, 336–339. [CrossRef] [PubMed]

17. Ma, Y.; Dobbins, S.E.; Sherborne, A.L.; Chubb, D.; Galbiati, M.; Cazzaniga, G.; Micalizzi, C.; Tearle, R.; Lloyd, A.L.; Hain, R.; et al. Developmental timing of mutations revealed by whole-genome sequencing of twins with acute lymphoblastic leukemia. *Proc. Natl. Acad. Sci. USA* **2013**, *110*, 7429–7433. [CrossRef] [PubMed]

18. Barnes, D.J.; Melo, J.V. Primitive, quiescent and difficult to kill: The role of non-proliferating stem cells in chronic myeloid leukemia. *Cell Cycle* **2006**, *5*, 2862–2866. [CrossRef] [PubMed]

19. Graham, S.M.; Jorgensen, H.G.; Allan, E.; Pearson, C.; Alcorn, M.J.; Richmond, L.; Holyoake, T.L. Primitive, quiescent, philadelphia-positive stem cells from patients with chronic myeloid leukemia are insensitive to sti571 In Vitro. *Blood* **2002**, *99*, 319–325. [CrossRef] [PubMed]

20. Romero-Camarero, I.; Jiang, X.; Natkunam, Y.; Lu, X.; Vicente-Duenas, C.; Gonzalez-Herrero, I.; Flores, T.; Luis Garcia, J.; McNamara, G.; Kunder, C.; et al. Germinal centre protein hgal promotes lymphoid hyperplasia and amyloidosis via bcr-mediated syk activation. *Nat. Commun.* **2013**, *4*, 1338. [CrossRef] [PubMed]

21. Vicente-Duenas, C.; Fontan, L.; Gonzalez-Herrero, I.; Romero-Camarero, I.; Segura, V.; Aznar, M.A.; Alonso-Escudero, E.; Campos-Sanchez, E.; Ruiz-Roca, L.; Barajas-Diego, M.; et al. Expression of malt1 oncogene in hematopoietic stem/progenitor cells recapitulates the pathogenesis of human lymphoma in mice. *Proc. Natl. Acad. Sci. USA* **2012**, *109*, 10534–10539. [CrossRef] [PubMed]

22. Vicente-Duenas, C.; Romero-Camarero, I.; Gonzalez-Herrero, I.; Alonso-Escudero, E.; Abollo-Jimenez, F.; Jiang, X.; Gutierrez, N.C.; Orfao, A.; Marin, N.; Villar, L.M.; et al. A novel molecular mechanism involved in multiple myeloma development revealed by targeting mafb to haematopoietic progenitors. *EMBO J.* **2012**, *31*, 3704–3717. [CrossRef] [PubMed]

23. Green, M.R.; Vicente-Duenas, C.; Romero-Camarero, I.; Long Liu, C.; Dai, B.; Gonzalez-Herrero, I.; Garcia-Ramirez, I.; Alonso-Escudero, E.; Iqbal, J.; Chan, W.C.; et al. Transient expression of bcl6 is sufficient for oncogenic function and induction of mature b-cell lymphoma. *Nat. Commun.* **2014**, *5*, 3904. [CrossRef] [PubMed]

24. Rodriguez-Hernandez, G.; Hauer, J.; Martin-Lorenzo, A.; Schafer, D.; Bartenhagen, C.; Garcia-Ramirez, I.; Auer, F.; Gonzalez-Herrero, I.; Ruiz-Roca, L.; Gombert, M.; et al. Infection exposure promotes etv6-runx1 precursor b-cell leukemia via impaired h3k4 demethylases. *Cancer Res.* **2017**, *77*, 4365–4377. [CrossRef] [PubMed]

25. Vicente-Duenas, C.; Romero-Camarero, I.; Garcia-Criado, F.J.; Cobaleda, C.; Sanchez-Garcia, I. The cellular architecture of multiple myeloma. *Cell Cycle* **2012**, *11*, 3715–3717. [CrossRef] [PubMed]

26. Vicente-Duenas, C.; Cobaleda, C.; Martinez-Climent, J.A.; Sanchez-Garcia, I. Malt lymphoma meets stem cells. *Cell Cycle* **2012**, *11*, 2961–2962. [CrossRef] [PubMed]

27. Riggi, N.; Suva, M.L.; De Vito, C.; Provero, P.; Stehle, J.C.; Baumer, K.; Cironi, L.; Janiszewska, M.; Petricevic, T.; Suva, D.; et al. Ews-fli-1 modulates $miRNA_{145}$ and SOX_2 expression to initiate mesenchymal stem cell reprogramming toward ewing sarcoma cancer stem cells. *Genes Dev.* **2010**, *24*, 916–932. [CrossRef] [PubMed]

28. Garcia, C.B.; Shaffer, C.M.; Alfaro, M.P.; Smith, A.L.; Sun, J.; Zhao, Z.; Young, P.P.; VanSaun, M.N.; Eid, J.E. Reprogramming of mesenchymal stem cells by the synovial sarcoma-associated oncogene SYT-SSX2. *Oncogene* **2012**, *31*, 2323–2334. [CrossRef] [PubMed]

29. Fang, D.; Gan, H.; Lee, J.H.; Han, J.; Wang, Z.; Riester, S.M.; Jin, L.; Chen, J.; Zhou, H.; Wang, J.; et al. The histone h3.k36m mutation reprograms the epigenome of chondroblastomas. *Science* **2016**, *352*, 1344–1348. [CrossRef] [PubMed]

30. Mainardi, S.; Mijimolle, N.; Francoz, S.; Vicente-Duenas, C.; Sanchez-Garcia, I.; Barbacid, M. Identification of cancer initiating cells in k-ras driven lung adenocarcinoma. *Proc. Natl. Acad. Sci. USA* **2014**, *111*, 255–260. [CrossRef] [PubMed]

31. Shin, K.; Lim, A.; Odegaard, J.I.; Honeycutt, J.D.; Kawano, S.; Hsieh, M.H.; Beachy, P.A. Cellular origin of bladder neoplasia and tissue dynamics of its progression to invasive carcinoma. *Nat. Cell Biol.* **2014**, *16*, 469–478. [CrossRef] [PubMed]

32. Iglesias-Bartolome, R.; Torres, D.; Marone, R.; Feng, X.; Martin, D.; Simaan, M.; Chen, M.; Weinstein, L.S.; Taylor, S.S.; Molinolo, A.A.; et al. Inactivation of a galpha(s)-pka tumour suppressor pathway in skin stem cells initiates basal-cell carcinogenesis. *Nat. Cell Biol.* **2015**, *17*, 793–803. [CrossRef] [PubMed]

33. Bartlett, T.E.; Chindera, K.; McDermott, J.; Breeze, C.E.; Cooke, W.R.; Jones, A.; Reisel, D.; Karegodar, S.T.; Arora, R.; Beck, S.; et al. Epigenetic reprogramming of fallopian tube fimbriae in brca mutation carriers defines early ovarian cancer evolution. *Nat. Commun.* **2016**, *7*, 11620. [CrossRef] [PubMed]

34. McDonald, O.G.; Li, X.; Saunders, T.; Tryggvadottir, R.; Mentch, S.J.; Warmoes, M.O.; Word, A.E.; Carrer, A.; Salz, T.H.; Natsume, S.; et al. Epigenomic reprogramming during pancreatic cancer progression links anabolic glucose metabolism to distant metastasis. *Nat. Genet.* **2017**, *49*, 367–376. [CrossRef] [PubMed]

35. Kozono, D.; Li, J.; Nitta, M.; Sampetrean, O.; Gonda, D.; Kushwaha, D.S.; Merzon, D.; Ramakrishnan, V.; Zhu, S.; Zhu, K.; et al. Dynamic epigenetic regulation of glioblastoma tumorigenicity through lsd1 modulation of myc expression. *Proc. Natl. Acad. Sci. USA* **2015**, *112*, E4055–E4064. [CrossRef] [PubMed]

36. Mazor, T.; Chesnelong, C.; Pankov, A.; Jalbert, L.E.; Hong, C.; Hayes, J.; Smirnov, I.V.; Marshall, R.; Souza, C.F.; Shen, Y.; et al. Clonal expansion and epigenetic reprogramming following deletion or amplification of mutant idh1. *Proc. Natl. Acad. Sci. USA* **2017**, *114*, 10743–10748. [CrossRef] [PubMed]

37. Liang, Y.; Ahmed, M.; Guo, H.; Soares, F.; Hua, J.T.; Gao, S.; Lu, C.; Poon, C.; Han, W.; Langstein, J.; et al. Lsd1-mediated epigenetic reprogramming drives cenpe expression and prostate cancer progression. *Cancer Res.* **2017**, *77*, 5479–5490. [CrossRef] [PubMed]

38. Kikushige, Y.; Ishikawa, F.; Miyamoto, T.; Shima, T.; Urata, S.; Yoshimoto, G.; Mori, Y.; Iino, T.; Yamauchi, T.; Eto, T.; et al. Self-renewing hematopoietic stem cell is the primary target in pathogenesis of human chronic lymphocytic leukemia. *Cancer Cell* **2011**, *20*, 246–259. [CrossRef] [PubMed]

39. Chomel, J.C.; Bonnet, M.L.; Sorel, N.; Bertrand, A.; Meunier, M.C.; Fichelson, S.; Melkus, M.; Bennaceur-Griscelli, A.; Guilhot, F.; Turhan, A.G. Leukemic stem cell persistence in chronic myeloid leukemia patients with sustained undetectable molecular residual disease. *Blood* **2011**, *118*, 3657–3660. [CrossRef] [PubMed]

40. Chu, S.; McDonald, T.; Lin, A.; Chakraborty, S.; Huang, Q.; Snyder, D.S.; Bhatia, R. Persistence of leukemia stem cells in chronic myelogenous leukemia patients in prolonged remission with imatinib treatment. *Blood* **2011**, *118*, 5565–5572. [CrossRef] [PubMed]

41. Corbin, A.S.; Agarwal, A.; Loriaux, M.; Cortes, J.; Deininger, M.W.; Druker, B.J. Human chronic myeloid leukemia stem cells are insensitive to imatinib despite inhibition of bcr-abl activity. *J. Clin. Investig.* **2011**, *121*, 396–409. [CrossRef] [PubMed]

42. Hamilton, A.; Helgason, G.V.; Schemionek, M.; Zhang, B.; Myssina, S.; Allan, E.K.; Nicolini, F.E.; Muller-Tidow, C.; Bhatia, R.; Brunton, V.G.; et al. Chronic myeloid leukemia stem cells are not dependent on bcr-abl kinase activity for their survival. *Blood* **2012**, *119*, 1501–1510. [CrossRef] [PubMed]

43. Kumari, A.; Brendel, C.; Hochhaus, A.; Neubauer, A.; Burchert, A. Low bcr-abl expression levels in hematopoietic precursor cells enable persistence of chronic myeloid leukemia under imatinib. *Blood* **2012**, *119*, 530–539. [CrossRef] [PubMed]

44. Velten, L.; Haas, S.F.; Raffel, S.; Blaszkiewicz, S.; Islam, S.; Hennig, B.P.; Hirche, C.; Lutz, C.; Buss, E.C.; Nowak, D.; et al. Human haematopoietic stem cell lineage commitment is a continuous process. *Nat. Cell Biol.* **2017**, *19*, 271–281. [CrossRef] [PubMed]

45. Jan, M.; Snyder, T.M.; Corces-Zimmerman, M.R.; Vyas, P.; Weissman, I.L.; Quake, S.R.; Majeti, R. Clonal evolution of preleukemic hematopoietic stem cells precedes human acute myeloid leukemia. *Sci. Trans. Med.* **2012**, *4*, 149ra118. [CrossRef] [PubMed]

46. Shlush, L.I.; Zandi, S.; Mitchell, A.; Chen, W.C.; Brandwein, J.M.; Gupta, V.; Kennedy, J.A.; Schimmer, A.D.; Schuh, A.C.; Yee, K.W.; et al. Identification of pre-leukaemic haematopoietic stem cells in acute leukaemia. *Nature* **2014**, *506*, 328–333. [CrossRef] [PubMed]

47. Stricker, S.H.; Feber, A.; Engstrom, P.G.; Caren, H.; Kurian, K.M.; Takashima, Y.; Watts, C.; Way, M.; Dirks, P.; Bertone, P.; et al. Widespread resetting of DNA methylation in glioblastoma-initiating cells suppresses malignant cellular behavior in a lineage-dependent manner. *Genes Dev.* **2013**, *27*, 654–669. [CrossRef] [PubMed]

48. McClellan, J.S.; Dove, C.; Gentles, A.J.; Ryan, C.E.; Majeti, R. Reprogramming of primary human philadelphia chromosome-positive B cell acute lymphoblastic leukemia cells into nonleukemic macrophages. *Proc. Natl. Acad. Sci. USA* **2015**, *112*, 4074–4079. [CrossRef] [PubMed]

49. Sanchez-Garcia, I. How tumour cell identity is established? *Semin. Cancer Biol.* **2015**, *32*, 1–2. [CrossRef] [PubMed]

50. Velasco-Hernandez, T.; Vicente-Duenas, C.; Sanchez-Garcia, I.; Martin-Zanca, D. P53 restoration kills primitive leukemia cells in vivo and increases survival of leukemic mice. *Cell Cycle* **2013**, *12*, 122–132. [CrossRef] [PubMed]

51. Martin-Lorenzo, A.; Hauer, J.; Vicente-Duenas, C.; Auer, F.; Gonzalez-Herrero, I.; Garcia-Ramirez, I.; Ginzel, S.; Thiele, R.; Constantinescu, S.N.; Bartenhagen, C.; et al. Infection exposure is a causal factor in b-cell precursor acute lymphoblastic leukemia as a result of pax5-inherited susceptibility. *Cancer Discov.* **2015**, *5*, 1328–1343. [CrossRef] [PubMed]

52. Swaminathan, S.; Klemm, L.; Park, E.; Papaemmanuil, E.; Ford, A.; Kweon, S.M.; Trageser, D.; Hasselfeld, B.; Henke, N.; Mooster, J.; et al. Mechanisms of clonal evolution in childhood acute lymphoblastic leukemia. *Nat. Immunol.* **2015**, *16*, 766–774. [CrossRef] [PubMed]

53. Pereira, H.M. A latitudinal gradient for genetic diversity. *Science* **2016**, *353*, 1494–1495. [CrossRef] [PubMed]

54. Lin, S.; Luo, R.T.; Ptasinska, A.; Kerry, J.; Assi, S.A.; Wunderlich, M.; Imamura, T.; Kaberlein, J.J.; Rayes, A.; Althoff, M.J.; et al. Instructive role of mll-fusion proteins revealed by a model of t(4;11) pro-b acute lymphoblastic leukemia. *Cancer Cell* **2016**, *30*, 737–749. [CrossRef] [PubMed]

55. Jacoby, E.; Nguyen, S.M.; Fountaine, T.J.; Welp, K.; Gryder, B.; Qin, H.; Yang, Y.; Chien, C.D.; Seif, A.E.; Lei, H.; et al. Cd19 car immune pressure induces b-precursor acute lymphoblastic leukaemia lineage switch exposing inherent leukaemic plasticity. *Nat. Commun.* **2016**, *7*, 12320. [CrossRef] [PubMed]

56. Bhatla, T.; Wang, J.; Morrison, D.J.; Raetz, E.A.; Burke, M.J.; Brown, P.; Carroll, W.L. Epigenetic reprogramming reverses the relapse-specific gene expression signature and restores chemosensitivity in childhood b-lymphoblastic leukemia. *Blood* **2012**, *119*, 5201–5210. [CrossRef] [PubMed]

57. Li, L.; Connelly, M.C.; Wetmore, C.; Curran, T.; Morgan, J.I. Mouse embryos cloned from brain tumors. *Cancer Res.* **2003**, *63*, 2733–2736. [PubMed]

58. Blelloch, R.H.; Hochedlinger, K.; Yamada, Y.; Brennan, C.; Kim, M.; Mintz, B.; Chin, L.; Jaenisch, R. Nuclear cloning of embryonal carcinoma cells. *Proc. Natl. Acad. Sci. USA* **2004**, *101*, 13985–13990. [PubMed]

59. Hochedlinger, K.; Blelloch, R.; Brennan, C.; Yamada, Y.; Kim, M.; Chin, L.; Jaenisch, R. Reprogramming of a melanoma genome by nuclear transplantation. *Genes Dev.* **2004**, *18*, 1875–1885. [CrossRef] [PubMed]

60. Scott, M.T.; Korfi, K.; Saffrey, P.; Hopcroft, L.E.; Kinstrie, R.; Pellicano, F.; Guenther, C.; Gallipoli, P.; Cruz, M.; Dunn, K.; et al. Epigenetic reprogramming sensitizes cml stem cells to combined ezh2 and tyrosine kinase inhibition. *Cancer Discov.* **2016**, *6*, 1248–1257. [CrossRef] [PubMed]

61. Chheda, M.G.; Gutmann, D.H. Using epigenetic reprogramming to treat pediatric brain cancer. *Cancer Cell* **2017**, *31*, 609–611. [CrossRef] [PubMed]

62. Nagaraja, S.; Vitanza, N.A.; Woo, P.J.; Taylor, K.R.; Liu, F.; Zhang, L.; Li, M.; Meng, W.; Ponnuswami, A.; Sun, W.; et al. Transcriptional dependencies in diffuse intrinsic pontine glioma. *Cancer Cell* **2017**, *31*, 635–652 e636. [CrossRef] [PubMed]

63. Li, R.; Grimm, S.A.; Chrysovergis, K.; Kosak, J.; Wang, X.; Du, Y.; Burkholder, A.; Janardhan, K.; Mav, D.; Shah, R.; et al. Obesity, rather than diet, drives epigenomic alterations in colonic epithelium resembling cancer progression. *Cell Metab.* **2014**, *19*, 702–711. [CrossRef] [PubMed]

64. Beyaz, S.; Mana, M.D.; Roper, J.; Kedrin, D.; Saadatpour, A.; Hong, S.J.; Bauer-Rowe, K.E.; Xifaras, M.E.; Akkad, A.; Arias, E.; et al. High-fat diet enhances stemness and tumorigenicity of intestinal progenitors. *Nature* **2016**, *531*, 53–58. [CrossRef] [PubMed]

65. Vaz, M.; Hwang, S.Y.; Kagiampakis, I.; Phallen, J.; Patil, A.; O'Hagan, H.M.; Murphy, L.; Zahnow, C.A.; Gabrielson, E.; Velculescu, V.E.; et al. Chronic cigarette smoke-induced epigenomic changes precede sensitization of bronchial epithelial cells to single-step transformation by kras mutations. *Cancer Cell* **2017**, *32*, 360–376.e6. [CrossRef] [PubMed]

66. Greathouse, K.L.; Bredfeldt, T.; Everitt, J.I.; Lin, K.; Berry, T.; Kannan, K.; Mittelstadt, M.L.; Ho, S.M.; Walker, C.L. Environmental estrogens differentially engage the histone methyltransferase ezh2 to increase risk of uterine tumorigenesis. *Mol. Cancer Res.* **2012**, *10*, 546–557. [CrossRef] [PubMed]

International Journal of
Molecular Sciences

MDPI

Review

Vitamins D: Relationship between Structure and Biological Activity

Andrzej Kutner [1] **and Geoffrey Brown** [2,*]

[1] Pharmaceutical Research Institute, 8 Rydygiera, Warsaw 01-793, Poland; a.kutner@ifarm.eu
[2] Institute of Clinical Sciences, Institute of Immunology and Immunotherapy, College of Medical and Dental Sciences, University of Birmingham, Edgbaston, Birmingham B15 2TT, UK
* Correspondence: g.brown@bham.ac.uk; Tel.: +44-(0)121-414-4082

Received: 3 July 2018; Accepted: 18 July 2018; Published: 20 July 2018

Abstract: The most active metabolite of vitamin D is 1α,25-dihydroxyvitamin D_3, which is a central regulator of mineral homeostasis: excessive administration leads to hypercalcemia. Additionally, 1α,25-dihydroxyvitamin D_3 is important to decision-making by cells, driving many cell types to growth arrest, differentiate and undergo apoptosis. 1α,25-Dihydroxyvitamin D_3 regulates gene transcription by binding to a single known receptor, the vitamin D receptor. Rapid intracellular signals are also elicited in vitro by 1α,25-dihydroxyvitamin D_3 that are independent of transcription. There are many aspects of the multiple actions of 1α,25-dihydroxyvitamin D_3 that we do not fully understand. These include how a single receptor and provoked rapid events relate to the different actions of 1α,25-dihydroxyvitamin D_3, its calcemic action per se, and whether a large number of genes are activated directly, via the vitamin D receptor, or indirectly. A strategy to resolving these issues has been to generate synthetic analogues of 1α,25-dihydroxyvitamin D_3: Some of these separate the anti-proliferative and calcemic actions of the parent hormone. Crystallography is important to understanding how differences between 1α,25-dihydroxyvitamin D_3- and analogue-provoked structural changes to the vitamin D receptor may underlie their different activity profiles. Current crystallographic resolution has not revealed such information. Studies of our new analogues have revealed the importance of the A-ring adopting the chair β-conformation upon interaction with the vitamin D receptor to receptor-affinity and biological activity. Vitamin D analogues are useful probes to providing a better understanding of the physiology of vitamin D.

Keywords: cell differentiation; vitamin D; vitamin D receptor; vitamin D analogues; crystallography

1. Introduction

1α,25-Dihydroxyvitamin D_3 (1,25D3, calcitriol, Figure 1) is the most active metabolite of vitamin D (vitD, cholecalciferol). This *seco*-steroid hormone has many biological roles [1]. A key physiological role of 1,25D3 is to regulate the absorption and transportation of essential minerals, particularly calcium, phosphorus, and magnesium, which are important to the maintenance of bone [2,3]. 1,25D3 also plays a key role in decision-making by cells because it can elicit events that result in many types of cells arresting their growth, differentiating, and undergoing apoptosis. The anti-proliferative action of 1,25D3 extends to malignant cells, leading to interest in its use for differentiation therapy of leukemia and other cancers. However, a substantial limitation to this use of 1,25D3 is that its calcemic action restricts achieving an effective therapeutic dose [4]. Many of the immune cell types express the receptor for vitamin D (VDR): 1,25D3 has an important role in their function and is, therefore, important to good health [5,6]. A related interest is the therapeutic use of 1,25D3 as an anti-inflammatory agent.

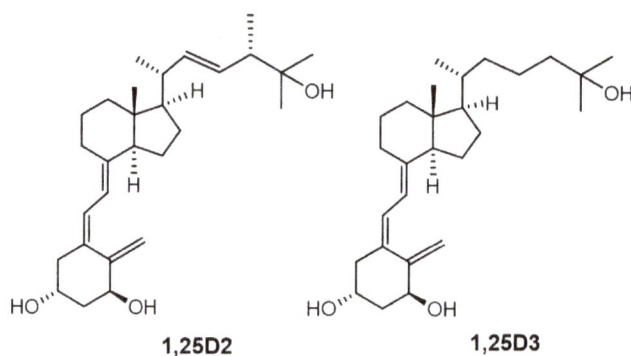

Figure 1. The structures of $1\alpha,25$-dihydroxyvitamin D_3 (1,25D3) and $1\alpha,25$-dihydroxyvitamin D_2 (1,25D2).

2. The Modes of Action of Vitamin D

Some of the biological effects of 1,25D3 result from the direct activation of target genes [7,8]. In this case, 1,25D3 binds to its single known receptor VDR. The VDR forms a heterodimer with the retinoid X receptor to regulate gene transcription. 1,25D3 regulates a large number of genes, including ones that link to the growth and differentiation of cells and diseases that include cancer, diabetes, and arthritis. At least 229 genes are subject to regulation by 1,25D3, as revealed by ChiP-seq analysis [9,10]. We do not understand whether the VDR activates all of the envisaged number of genes directly or indirectly.

For some cells, 1,25D3 treatment leads to the activation of non-genomic signaling pathways. The delineation of these events is from in vitro studies, and they occur within seconds or minutes after treating cells with 1,25D3 [11,12]. In this scenario, 1,25D3 initiates signals at the plasma membrane or in the cytosol. The mitogen-activated protein kinases and phosphatidylinositol 3-kinase signaling cascades mediate this rapid response to some extent [13–15]. There are putative receptors for 1,25D3 at the cell membrane that include the membrane-associated rapid response steroid-binding protein (MARRS, ERp57, or PDIA3), megalin, and cubilin, and the latter two transport vitamin D, complexed to its serum binding protein [16]. Additionally, 1,25D3 binds to the VDR in an alternative VDR-binding pocket through a 6-*cis*-1,25D conformation as proposed by Mizwicki and Norman [17]. Endocytosis of this and other putative membrane receptors can generate immediate intracellular signals *via* the Rab5/PI3-kinase pathway. A shift in the balance between VDR-provoked gene transcription and rapid signaling events might underlie the anti-proliferative versus calcemic actions of 1,25D3. However, the structure-calcemic activity relationship for most of the known vitamin D analogues is not clear to date.

3. Vitamin D Analogues

Over recent years, investigators have generated and studied hundreds of vitD analogues and several metabolites. Their structures are important to biological activity. A total of 17 crystal structures of 1α-hydroxylated vitDs are at the Cambridge Structural Database, and there are structures of 63 vitD analogues bound to the engineered VDR. Despite all of this, we still do not understand the molecular events that force an analogue to adopt the A-ring slightly distorted chair β-conformation *per se* and when bound to the VDR. It remains a mystery as to why the three hydroxyls (1, 3, and 25) that mediate analogue binding to the VDR are almost overlapping and why analogues have very different structures and activities.

The most important parts of vitD and analogues regarding the affinity for the VDR, and consequently activity profile, are the A-ring, the side-chain, and the CD-ring system.

Recently, we divided double point modified (DPM) analogues of vitDs [18] into structural groups. We introduced the new classification system for *semi*-quantifying the biological activities of DPM analogues and outlined new directions for structural modifications. Based on this overview, we designed new analogues of 1α,25-dihydroxyvitamin D$_2$ (Figure 1, 1,25D2) to study the active conformation of the A-ring. We used a panel of closely related analogues of 1,25D2, [19] with solved crystal structures [20], rather than just a single analogue. For reference, we used a new compound (not shown) with all the functional groups (C-25 carboxy and 1- and 3-hydroxyl) protected and, therefore, deprived of electrostatic interactions [20]. The correlation of the structures of the new analogues with their biological activities [21,22] allowed us to solve a long-lasting enigma [23,24] surrounding the 1α-hydroxylated analogues of vitD. We have proposed a new general rule (see below), confirmed by theoretical calculations, which applies to all of our crystallographic data and which has been obtained for 1α -hydroxylated analogues of vitD over the last 25 years.

We investigated the side-chain structure by generating (24Z) geometric isomers [25] of our hypocalcemic analogues of 1,25D2. We examined their binding affinities for the VDR, the potency regarding inhibition of cell growth, and the regulation of stem cell-related gene expression in colon cancer cells [26]. Resistance to *hCYP24A1*-metabolism, which deactivates 1,25D3 by enzymatic hydroxylation [25], relates to a preferred side-chain geometry. An extended and rigidified side-chain of 1,25D2 is responsible for the longer-term biological effects of vitD (see below). We used human AML cell lines to examine the correlation between the side-chain geometry of 1,25D2 and potency in arresting cell growth and inducing cell differentiation [27].

The original principle of CD-ring modified analogues [28] was that the entire CD-ring system of an analogue was not required for activity [29]. The hydrindane CD-ring moiety originates from the biotransformation of sterols into 9,10-*seco*-steroids (vitD), and it was not specifically biosynthesized for vitD. Additionally, the CD-ring is the only part of the vitD molecule that does not participate in the metabolic transformations of vitD. Moreover, molecular modeling revealed that *des*-C,D analogues relate structurally to the biotransformation product [29] of all-*trans*-retinoic acid (ATRA), a potent differentiating agent. *Des*-C,D analogues combine aspects of both the vitD and retinoid structures, and we call them retiferols. Our original concept of retiferols led to the synthesis of a substantial variety of CD-ring modified analogues, including other *des*-C,D analogues [30], *des*-D,C-ring analogues [31], and *des*-C,D-ring homo analogues [32], and their biological activity was evaluated [30–32]. Our first retiferol (Figure 2) RAD$_2$ became a synthetic target [33]. Roche researchers [30] obtained 19-*nor* RAD$_2$ and reported that this analogue was useful in the treatment of hyper-proliferative skin diseases in vivo. Here, we discuss how CD-ring modifications affect activity.

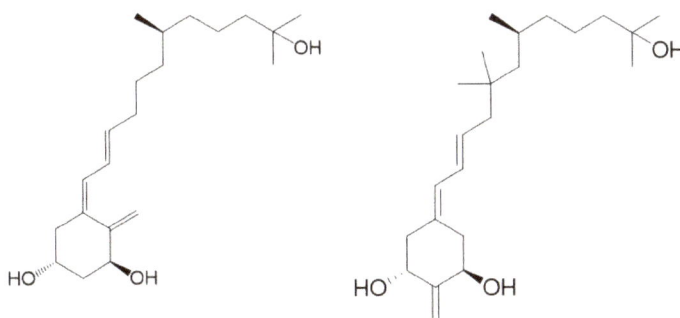

Figure 2. The structure of RAD$_2$, the first *des*-C,D analogue of 1α,25-dihydroxyvitamin D$_3$ and of (20S)-13,13-dimethyl-*des*-C,D-1,25-dihydroxy-2-methylene-19-*nor*-vitamin D$_3$.

Below, we examine the use of various classes of new analogues of vitD to unravel some of the issues relating to structure and biological activity. Of particular importance is that new analogues separate the anti-proliferative and calcemic actions of the parent hormone more effectively than previous analogues (see below), and how might this be the case?

4. The Vitamin D A-Ring Conformation

To investigate the correlation between the A-ring conformation and activity, we conceived and synthesized, by a novel convergent strategy, a panel of DPM analogues of 1,25D2 coded PRI-1730, PRI-1731, PRI-1732, PRI-1733, and PRI-1734 (Figure 3) [19]. Our modifications included new 5,6-*trans* (5*E*,7*E*) geometry of the A-ring *per se* or combined with the further modifications in the side chain. These included an additional (22*S*)-hydroxyl, 22,23-single bond, and a reversed absolute configuration (24-*epi*) at C-24. All our analogues induced differentiation of the VDR positive A375 and VDR negative SK-MEL 188b human malignant melanoma cell lines [22]. As expected, 5,6-*trans* modification of the A-ring was advantageous to enhancing the anti-proliferative activity of the analogues but not as a single point modification. Very unexpectedly, the additional 22-hydroxyl in the side-chain, conceived to enhance VDR binding, reduced significantly the anti-proliferative activity of both the natural and 5,6-*trans* series of analogues [21].

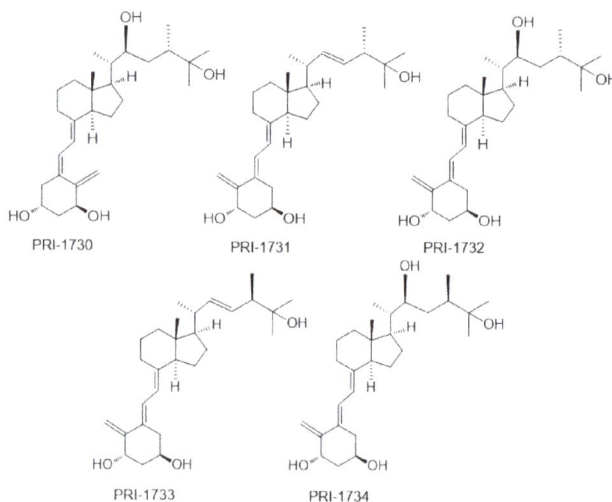

Figure 3. The structures of double point modified analogues of 1α,25-dihydroxyvitamin D$_2$.

PRI-1731 and PRI-1733 increased translocation of the VDR to the nucleus of HL60 cells but to a lesser extent than provoked by 1,25D2 and 1,25D3. 5,6-*Trans* modification contributed substantially to the increased stability of the PRI-1731 and PRI-1733 against enzymatic hydroxylation by *h*CYP24A1, produced by expressing recombinant protein in *Escherichia coli*. Unexpectedly, reversing the chirality at C-24 from the natural (24*S*) in PRI-1731 to the (24*R*) in PRI-1733 did not affect metabolic resistance. The conversion remained at the high level of only 12% for both analogues, compared with 44% and 35% for 1,25D3 and 1,25D2, respectively. The addition of 22-hydroxyl and the saturation of the 22,23-double bond resulted in a dramatic loss of metabolic stability from 12% for PRI-1731 to a 52% conversion for PRI-1732. We used fluorescence polarization-based competition assay to measure the binding affinity of our analogues for the VDR. Only the 5,6-*trans* analogue of 1,25D2 (PRI-1731) showed a binding affinity comparable to that of both 1,25D2 and 1,25D3. Very intriguingly, a combination of all four structural modifications resulted in a complete loss of activity in the case of PRI-1734. This analogue showed weak binding to the VDR [21] and failed to agonize the VDR. However, its structure might be a good starting point for the design of a vitD antagonist, once the binding is improved [21]. The modifications introduced have not led to an increase in differentiation-inducing potency for the above new panel of analogues. However, they have resulted in a very divergent group of analogues that have provided very important data regarding structure versus activity relationships.

VitD analogues are resistant to crystallization due to a high flexibility over the number of rotated single bonds in the side-chain and in the triene system. Therefore, we were very fortunate to obtain single crystals of as many as three analogues (PRI-1730, PRI-1731, and PRI-1732), out of a panel of our five analogues [19], suitable for X-ray diffraction. For our structure–activity relationship, it was also of key importance to obtain a single crystal of the synthetic intermediate with all the functional groups (1,3, and 25-hydroxyl and 25-carboxyl) protected and, therefore, deprived of electrostatic interactions [25]. Very interestingly, we observed that the A-ring of PRI-1730 and PRI-1731 exists in a crystal state in a chair β-conformation, and that of PRI-1732 and of the totally protected synthetic intermediate in a chair α-conformation [20]. Using this unusually large collection of solid-state structures, our crystallography study revealed the new general rule regarding the solid-state A-ring conformation. We concluded that the direct hydrogen bond between 1-hydroxyl and 3-hydroxyl of the neighboring molecule forces the A-ring of 1α-hydroxylated analogues of vitamin D to adopt the chair β-conformation. According to our rule, the analogues adopting an A-ring chair β-conformation exhibit higher biological activity than the structurally related analogues existing in the chair α-conformation. This is due to the possibility of stronger electrostatic interactions of the chair β-conformation with the VDR. This explains why PRI-1730, which adopts the chair β-conformation, shows much higher activity (e.g., in nuclear translocation of the VDR) than PRI-1732, which has the chair α-conformation, and why PRI-1730 shows a much higher metabolic stability (31% conversion) than PRI-1732 (52% conversion). Contrary to the common understanding, our crystallographic studies demonstrated that the structure of an analogue in a solid state very much relates to its structure in a solution and when interacting with the VDR, predicts its biological activity as high or low. We should consider testing only the analogues existing in a solid state in A-ring chair β-conformation for biological activity.

5. Modifications to the Vitamin D Side-Chain

In our studies of the relationship between the side-chain structure and activity, we developed another new convergent strategy to modify our leading 1,25D2 analogues, PRI-1906 and PRI-1907 (Figure 4). We extended and rigidified the side-chain and obtained new analogues, PRI-1916 and PRI-1917, with the previously unknown geometry at C-24 [(24Z), instead of (24E)]. The binding affinity of PRI-1916, with two methyl at the terminus of the side-chain at C-25, for the full-length human VDR in a fluorescence polarization assay was substantially higher than that of the previously obtained PRI-1906. However, the affinity of PRI-1917, with two ethyl at C-25, was much lower than that of the parent PRI-1907. This finding indicated that terminal alkyls at C-25 strongly influence the binding affinity of analogues for the VDR. Our PRI-1906 and PRI-1907 have a very high resistance to metabolic conversion (2.3% and 0.8% for PRI-1906 and PRI-1907, respectively, compared with 44% and 35% for 1,25D3 and 1,25D2, respectively). Our new PRI-1916 and PRI-1917 showed a somewhat lower resistance to conversion, although still higher than that of 1,25D3 and 1,25D2. From these findings we proved that a rigid and straight (24E) geometry of the side-chain is preferred for metabolic resistance, and in keeping (24E), (24-*trans*) analogues elicit long-term biological effects against cancer cells [25]. (24Z) Modification of the side-chain of 1,25D2 analogues has a contrasting effect on the differentiating activity of PRI-1906 and PRI-1907. Although the VDR affinity of the (24Z) analogue PRI-1916 was lower than that of (24E) analogue PRI-1906, the potency of PRI-1916 was slightly higher than that of PRI-1906 when tested against the human AML cell lines KG-1a, HL60, U937, and MOLM-13, which typify different stages of myeloid maturation [27] However, PRI-1917 was significantly less potent than PRI-1907. We, therefore, finally concluded that the (24E) side-chain geometry combined with selected modifications of the A-ring is preferable in terms of generating potent anti-proliferative and differentiating vitDs. Evaluation of the differentiating activity and calcemic action of newer analogues of 1,25D2 has shown they are more potent differentiating agents than 1,25D3 and have a reduced calcemic action (see Table 1).

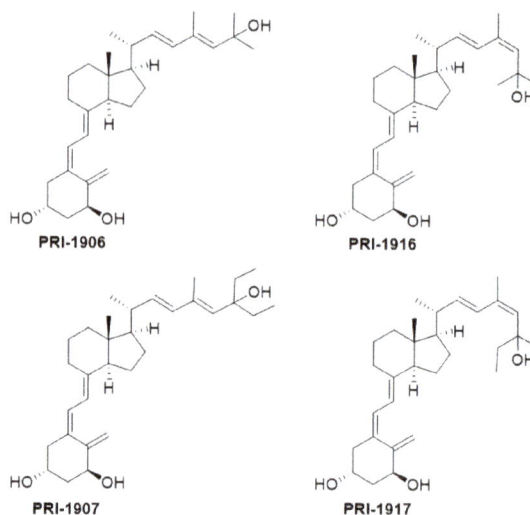

Figure 4. The side-chain extended and rigidified analogues of $1\alpha,25$-dihydroxyvitamin D_2 (PRI-1906 and PRI-1907) and their (24Z) geometric isomers (PRI-1916 and PRI-1917).

Table 1. Differentiating potency, calcemic action, and binding to the receptor for vitamin D (VDR) of PRI analogues of $1\alpha,25$-dihydroxyvitamin D_2.

Analogue	Differentiation $EC_{50}\ m \times 10^{-11}$	Calcaemic Action Ca^{2+} (Serum) $m \times 10^{-6}$	VDR Binding $IC_{50}\ m \times 10^{-10}$
1,25D3	53	107	23
PRI-1907	6	75	62
PRI-5100	113	86	6
PRI-5101	118	90	5
PRI-5201	3	93	11
PRI-5202	2	81	36

The values obtained for $1\alpha,25$-dihydroxyvitamin D_3 (1,24D3) are shown for comparison. The EC_{50} for differentiating potency is the value that drives half-maximal differentiation of the promyeloid cell line HL60 towards neutrophils. Calcium levels are the mean of the values obtained for five mice treated with 0.3 µg/kg of 1,25D3 or an analogue every other day for 3 weeks and measured on day 21. The serum calcium level in ethanol-treated mice was 62 µM. The analogues PRI-1907, PRI-5201, and PRI-5202 are more potent differentiating agents than 1,25D3 and have a lower calcemic action.

Quite unexpectedly, both (24E) and the new (24Z) analogues are equipotent in decreasing the cloning capacity and the proliferative activity of the human colon cancer cells HT-29 [26]. These cells are refractory to the anti-proliferative action of the chemotherapeutic agent 5-flurouracil. Both of the C-25 diethyl analogues, PRI-1907 and PRI-1917, decreased the level of expression of stemness-related genes, while both dimethyl analogues, PRI-1906 and PRI-1916, were not able to downregulate these genes. Therefore, the new geometric analogues, PRI-1907 and PRI-1917, are good candidates for further studies to examine the benefit to preventing cancer relapse. In this regard, they act to decrease the proliferative capacity of cells that initiate tumor regrowth, by virtue of downregulating stemness-related genes. The analogues might be useful in post-treatment of cancer patients with conventional reductive chemotherapy, particularly regarding the reduced calcemic action of PRI-1907 and the newer analogues of 1,25D2 (see Table 1) [26].

6. CD-Ring Modifications of Vitamin D

There is evidence to support the viewpoint that methyl substituents at C-13 of 13,13-dimethyl-*des*-C,D-1,25-dihydroxy-2-methylene-19-*nor*-vitamin D3 mimic the C-ring of 1,25D3. The 13,13-dimethyl *des*-C,D-analogue of (20S)-1α,25-dihydroxy-2-methylene-19-*nor* vitamin D$_3$ (Figure 2) still retains much of the activity of vitD [34]. As predicted [29] and confirmed recently [35], even an extensive modification of the CD-ring does not affect the functional activity of vitD as long as the three-dimensional (3D) arrangement of the three 1α, 3, and 25 hydroxyls, responsible for VDR binding, is preserved. As expected, a number of CD-ring modifications resulted in a lowering or complete loss of the undesired calcemic action regarding the development of a potent agent that selectively drives differentiation [30]. A lowering of calcemic activity is also the case for a novel class of five *des*-C analogues with the CD-ring fragment partially replaced by an alkyl chain to mimic the C-ring and an aromatic *m*-phenylene ring replacing the D-ring (Figure 5). Docking studies to the human ligand-binding domain of the VDR led to the design of these analogues. The analogue with an ethyl substituent to mimic the missing C-ring is active against breast cancer cells and, as expected, has negligible calcemic action [36]. These analogues provide important information about the structure–activity relationships regarding CD-ring modifications. This novel class of aromatic-based vitamin D analogues also confirmed that retaining the network of hydrogen bonds of 1,25D3 is crucial for transactivation activity and that aromatic modification of the CD-ring fragment abolishes calcemic activity.

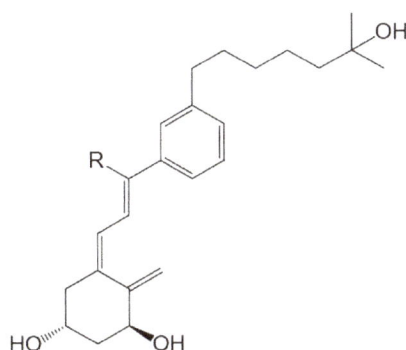

Figure 5. The *des*-C-*m*-phenylene-D-21-*nor*-analogues of 1α,25-dihydroxyvitamin D$_3$ (R = Et, *n*-Pr, *n*-Bu, *n*-Hex, *n*-Hept).

7. Future Directions

As described, various modifications to vitD underlie the search for a new drug candidate. Beneficial modifications are those to the A-ring, the triene system, the CD-ring fragment, and the side-chain. An old viewpoint was to make very subtle and single modifications. The current understanding to obtaining the most potent VDR agonist is to combine several modifications at various parts of the molecule. However, the number of possible combinations is numerous. The route to the best combinations remains elusive, because the ultimate outcome of combinations is quite unpredictable. They might result in enhancement or in a complete loss of activity. Even so, several vitamin D super-agonists have been rationally designed and synthesized.

Intuition has mostly driven the design of new vitamin D analogues, coupled with trial and error. A crystal of the full-length native VDR that is suitable for X-ray diffraction study is presently not available. In this case, a classical model [37] that makes use of a truncated VDR protein remains the only available approach to examining the interactions of vitDs with the VDR. This artificial VDR protein has the flexible hinge region cut off to improve the ability of the residual VDR to crystallize.

The accumulating of more and more data leads to the viewpoint that the standing model might be of limited significance, because the resulting solid-state structures of various analogues show overlapping of the functional hydroxyls. Very recently, a study of a panel of five analogues of very different structures recorded no detectable differences regarding binding to the VDR [36].

A common understanding is that crystallography plays the major role as the only direct method for studying analogues *per se* and their protein complexes to reveal structure versus activity relationships. Relative spectroscopic methods, such as high frequency nuclear magnetic resonance (900 MHz) [38], are still of a very limited use, as not enough data have been accumulated for a complete signal assignment. Up to now and for complexes of vitDs and the VDR, there is only low-resolution crystallographic data, usually well above 2 Å. We might expect that high-resolution crystallographic data, below 0.5 Å, will give a much better insight into the structural differences between the VDR complexed with various analogues.

However, to make substantial progress and ensure the design of analogues in a truly rational manner an entirely new crystallographic approach is required. It is highly desirable to obtain single crystal high-resolution data of the native VDR, other than for a truncated model protein. Considering the recent developments of crystallography, the time is right to make a concerted effort to obtain the crystal structure of the full-length human VDR *per se* and complexed with vitDs. Automated high-throughput robots now crystallize proteins up to three times larger than the VDR. Additionally, consideration of modern cryo-electron microscopy as a tool for examining a native VDR-retinoid X receptor (RXR) complex is worthwhile. Success in the above directions will open a new era that should reveal the real relationships between the structure and the biological activity of vitDs.

To understand the various biological roles of the VDR there is a need to develop an antagonist of the VDR. The rational design of a potent antagonist is much more difficult than designing an agonist, because there are only few antagonistic structures for use to examine correlations between structure and function. Additionally, computerized docking approaches are especially useful for designing agonists, and the same method is useful to the design of antagonists. The crystal structure of the full-length VDR will be valuable to the design of an antagonist.

8. Concluding Remarks

As outlined, the precise nature of modifications to the structure of vitD is important to the gain and loss of particular biological actions. The modified analogues, in turn, are important to developing a better understanding of how vitD can mediate a variety of different biological roles. Of particular interest is the generation of vitDs that are potent differentiating agents and lack calcemic action, for use in studies of cell differentiation and as potential anti-cancer agents. Work over a number of years has generated a range of analogues with very different modifications to the structure of vitD and has correlated their structures with biological actions. A number of important structure–activity rules have emerged (see Figure 6). Analogues with an aromatic modification of the CD-ring fragment have a reduced calcemic action. The hydrogen bonds of the natural 1,25D3 are crucial for transactivation activity, and vitDs with the (24E) side-chain geometry combined A-ring modifications are potent anti-proliferative and differentiating agents. Too many structural modifications, a combination of four in the case of PRI-1734, led to a complete loss of biological activity. A rigid and straight (24E) geometry of the side-chain confers resistance to catabolism *via* enzymatic hydroxylation by *h*CYP24A1. The provision of a better understanding of how vitD and analogues bind to VDR is also essential to unraveling function. In this regard, the existence of the vitD A-ring in the β-chair conformation, rather than the α-chair conformation, is essential to the differentiating activity of analogues. The application of the rules to activity gained so far should allow analogues to be refined further regarding potency and selectivity.

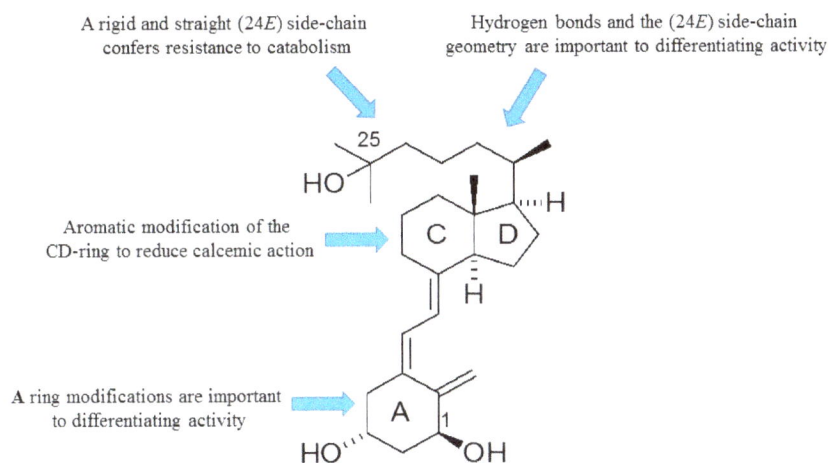

Figure 6. Map of the relationships between structural modifications to 1α,25-dihydroxyvitamin D$_2$ and biological actions.

Author Contributions: A.K. wrote the chemistry sections and G.B. about the biological aspects. Both authors then revised the manuscript.

Acknowledgments: This project received funding from the European Union's Seventh Framework Programme for research, technological development, and demonstration under grant agreement no 315902. Andrzej Kutner and Geoffrey Brown were partners within the Marie Curie Initial Training Network DECIDE (Decision-making within cells and differentiation entity therapies). The authors dedicate this article to their friend, colleague, and mentor, Antonius G. Rolink, whose incisive comments greatly influenced our way of viewing hematopoiesis and studies of the actions of vitamins D.

Conflicts of Interest: The authors declare no conflict of interest.

References

1. Norman, A.W. The history of the discovery of vitamin D and its daughter steroid hormone. *Ann. Nutr. Metab.* **2012**, *61*, 199–206. [CrossRef] [PubMed]
2. Lieben, L.; Carmeliet, G. Vitamin D signalling in osteocytes: effects on bone and mineral homeostasis. *Bone* **2013**, *54*, 237–243. [CrossRef] [PubMed]
3. Turner, A.G.; Anderson, P.H.; Morris, H.A. Vitamin D and bone health. *Scand. J. Clin. Lab. Invest. Suppl.* **2012**, *243*, 65–72. [PubMed]
4. Brown, G.; Kutner, A.; Marcinkowska, E. Vitamin D and leukaemia. In *Extraskeletal Effects of Vitamin D*; Liao, E.P., Ed.; Springer: New York, NY, USA, 2018; pp. 115–134.
5. Norman, A.W. From vitamin D to hormone D: Fundamentals if the vitamin D endocrine system essential for good health. *Am. J. Clin. Nutr.* **2008**, *88*, 491S–499S. [CrossRef] [PubMed]
6. Hewison, M. Vitamin D and immune function: autocrine, paracrine or endocrine. *Scand. J. Clin. Lab. Invest. Suppl.* **2012**, *243*, 92–102. [PubMed]
7. Haussler, M.R.; Jurutka, P.W.; Mizwicki, M.; Norman, A.W. Vitamin D receptor (VDR)-mediated actions of 1α,25(OH)$_2$vitamin D$_3$: Genomic and non-genomic mechanisms. *Best Pract. Res. Clin. Endocrinol. Metab.* **2011**, *25*, 543–549. [CrossRef] [PubMed]
8. Aranda, A.; Pascual, A. Nucelar hormone receptors and gene expression. *Physiol. Rev.* **2001**, *81*, 1269–1304. [CrossRef] [PubMed]
9. Ramagopalan, S.V.; Heger, A.; Berlanga, A.J.; Maugeri, N.J.; Lincoln, M.R.; Burrell, A.; Handunnetthi, L.; Handel, A.E.; Disanto, G.; Orton, S.M.; et al. A ChIP-seq defined genome-wide map of vitamin D receptor binding: Associations with disease and evolution. *Genome Res.* **2010**, *20*, 1352–1360. [CrossRef] [PubMed]
10. Carlberg, C. Endocrine functions of vitamin D. *Mol. Cell. Endocrinol.* **2017**, *453*, 1–2. [CrossRef] [PubMed]

11. Norman, A.W.; Bishop, J.E.; Bula, C.M.; Olivera, C.J.; Mizwicki, M.T.; Zanello, L.P.; Ishida, H.; Okamura, W.H. Molecular tools for study of genomic and rapid signal transduction responses initiated by 1α,25(OH)$_{(2)}$-vitamin D$_{(3)}$. *Steroids* **2002**, *67*, 457–466. [CrossRef]

12. Gniadecki, R. Nongenomic signalling by vitamin D: A new face of Src. *Biochem. Pharmacol.* **1998**, *56*, 1273–1277. [CrossRef]

13. Hughes, P.J.; Brown, G. 1Alpha,25-dihydroxyvitamin D3-mediated stimulation of steroid sulphatase activity in myeloid leukaemic cell lines requires VDRnuc-mediated activation of the RAS/RAF/ERK-MAP kinase signalling pathway. *J. Cell. Biochem.* **2006**, *98*, 590–617. [CrossRef] [PubMed]

14. Hughes, P.J.; Lee, J.S.; Reiner, N.E.; Brown, G. The vitamin D receptor-mediated activation of phosphatidylinositol 3-kinase (PI3Kalpha) plays a role in the 1α,25-dihydroxyvitamin D3-stimulated increase in steroid sulphatase activity in myeloid leukaemic cell lines. *J. Cell. Biochem.* **2008**, *103*, 1551–1572. [CrossRef] [PubMed]

15. Marcinkowska, E.; Wiedlocha, A.; Radzikowski, C. 1,25-Dihydroxyvitamin D3 induced activation and subsequent nuclear translocation of MAPK is upstream regulated by PKC in HL-60 cells. *Biochem. Biophys. Res. Commun.* **1997**, *241*, 419–426. [CrossRef] [PubMed]

16. Nykjaer, A.; Fyfe, J.C.; Kozyraki, R.; Leheste, J.R.; Jacobsen, C.; Nielsen, M.S.; Verroust, P.J.; Aminoff, M.; de la Chapelle, A.; Moestrup, S.K.; et al. Cubilin dysfunction causes abnormal metabolism of the steroid hormone 25(OH) vitamin D(3). *Proc. Natl. Acad. Sci. USA* **2001**, *98*, 13895–13900. [CrossRef] [PubMed]

17. Mizwicki, M.T.; Norman, A.W. Vitamin D Sterol/Vitamin D Receptor Conformational Dynamics and Nongenomic Actions. In *Vitamin D*, 4th ed.; Feldmann, D., Pike, J., Bouillon, R., Giovannucci, E., Goltzman, D., Hewison, M., Eds.; Academic Press: San Diego, CA, USA, 2018; Volume 1, pp. 69–294.

18. Nadkarni, S.; Chodyński, M.; Corcoran, A.; Marcinkowska, E.; Brown, G.; Kutner, A. Double point modified analogs of vitamin D as potent activators of vitamin D receptor. *Curr. Pharm. Design* **2015**, *21*, 1741–1763. [CrossRef]

19. Nadkarni, S.; Chodyński, M.; Krajewski, K.; Cmoch, P.; Marcinkowska, E.; Brown, G.; Kutner, A. Convergent synthesis of double point modified analogs of 1α,25-dihydroxyvitamin D$_2$ for biological evaluation. *J. Steroid. Biochem. Mol. Biol.* **2016**, *164*, 45–49. [CrossRef] [PubMed]

20. Wanat, M.; Malinska, M.; Kutner, A.; Wozniak, K. Effect of vitamin D conformation on interactions and packing in the crystal lattice. *Cryst. Growth Des.* **2018**. [CrossRef]

21. Corcoran, A.; Nadkarni, S.; Yasuda, K.; Sakaki, T.; Brown, G.; Kutner, A.; Marcinkowska, E. Biological evaluation of double point modified analogues of 1,25-dihydroxyvitamin D$_2$ as potential anti-leukemic agents. *Int. J. Mol. Sci.* **2016**, *17*, E91. [CrossRef] [PubMed]

22. Piotrowska, A.; Wierzbicka, J.; Nadkarni, S.; Brown, G.; Kutner, A.; Żmijewski, M.A. Antiproliferative activity of double point modified analogs of 1,25-dihydroxyvitamin D$_2$ against human malignant melanoma cell lines: VDR positive A375 and VDR negative SK-MEL 188b. *Int. J. Mol. Sci.* **2016**, *17*, E76. [CrossRef] [PubMed]

23. Larsen, S.; Hansen, E.T.; Hoffmeyer, L.; Rastrup-Andersen, N. Structure and absolute configuration of calcipotriol. *Acta Cryst.* **1993**, *C49*, 618–621.

24. Suwińska, K.; Kutner, A. Crystal and Molecular structure of 1,25-dihydroxycholecalciferol. *Acta Cryst.* **1996**, *B52*, 550–554. [CrossRef]

25. Bolla, N.R.; Corcoran, A.; Yasuda, K.; Chodyński, M.; Krajewski, K.; Cmoch, P.; Marcinkowska, E.; Brown, G.; Sakaki, T.; Kutner, A. Synthesis and evaluation of geometric analogs of 1α,25-dihydroxyvitamin D$_2$ as potential therapeutics. *J. Steroid Biochem. Mol. Biol.* **2016**, *164*, 50–55. [CrossRef] [PubMed]

26. Neska, J.; Swoboda, P.; Przybyszewska, M.; Kotlarz, A.; Bolla, N.R.; Miłoszewska, J.; Grygorowicz, M.A.; Kutner, A.; Markowicz, S. The effect of analogues of 1α,25-dihydroxyvitamin D$_2$ on the regrowth and gene expression of human colon cancer cells refractory to 5-fluorouracil. *Int. J. Mol. Sci.* **2016**, *17*, E903. [CrossRef] [PubMed]

27. Nachliely, M.; Sharony, E.; Bolla, N.R.; Kutner, A.; Danilenko, M. Prodifferentiation activity of novel vitamin D$_2$ analogs PRI-1916 and PRI-1917 and their combinations with a plant polyphenol toward acute myeloid leukemia cells. *Int. J. Mol. Sci.* **2016**, *17*, E1068. [CrossRef] [PubMed]

28. Kutner, A.; Zhao, H.; Podosenin, A.; Fitak, H.; Chodyński, M.; Halkes, S.J.; Wilson, S.R. Retiferols. Design and Synthesis of a New Class of Vitamin D Analogs. In *Vitamin D A Pluripotent Steroid Hormone: Structural Studies, Molecular Endocrinology and Clinical Applications*; Norman, A.W., Bouillon, R., Thomaset, M., Eds.; de Gruyter & Co.: Berlin, Germany, 1994; pp. 29–30.

29. Kutner, A.; Zhao, H.; Fitak, H.; Wilson, S.R. Synthesis of retiferol RAD$_1$ and RAD$_2$, the lead representatives of new class of *des*-C,D analogs of cholecalciferol. *Bioorg. Chem.* **1995**, *23*, 22–32. [CrossRef]

30. Hilpert, H.; Wirtz, B. Novel versatile approach to an enantiopure 19-*nor*, *des*-C,D vitamin D$_3$ derivative. *Tetrahedron* **2001**, *57*, 681–694. [CrossRef]

31. Zhu, G.D.; Chen, Y.J.; Zhou, X.M.; Vandewalle, M.; de Clercq, P.J.; Versstuyf, A. Synthesis of CD-ring modified 1α,25-dihydroxy vitamin D analogues: C-ring analogues. *Bioorg. Med. Chem. Lett.* **1996**, *6*, 1703–1708. [CrossRef]

32. Yong, W.; Ling, S.D.; Halleweyn, C.; Van Harver, D.; DeClercq, P.; Vandewalle, M.; Bouyllion, R.; Versuyf, A. Synthesis of CD-ring modified 1α,25-dihydroxy vitamin D analogues: five-membered D-ring analogues. *Biorg. Med. Chem. Lett.* **1997**, *7*, 923–928. [CrossRef]

33. Hanazawa, T.; Koyama, A.; Nakat, K.; Okamoto, S.; Sato, F. New convergent synthesis of 1α,25-dihydroxyvitamin D$_3$ and its analogues by Suzuki-Miyaura coupling between A-ring and C,D-ring parts. *J. Org. Chem.* **2003**, *68*, 9767–9772. [CrossRef] [PubMed]

34. Plonska-Ocypa, K.; Sibilska, I.; Sicinski, R.R.; Sicinska, W.; Plum, L.A.; DeLuca, H.F. 13,13-Dimethyl-*des*-C,D analogues of (20*S*)-1α,25-dihydroxy-2-methylene-19-*nor*-vitamin D$_3$ (2MD): Total synthesis, docking to the VDR, and biological evaluation. *Bioorg. Med. Chem.* **2011**, *19*, 7205–7220. [CrossRef] [PubMed]

35. Eelen, G.; Verlinden, L.; Bouillon, R.; De Clercq, P.; Munoz, A.; Verstuyf, A. CD-ring modified vitamin D3 analogs and their superagonistic action. *J. Steroid. Biochem. Mol. Biol.* **2010**, *121*, 417–419. [CrossRef] [PubMed]

36. Gogoi, P.; Seoane, S.; Sihueiro, R.; Guiberteau, T.; Maestro, M.A.; Perez-Fernandez, R.; Rochel, N.; Mourino, A. Aromatic-based design of highly active and noncalcemic vitamin D receptor agonists. *J. Med. Chem.* **2018**. [CrossRef] [PubMed]

37. Tocchini-Valentini, G.; Rochel, N.; Wurtz, J.M.; Mitschler, A.; Moras, D. Crystal structures of the vitamin D receptor complexed to superagonists 20-*epi* ligands. *Proc. Natl. Acad. Sci. USA* **2001**, *98*, 5491–5496. [CrossRef] [PubMed]

38. Singarapu, K.K.; Zhu, J.; Tonelli, M.; Rao, H.; Assadi-Porter, F.M.; Westler, W.M.; DeLuca, H.F.; Markley, J.L. Ligand-specific structural changes in the vitamin D receptor in solution. *Biochemistry* **2011**, *50*, 11025–11033. [CrossRef] [PubMed]

International Journal of
Molecular Sciences

MDPI

Article

Regulation of Expression of *CEBP* Genes by Variably Expressed Vitamin D Receptor and Retinoic Acid Receptor *α* in Human Acute Myeloid Leukemia Cell Lines

Aleksandra Marchwicka and Ewa Marcinkowska *

Department of Biotechnology, University of Wroclaw, Joliot-Curie 14a, 50-383 Wroclaw, Poland;
alexandramarchwicka@interia.pl
* Correspondence: ema@cs.uni.wroc.pl; Tel.: +48-71-375-29-29

Received: 5 June 2018; Accepted: 28 June 2018; Published: 29 June 2018

Abstract: All-*trans*-retinoic acid (ATRA) and 1α,25-dihydroxyvitamin D (1,25D) are potent inducers of differentiation of myeloid leukemia cells. During myeloid differentiation specific transcription factors are expressed at crucial developmental stages. However, precise mechanism controlling the diversification of myeloid progenitors is largely unknown, CCAAT/enhancer-binding protein (C/EBP) transcription factors have been characterized as key regulators of the development and function of the myeloid system. Past data point at functional redundancy among C/EBP family members during myeloid differentiation. In this study, we show that in acute myeloid leukemia (AML) cells, high expression of vitamin D receptor gene (*VDR*) is needed for strong and sustained upregulation of *CEBPB* gene, while the moderate expression of *VDR* is sufficient for upregulation of *CEBPD* in response to 1,25D. The high expression level of the gene encoding for retinoic acid receptor α (*RARA*) allows for high and sustained expression of *CEBPB*, which becomes decreased along with a decrease of *RARA* expression. Expression of *CEBPB* induced by ATRA is accompanied by upregulated expression of *CEBPE* with similar kinetics. Our results suggest that *CEBPB* is the major VDR and RARA-responsive gene among the *CEBP* family, necessary for expression of genes connected with myeloid functions.

Keywords: nuclear receptors; CCAAT-enhancer-binding proteins; *CEBP* genes; vitamin D receptor; retinoic acid receptor α; differentiation; acute myeloid leukemia

---

1. Introduction

CCAAT/enhancer-binding proteins (C/EBPs) are transcription factors that activate the expression of target genes through interaction with response elements within their promoters [1]. There are six members of C/EBP family, and they regulate differentiation process in various tissues [2]. The first transcription factor in this family, C/EBPα, was isolated from the rat liver and it appeared to be important for adipocyte differentiation [3]. C/EBPs are modular proteins consisting of an activation domain, a DNA binding domain, and a leucine-rich dimerization domain that is responsible for forming dimers with other members of the family [4]. In order to activate transcription, the C/EBP dimers bind to the consensus sequence 5'-TT/GNNGNAAT/G-3' in promoter regions of target genes. For three out of six genes encoding C/EBP family members, alternative protein products are translated, due to a leaky ribosomal scanning mechanism. Some of them lack the N-terminal activation domains and exert inhibitory functions, presumably by a dominant negative mechanism [5].

Hematopoiesis is a process in which all blood elements are formed from multipotential hematopoietic stem cells (HSCs). In the process of hematopoiesis, the HSCs and their progeny interact with the bone marrow stromal cells and they are stimulated by the numerous growth factors that are secreted in the bone marrow environment. The eventual cell fate during hematopoiesis is governed by spatiotemporal fluctuations in transcription factor concentrations, which either cooperate or compete in driving target gene expression [6]. Some members of C/EBP family of transcription factors are important at certain steps of hematopoiesis [7]. C/EBPα appears in differentiating cells at the stage of early progenitors with lymphoid and myeloid potential and then reappears only in the cells that are differentiating into granulocytes [8]. C/EBPα-deficient mice show disturbances in monocyte and neutrophil development [9,10]. High level of C/EBPβ leads to monocyte and macrophage development [11,12], while high level of C/EBPε leads to neutrophil differentiation [13]. The role of C/EBPδ in blood cells development is less defined, since *CEBPD*−/− mice did not reveal any blood disturbances [14]. It has been documented that C/EBPδ regulates expression of genes important for granulocyte function [15].

However, the most important factors that drive blood cells development are cytokines [16], some ligands for nuclear receptors can also modulate cell fate during hematopoiesis [7]. The best described in this respect are ligands for retinoic acid receptors (RARs). Active metabolites of vitamin A are natural ligands for RARs. A dominating retinoic acid (RA) metabolite is all-*trans*-RA (ATRA), which binds with high affinity to all RARs (α, β, and γ) [17]. During embryogenesis, ATRA causes the appearance of hematopoietic progenitors from the hemogenic endothelium [18], while in adults, it is important for the differentiation of granulocytes, as well as B and T lymphocytes [19]. This activity of ATRA has been used in clinics. The most clinically significant application of ATRA is to treat a rare subtype of an acute myeloid leukemia (AML), called acute promyelocytic leukemia (APL). At the first description this subtype was considered the most difficult to treat [20], while it is now considered as highly curable using the combination of ATRA and anthracycline-based chemotherapy [21]. Another ligand for the nuclear receptor which influences hematopoiesis is an active metabolite of vitamin D. The correct physiological concentrations of 1,25-dihydroxyvitamin D (1,25D), which is a natural ligand for vitamin D receptor (VDR), are necessary to induce markers of monocytic differentiation in HSCs [22]. The expression of the *VDR* gene is higher at the early steps of hematopoiesis than at later stages and in mature blood cells [23]. However, both these ligands do not seem indispensable for blood cells development since RARα-deficient and VDR-deficient mice show no defects in hematopoiesis [24,25]. The possibility that these nuclear receptors can, in some aspects, functionally compensate each other should be considered.

It has been documented in the past that members of C/EBP family of transcription factors can be upregulated in blood cells by an exposure to RA, 1,25D, or to their active analogs. For example, the expression of C/EBPε mRNA and protein increases in AML cells exposed to 9-*cis*–RA or 1,25D analog (KH1060) [26]. The gene encoding C/EBPβ has been shown to be a target for VDR regulation [27] and all isoforms of this transcription factor are increased in AML cells exposed to 1,25D or to analogs of 1,25D [11,28]. This gene is also strongly upregulated in AML cells that were exposed to ATRA [29]. 1,25D induces a transient increase of C/EBPα [11], which also participates in the ATRA-induced differentiation of AML cells [30].

In this study, we addressed a question of whether the lack of one of the nuclear receptors mentioned above could be compensated by the other in terms of *CEBP* activation. Therefore, we used four cell lines in our study, with different expression of retinoic acid receptor α (*RARA*) or *VDR*. In HL60 cells, *VDR* expression is on a high level and *RARA* is moderate [31]. For the purpose of this study, we silenced the expression of *VDR* in HL60 cells using shRNA. In contrast to HL60 cells, KG1 cell express high levels of *RARA*, but low of *VDR* [31]. The effects of *RARA* silencing were studied using a sub-line KG1-RARα(−).

2. Results

2.1. Activation of Expression of CEBP Transcription Factor's Genes in AML Cells with High Level of VDR and Low Level of RARα

In previous studies, we have shown that different AML cell lines have variable sensitivity to 1,25D– and ATRA-induced differentiation [32]. HL60 cell line responded to 1,25D with robust monocytic differentiation and to ATRA with moderate granulocytic differentiation. That corresponded to high basal level of expression of *VDR* and low basal level of expression of *RARA* [31]. In view of demonstrated regulation of differentiation of myeloid leukemia cells by these two compounds, it was of interest to determine the expression profiles of *CEBP* genes in response to 1,25D and ATRA in HL60 cells. Therefore, the expression of *CEBPA, CEBPB, CEBPD, CEBPE,* and *CEBPG* in HL60 cells that were exposed to 1 µM ATRA or to 10 nM 1,25D for different time periods was tested. As depicted in Figure 1a, transient upregulation of *CEBPA* was detected in HL60 cells stimulated with 1,25D, followed by fast decline. This was in concordance with an earlier observed transient upregulation of C/EBPα protein in HL60 cells after exposure to 1,25D [11]. The increase in expression of *CEBPB* was more sustained, with a peak at 24 h and more gradual decline. As presented before, protein level of C/EBPβ follows this sustained expression pattern and it peaks between two and three days of exposure to 1,25D [11]. *CEBPD, CEBPE,* and *CEBPG* were not upregulated in response to 1,25D exposure of HL60 cells. As presented in Figure 1b in HL60 cells that were exposed to ATRA, *CEBPA* was upregulated weakly and transiently. Expression of *CEBPB* and *CEBPE* was stimulated by *ATRA* stronger and in a sustained manner. Modest upregulation of *CEBPD* was observed at 96 h from exposure to ATRA. Again, no stimulation of *CEBPG* was observed. Values of mRNA expression obtained using comparative quantification algorithm are presented in Table A1.

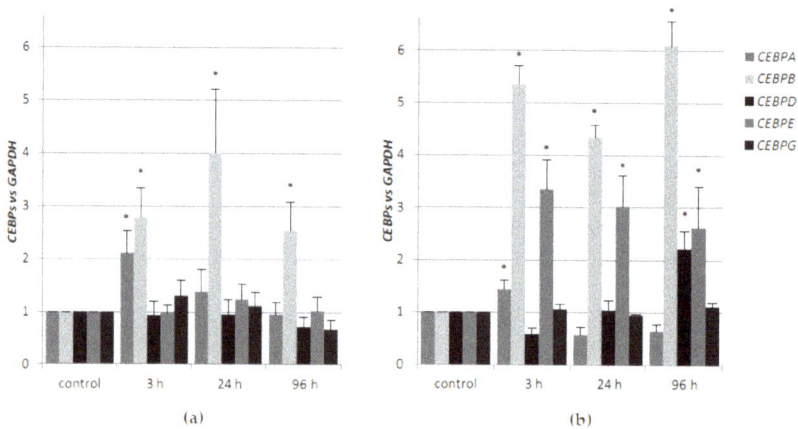

Figure 1. Regulation of *CEBP* genes in HL60 exposed to 1,25D or all-*trans*-retinoic acid (ATRA). HL60 cells were exposed to 10 nM 1,25D (**a**) or 1 µM ATRA (**b**) and after desired time the expression of *CEBP* genes was measured by Real-time PCR. The bars represent mean values (±standard error of the mean (SEM)) of the fold changes in mRNA levels relative to glyceraldehyde 3-phosphate dehydrogenase (*GAPDH*) mRNA levels. Expressions in control cells were treated as calibrators. Values significantly different from those obtained from respective controls cells are marked with an asterisk (* $p < 0.05$).

2.2. Activation of Expression of CEBP Transcription Factor's Genes in AML Cells with Low Level of VDR and High Level of RARα

In contrast to HL60 cells, KG1 cells are not responsive to 1,25D and they have a low level of VDR protein, whilst being susceptible to ATRA-driven granulocytic differentiation [33]. This corresponds with the high basal level of expression of *RARA* gene and high constitutive content of RARα protein [31]. In the next series of experiments, KG1 cells were exposed to 1 μM ATRA or to 10 nM 1,25D for different time periods. In KG1 cells, the transcript levels of *CEBP* genes remained unchanged after exposure to 1,25D (Figure 2a). In contrast, significant changes in expression of *CEBP* genes after exposure of KG1 cells to ATRA were observed. Modest upregulation of *CEBPA* was detected at 3 h and 96 h from exposure. *CEBPB* was the most responsive to ATRA out of the genes studied, the expression upregulation was fast and long-lasting. The second ATRA-responsive gene was *CEBPE*, where the expression peaked at 24 h. The expression of *CEBPD* and *CEBPG* was modest with a peak at 96 h (Figure 2b). Values of mRNA expression that were obtained using comparative quantification algorithm are presented in Table A2.

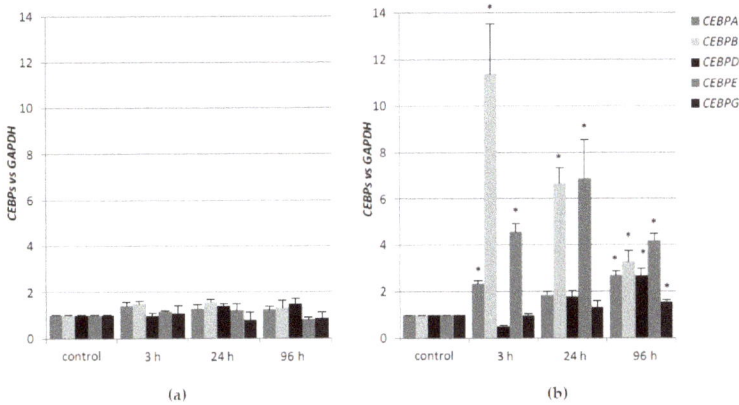

Figure 2. Regulation of *CEBP* genes in KG1 exposed to 1,25D (**a**) or to ATRA (**b**). KG1 cells were exposed to 10 nM 1,25D or 1 μM ATRA and after desired time the expression of *CEBP* genes was measured by Real-time PCR. The bars represent mean values of the fold changes (±SEM) in mRNA levels relative to *GAPDH* mRNA levels. Expressions in control cells were treated as calibrators. Values significantly different from those that were obtained from respective controls cells are marked with asterisk (* $p < 0.05$).

2.3. Effects of Silencing High RARA on Expression of CEBP Transcription Factor's Genes in KG1 Cells

In an attempt to elucidate whether the lack of one of the nuclear receptors VDR and RARα could be compensated by the other in terms of *CEBP* activation, we used KG1 sublines with silenced *RARA* gene (KG1-RARα(−)) and KG1 control cells (KG1-CTR), which were obtained before [31]. These cells have substantially reduced level of *RARA* gene expression and RARα protein, but also exhibit the increased expression of *VDR* gene and VDR protein, when compared to wild-type KG1 and KG1-CTR [31]. It should be noted that the expression of *VDR* gene in KG1-RARα(−) is still lower than in HL60 cells (Figure 3a), and it was not sufficient to induce antigen CD14 typical for monocytes (Figure 3b). The KG1 sublines were stimulated with 1,25D or ATRA in a similar manner as before. As presented in Figure 3e, KG1-RARα(−) cells started to be responsive to 1,25D, however in a manner that was different from HL60 cells. Only *CEBPD* gene became responsive to 1,25D in RARA silenced KG1 cells, and the expression of *CEBPB* remained at a control level. As expected, expression levels of *CEBPA*, *CEBPB*, and *CEBPE* were reduced in KG1-RARα(−) when compared to KG1-CTR cells after ATRA stimulation, especially at the early hours from stimulation (Figure 3c–f). Values of mRNA expression that were obtained using comparative quantification algorithm are presented in Table A3.

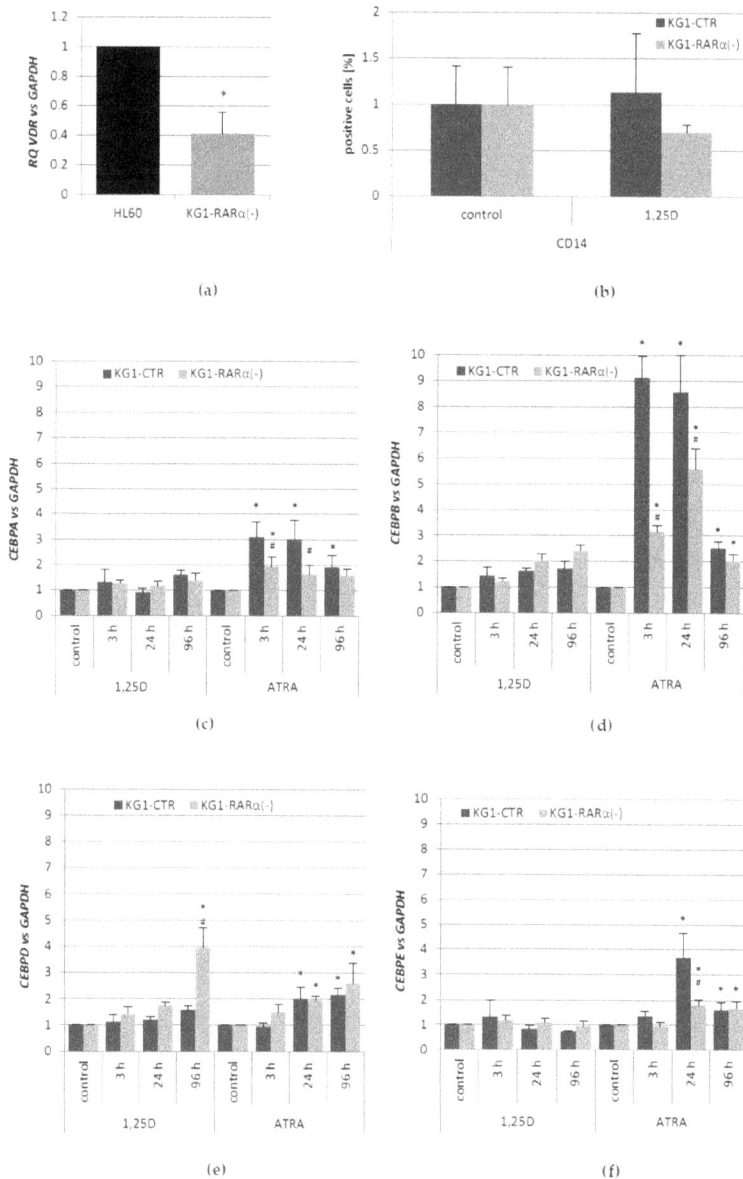

Figure 3. Responses to 1,25D and to ATRA in KG1 cells with silenced retinoic acid receptor α (*RARA*) gene. Expression of vitamin D receptor (*VDR*) gene in KG1-RARα(−) cells compared to HL60 cells, which were treated as calibrator (**a**). Differentiation of KG1 sublines in response to 1,25D. KG1-CTR and KG1-RARα(−) cells were stimulated with 10 nM 1,25D for 96 h and then the expression of CD14 differentiation marker was detected using flow cytometry (**b**). Expression of *CEBPA* (**c**), *CEBPB* (**d**), *CEBPD* (**e**), and *CEBPE* (**f**) genes in KG1-CTR and KG1-RARα(−). Cells were stimulated with 10 nM 1,25D or 1 μM ATRA and after desired time the expression of *CEBP* genes was measured by Real-time PCR. The bars represent mean values of the fold changes (±SEM) in mRNA levels relative to *GAPDH* mRNA levels. Expressions in control cells were treated as calibrators. Values that are significantly different from those obtained from respective controls cells are marked with asterisk (* $p < 0.05$); values that differ significantly from those obtained from respective KG1-CTR control cells are marked with hash (# $p < 0.05$).

2.4. Effects of Silencing High VDR on Expression of CEBP Transcription Factor's Genes in HL60 Cells

Having shown that KG1-RARα(−) cells demonstrate an altered *CEBP* expression profile, we decided to silence the expression of VDR gene in HL60 cells. The gene silencing was performed using shRNA plasmid and the scrambled shRNA plasmid, as described before [34]. This way, two HL60 sublines were generated: HL60-VDR(−) and HL60-CtrA. In order to validate whether the expression of *VDR* gene was indeed efficiently knocked down in HL60-VDR(−) cells, the mRNA and protein levels were compared to HL60-CtrA cells. Unfortunately, the silencing was far from complete and mRNA level was reduced to approximately 80% of the initial level (Figure 4a). In order to verify whether this reduction would lead to VDR-dependent effects, the expression of the gene that encodes 24-hydroxylase of 1,25D (CYP24A1) was tested in both HL60 sublines exposed to 10 nM 1,25D. CYP24A1 is the most strongly regulated out of all 1,25D-target genes and is the best measure of VDR's activity [35]. 1,25D-induced expression of *CYP24A1* was significantly reduced in HL60-VDR(−) cells when compared to HL60-CtrA cells. Figure 4c shows that VDR protein content was also significantly reduced in the nuclei of HL60-VDR(−) cells to 77% in control cells, and to 39% after 1,25D treatment when compared to HL60-CtrA cells.

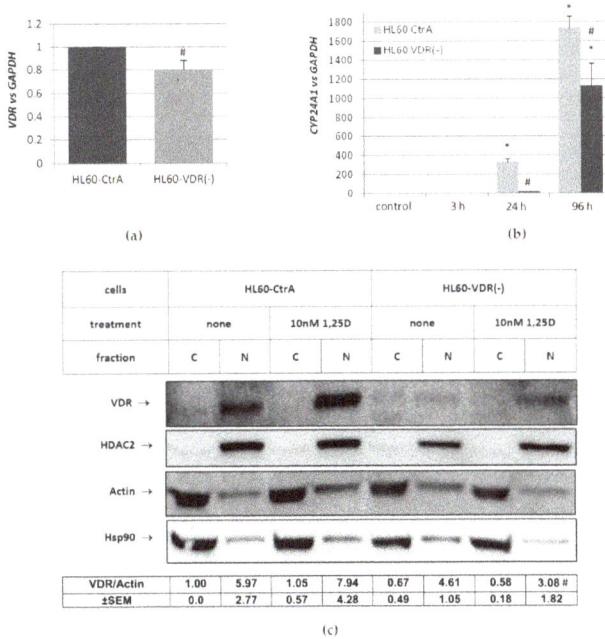

(a)

(b)

cells	HL60-CtrA				HL60-VDR(-)			
treatment	none		10nM 1,25D		none		10nM 1,25D	
fraction	C	N	C	N	C	N	C	N
VDR →								
HDAC2 →								
Actin →								
Hsp90 →								
VDR/Actin	1.00	5.97	1.05	7.94	0.67	4.61	0.58	3.08 #
±SEM	0.0	2.77	0.57	4.28	0.49	1.05	0.18	1.82

(c)

Figure 4. Generation of HL60 cells with reduced *VDR* expression. Constitutive expression of *VDR* gene in HL60-CtrA and HL60-VDR(−) cells. Expression in HL60-CtrA was treated as calibrator (**a**). Expression of *CYP24A1* was measured in both cell lines exposed to 10 nM 1,25D for different times (**b**). The levels of VDR protein were determined in the cytosol and nuclei of HL60-CtrA and HL60-VDR(−) cells by western blots after 10 nM 1,25D stimulation for 24h (**c**). Cell lysates were tested while using anti-VDR. Proper cell fractionation was revealed using anti-histone deacetylase 2 (anti-HDAC2), while proper lane loading using anti-Hsp90 and anti-actin. Values below the blots are means (±SEM), as obtained from five experiments. The bars represent mean values of the fold changes (±SEM) in mRNA levels relative to *GAPDH* mRNA levels. Values that are significantly different from those obtained from respective controls cells are marked with asterisk (* $p < 0.05$); values that differ significantly from those obtained from respective HL60-CtrA cells are marked with hash (# $p < 0.05$).

Even though HL60-VDR(−) subline was not entirely devoid of VDR, the cells were exposed 1 μM ATRA or 10 nM 1,25D in order to examine the expression of selected *CEBP* genes. *CEBPA* and *CEBPB* have been selected, since they are direct targets of VDR-dependent transcriptional regulation [36,37]. As presented in Figure 5a,b, the limited decrease of *VDR* expression level resulted in a reduced *CEBPA* and *CEBPB* expression levels in response to 1,25D when compared to HL60-CtrA cells. Interestingly, the response to ATRA in HL60-VDR(−) cells was different than in HL60-CtrA cells, and upregulated regarding *CEBPA* expression, while downregulated regarding *CEBPB* expression. Values of mRNA expression that were obtained using comparative quantification algorithm are presented in Table A4.

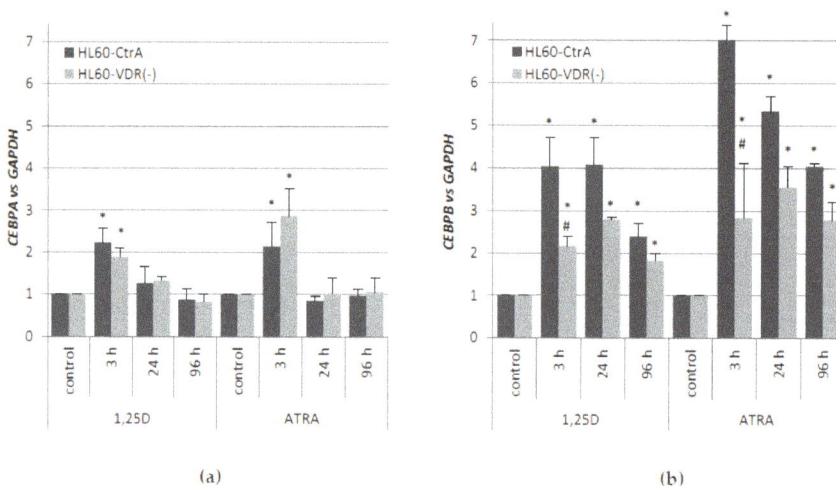

(a)   (b)

Figure 5. Expression of *CEBPA* (**a**) and *CEBPB* (**b**) genes in HL60-CtrA and HL60-VDR(−). Cells were exposed to 10 nM 1,25D or 1 μM ATRA and after desired time the expression of *CEBP* genes was measured by Real-time PCR. The bars represent mean values of the fold changes (±SEM) in mRNA levels relative to *GAPDH* mRNA levels. Expressions in control cells were treated as calibrators. Values that are significantly different from those obtained from respective controls cells are marked with an asterisk (* $p < 0.05$); values that differ significantly from those obtained from respective HL60-CtrA cells are marked with hash (# $p < 0.05$).

3. Discussion

The process of hematopoiesis leads to the acquisition of immune functions by terminally differentiated cells. Lineage selection within hematopoiesis depends on the appropriate levels of key transcription factors, which are regulated in response to numerous hematopoietic cytokines and interactions with bone marrow environment [38]. Transcription factors C/EBPα, C/EBPβ, and C/EBPδ have been demonstrated in granulocytes, monocytes, and eosinophils, as well as in myeloid progenitor cells [39]. C/EBPε has been identified as a critical regulator of terminal granulopoiesis [13]. Many genes that are important for myeloid functions contain in their promoters binding sites for C/EBP transcription factors [5]. In normal hematopoiesis, C/EBP transcription factors are produced in response to coordinated actions of cytokines and upstream transcription factors, and their activity is further modulated by posttranslational modifications [5].

C/EBPα seems to be the most important for normal blood development, since mutations in *CEBPA* gene lead to AML. *CEBPA* is mutated in around 13% of all AML patients [40], and mutations in this gene appear early, indicating at the driver role in leukemogenesis [41]. This is why the expression of *CEBPA* in AML patients has been extensively studied, and it has been shown to be downregulated by another driver of leukemogenesis, namely the AML-ETO fusion protein, which is present in 5–10%

of patients with AML [42]. Later studies have documented that the downregulation of *CEBPA* also accompanies AML cases with inv(16), which creates *CBFB-MYH11* gene fusion and occurs in about 10% of AML patients [43]. All together, the above data show that more than 30% of patients with AML exhibit disturbances in expression of *CEBPA*.

It has been shown that both 1,25D and ATRA are able to upregulate expression of C/EBP factors without the addition of hematopoietic cytokines [11,26,27,29]. Whether all the *CEBP* genes are direct targets for either RARα or VDR is not clear. Retinoic acid response elements (RAREs) have been found in the promoter of *CEBPE* gene, and not in other genes of this family [44], but at present, we know that RARα can bind to big variety of RAREs [45], which are sometimes located in a long distance from the transcription start [44]. *CEBPA* and *CEBPB* are direct targets of VDR-dependent transcriptional regulation [36,37], but it is not sure whether such a mechanism occurs also for *CEBPD*.

We thus wanted to find out how variable levels of VDR and RARα proteins affect the expression of *CEBP* genes, and whether the lack of one of the nuclear receptors that are mentioned above, could be compensated by the other. In the first place, we determined the expression profiles of these genes after stimulation with 1,25D or ATRA in HL60 and KG1 cells. These cells differ in basal levels of expression of VDR and RARα, and in susceptibility to 1,25D-induced differentiation. The basal levels of *VDR* and *RARA* mRNA expression in all of the cell lines that were used for the purpose of this research is presented in Table A5. HL60 cells, which have high level of VDR protein, respond to 1,25D with transient upregulation of *CEBPA*, and strong and sustained upregulation of *CEBPB*. It appeared that KG1 cells that have low level of VDR protein do not express *CEBP* genes in response to 1,25D at all. After the silencing of the *RARA* gene in KG1, these cells reduced the responsiveness to ATRA, but started to be responsive to 1,25D, most probably because of an increased expression of VDR [31]. The restored *VDR* expression level was not high enough to upregulate the expression of *CEBPB*, but it was sufficient to upregulate *CEBPD*. As presented here, the upregulation of *CEBPD* alone was not sufficient to complete the myeloid differentiation process. KG1 cells and HL60 cells are both responsive to ATRA, however, due to higher basal expression of *RARA*, KG1 cells respond stronger than HL60. In KG1 cells, the upregulation of *CEBPB* and *CEBPE* is approximately two times higher than in HL60 cells in response to ATRA; however, the kinetics of expression is similar.

Our results suggest that the ability of 1,25D or ATRA to effectively force the final myeloid differentiation of AML cells strongly depends on effective levels of nuclear receptors for these compounds. It also seems that expression of *CEBPB* is indispensable for the final effect of myeloid differentiation, and that VDR and RARα do not compensate each other in terms of the induction of *CEBP* expression. Our data are in agreement with the earlier findings that strong and sustained expression of *CEBPB*, when accompanied by transient expression of *CEBPA* leads to the differentiation towards monocytes [11], while, when accompanied by the sustained expression of *CEBPE*, it leads the differentiation process to granulocytes [46].

4. Materials and Methods

4.1. Cell Lines and Cultures

HL60 cells were from the local cell bank at the Institute of Immunology and Experimental Therapy in Wrocław, and KG1 cells were purchased from the German Resource Center for Biological Material (DSMZ GmbH, Braunschweig, Germany). The cells were grown in RPMI-1640 medium with 10% fetal bovine serum, 100 units/ml penicillin, and 100 µg/mL streptomycin (Sigma, St. Louis, MO, USA), and maintained at standard cell culture conditions.

4.2. Chemicals and Antibodies

1,25D was purchased from Cayman Europe (Tallinn, Estonia) and ATRA was from Sigma-Aldrich (St. Louis, MO, USA). The compounds were dissolved in an absolute ethanol to 1000× final concentrations, and subsequently, diluted in the culture medium to the required concentration.

4.3. cDNA Synthesis and Real-Time PCR

Total RNA was extracted from the cells treated with 1 µM ATRA or 10 nM 1,25D for different time points (3 h, 24 h, 96 h). Briefly, the isolation of total RNA, reverse transcription into cDNA, and Real-time PCR reactions were performed as published before [33], using CFX Real-time PCR System (Bio-Rad Laboratories Inc., Hercules, CA, USA). The sequences of *GAPDH*, *CYP24A1*, *VDR*, and *RARA* primers, and the reaction conditions were described previously [31,47]. The *CEBPA*, *CEBPB*, *CEBPD*, *CEBPE*, and *CEBPG* primers were obtained from RealTimePrimers.com (Real Time Primers, LLC, PA, USA). Their sequences are as follows: *CEBPA*: forward 5'-TTGGTGCGTCTAAGATGAGG-3', reverse 5'-GGCAGGAAACCTCCAAATAA-3'; *CEBPB*: forward 5'-AACTCTCTGCTTCTCCCTCTG-3', reverse 5'-AAGCCCGTAGGAACATCTTT-3'; *CEBPD*: forward 5'-ATCGACTTCAGCGCCTACAT-3', reverse 5'-GCCTTGTGATTGCTGTTGAA-3'; *CEBPE*: forward 5'-GAGGAGGTTGCTCAGAGTGG-3', reverse 5'-TCCTGGCCTATTCAGCAGTT-3'; *CEBPG*: forward 5'-GAACAACCCATTTTGCACTC-3', reverse 5'-TGAAAGCCAGGAACAAAAAG-3'; *APDH*: forward 5'-CATGAGAAGTATGACAACAGCCT-3', reverse 5'-AGTCCTTCCACGATACCAAAGT-3'. Quantification of gene expression was analyzed with either the ΔCq (to present comparative quantification of expression levels) or with the ΔΔCq (to present changes in expression induced by treatment) methods using *GAPDH* as the endogenous control Primers efficiencies were measured in all of the cell lines using Real-time PCR reaction based on the slope of the standard curve. The results were normalized to primer efficiencies to compare gene expression in different cell lines [48]. Real-time PCR assays were performed at least in triplicate.

4.4. Flow Cytometry

The expression of CD14 was determined by flow cytometry. The cells were incubated with 10 nM 1,25D for 96 h, then washed, and stained with 1 µL of Phycoerythrin labeled antibody (or the appropriate control immunoglobulins; both from ImmunoTools, Friesoythe, Germany) for 1 h on ice. Next, they were washed with ice-cold PBS supplemented with 0.1% BSA and suspended in 0.5 mL of PBS supplemented with 0.1% BSA prior to analysis on FACS Calibur flow cytometer (Becton–Dickinson, San Jose, CA, USA). Experiments were repeated at least three times. The acquisition parameters were set for an isotype control. Data analysis was performed with the use of WinMDI 2.8 software (freeware by Joseph Trotter).

4.5. Western Blotting

In order to obtain cytosolic and nuclear extracts, 5×10^6 cells/sample were washed and lysed using NE-PER Nuclear and Cytoplasmic Extraction Reagents (Thermo Fisher Scientific Inc., Worcester, MA, USA), according to the user's manual. Lysates were denatured by adding 5× sample buffer (1/4 volume of the lysate) and boiled for 5 min. 25 µL of each lysate were separated in SDS-PAGE and then electroblotted to PVDF membrane. The membranes were then dried and incubated sequentially with primary and a horseradish peroxidase-conjugated secondary antibody. The protein bands were visualized with a chemiluminescence. Then, the membranes were stripped, dried again, and probed with subsequent antibodies. Western blots were repeated five times.

4.6. Gene Silencing Reagents and Procedure

The *RARA* gene silencing in KG1 cells was described before [31]. The *VDR* gene silencing in HL60 cells was performed using shRNA plasmids and Neon® Transfection System (Invitrogen™, Carlsbad, CA, USA) using control shRNA plasmid-A (sc-108060) and the *VDR* shRNA plasmid (sc-106692-SH; both from Santa Cruz Biotechnology, Inc., Dallas, TX, USA). The procedure of electrotransfection by Neon® Transfection System was described before [49].

4.7. Statistical Analysis

For statistical analysis one-way ANOVA was used to test the null hypothesis that the samples in two or more groups are drawn from populations with the same mean values. When the ANOVA test had shown that the null hypothesis is not true, the Student's *t*-test for independent samples was used to analyze the differences between the pairs of groups (Excel, Microsoft Office and free ANOVA Calculator: http://www.danielsoper.com/statcalc3/calc.aspx?id=43).

Author Contributions: Both authors conceived and designed the experiments; analyzed the data; provided financial support and wrote the paper. E.M. conceived the general idea of the research and performed one western blot experiment. A.M. performed most of the experiments.

Funding: The research was funded by a grant PRELUDIUM No 2013/11/N/NZ3/00197 (A.M.), and grant OPUS No 2016/23/B/NZ5/00065 (E.M.), both from the National Science Centre in Poland. Publication cost was supported by Wroclaw Center of Biotechnology Program, The Leading National Research Center (KNOW) for years 2014–2018.

Conflicts of Interest: The authors declare no conflict of interest.

Abbreviations

ATRA	all-*trans*-retinoic acid
1,25D	1α,25-dihydroxyvitamin D
C/EBP	CCAAT/enhancer-binding protein
RARs	retinoic acid receptors
AML	acute myeloid leukemia
HSCs	hematopoietic stem cells
APL	acute promyelocytic leukemia
VDR	vitamin D receptor
GAPDH	glyceraldehyde 3-phosphate dehydrogenase
SEM	standard error of the mean

Appendix A

Table A1. Comparative quantification of *CEBP* genes related to a reference *GAPDH* gene in HL60 cells.

HL60		*CEBPA*	SEM	*CEBPB*	SEM	*CEBPD*	SEM	*CEBPE*	SEM	*CEBPG*	SEM
control		0.16	0.06	0.07	0.02	0.004	0.001	0.004	0.002	0.068	0.026
1,25D	3 h	0.33	0.07	0.19	0.03	0.003	0.001	0.003	0.002	0.085	0.051
	24 h	0.22	0.03	0.27	0.03	0.003	0.001	0.004	0.002	0.072	0.042
	96 h	0.15	0.05	0.18	0.02	0.003	0.001	0.004	0.001	0.046	0.030
ATRA	3 h	0.18	0.04	0.06	0.03	0.003	0.001	0.002	0.002	0.058	0.014
	24 h	0.25	0.07	0.30	0.14	0.002	0.000	0.007	0.008	0.061	0.020
	96 h	0.13	0.04	0.25	0.07	0.004	0.001	0.006	0.0071	0.055	0.012

Table A2. Comparative quantification of *CEBP* genes related to a reference *GAPDH* gene in KG1 cells.

KG1		*CEBPA*	SEM	*CEBPB*	SEM	*CEBPD*	SEM	*CEBPE*	SEM	*CEBPG*	SEM
control		0.025	0.016	0.039	0.000	0.0006	0.000	0.00003	0.0000	0.0037	0.0009
1,25D	3 h	0.035	0.022	0.060	0.000	0.0006	0.000	0.00003	0.0000	0.0041	0.0010
	24 h	0.032	0.015	0.063	0.000	0.0008	0.000	0.00003	0.0000	0.0031	0.0009
	96 h	0.031	0.058	0.053	0.000	0.0009	0.002	0.00002	0.0000	0.0034	0.0018
ATRA	3 h	0.026	0.011	0.039	0.000	0.0006	0.000	0.00002	0.0000	0.0023	0.0011
	24 h	0.063	0.030	0.496	0.000	0.0003	0.000	0.00009	0.0000	0.0023	0.0012
	96 h	0.050	0.027	0.284	0.000	0.0011	0.000	0.00013	0.0001	0.0031	0.0018

Table A3. Comparative quantification of *CEBP* genes related to a reference *GAPDH* gene in KG1-CTR and KG1-RARα(−) cells.

KG1 CTR		*CEBPA*	SEM	*CEBPB*	SEM	*CEBPD*	SEM	*CEBPE*	SEM
control		0.0061	0.0022	0.0199	0.0016	0.00002	0.0000	0.00003	0.0022
1,25D	3 h	0.0056	0.0019	0.0197	0.0013	0.00002	0.0000	0.00003	0.0018
	24 h	0.0048	0.0011	0.0282	0.0008	0.00002	0.0000	0.00002	0.0012
	96 h	0.0106	0.0017	0.0373	0.0012	0.00003	0.0000	0.00002	0.0016
ATRA	3 h	0.0153	0.0056	0.1484	0.0040	0.00001	0.0000	0.00003	0.0056
	24 h	0.0133	0.0080	0.1242	0.0056	0.00002	0.0000	0.00008	0.0079
	96 h	0.0134	0.0007	0.0577	0.0005	0.00003	0.0000	0.00005	0.0007
KG1 RARα(−)		*CEBPA*	SEM	*CEBPB*	SEM	*CEBPD*	SEM	*CEBPE*	SEM
control		0.0082	0.0023	0.0261	0.0016	0.00003	0.0000	0.00002	0.0022
1,25D	3 h	0.0123	0.0034	0.0377	0.0024	0.00003	0.0000	0.00003	0.0033
	24 h	0.0102	0.0034	0.0548	0.0024	0.00003	0.0000	0.00003	0.0033
	96 h	0.0120	0.0042	0.0660	0.0030	0.00007	0.0000	0.00002	0.0041
ATRA	3 h	0.0140	0.0038	0.0729	0.0027	0.00002	0.0000	0.00002	0.0037
	24 h	0.0135	0.0025	0.1457	0.0018	0.00003	0.0000	0.00004	0.0025
	96 h	0.0154	0.0036	0.0619	0.0025	0.00004	0.0000	0.00005	0.0034

Table A4. Comparative quantification of *CEBP* genes related to a reference *GAPDH* gene in HL60-CtrA and HL60-VDR(−) cells.

HL60 CtrA		*CEBPA*	SEM	*CEBPB*	SEM	HL60 VDR(−)		*CEBPA*	SEM	*CEBPB*	SEM
control		0.06228	0.0214	0.0753	0.0125	control		0.09500	0.0318	0.08958	0.0099
1,25D	3 h	0.05003	0.0089	0.3774	0.2312	1,25D	3 h	0.07767	0.0227	0.28449	0.0147
	24 h	0.07403	0.0172	0.2351	0.0788		24 h	0.12883	0.0455	0.41116	0.2031
	96 h	0.14063	0.0437	0.1749	0.0622		96 h	0.17163	0.0468	0.24263	0.0734
ATRA	3 h	0.06025	0.0180	0.1710	0.0118	ATRA	3 h	0.09056	0.0082	0.24728	0.0580
	24 h	0.05366	0.0167	0.2221	0.0188		24 h	0.09029	0.0166	0.40953	0.0871
	96 h	0.13521	0.0491	0.6720	0.0612		96 h	0.24610	0.0229	0.358334	0.2188

Table A5. Comparative quantification of *VDR* and *RARA* genes related to a reference *GAPDH* gene in all cell lines used in this study.

Cell Line	VDR	SEM	RARA	SEM
HL60	0.00048	0.00001	0.0679	0.001
HL60 CtrA	0.00045	0.00003	0.0600	0.003
HL60 VDR(−)	0.00035	0.00006	0.0638	0.003
KG1	0.00002	0.00001	0.1241	0.008
KG1CTR	0.00005	0.00000	0.1792	0.002
KG1 RARα(−)	0.00018	0.00003	0.0563	0.003

References

1. Nerlov, C. The C/EBP family of transcription factors: A paradigm for interaction between gene expression and proliferation control. *Trends Cell Biol.* **2007**, *17*, 318–324. [CrossRef] [PubMed]
2. Tsukada, J.; Yoshida, Y.; Kominato, Y.; Auron, P. The CCAAT/enhancer (C/EBP) family of basic-leucine zipper (bZIP) transcription factors is a multifaceted highly-regulated system for gene regulation. *Cytokine* **2011**, *54*, 6–19. [CrossRef] [PubMed]
3. Johnson, P.; Landschulz, W.; Graves, B.; McKnight, S. Identification of a rat liver nuclear protein that binds to the enhancer core element of three animal viruses. *Genes Dev. Biol.* **1987**, *2*, 133–146. [CrossRef]

4. Landshulz, W.; Johnson, P.; McKnight, S. The DNA binding domain of the rat liver nuclear protein C/EBP is bipartite. *Science* **1989**, *243*, 1681–1688. [CrossRef]

5. Ramji, D.; Foka, P. CCAAT/enhancer-binding proteins: structure, function and regulation. *Biochem. J.* **2002**, *365*, 561–575. [CrossRef] [PubMed]

6. Brown, G.; Hughes, P.; Michell, R.; Rolink, A.; Ceredig, R. The sequential determination model of hematopoiesis. *Trends Immunol.* **2007**, *28*, 442–448. [CrossRef] [PubMed]

7. Studzinski, G.; Marcinkowska, E. Intracellular signaling for granulocytic and monocytic differentiation. In *Cell Determination during Hematopoiesis*; Brown, G., Ceredig, R., Eds.; Nova Science Publishers: Hauppauge, NY, USA, 2009; pp. 53–77.

8. Friedman, A. Transcriptional control of granulocyte and monocyte development. *Oncogene* **2007**, *26*, 6816–6828. [CrossRef] [PubMed]

9. Heath, V.; Suh, H.; Holman, M.; Renn, K.; Gooya, J.; Parkin, S.; Klarmann, K.; Ortiz, M.; Johnson, P.; Keller, J. C/EBPα deficiency results in hyperproliferation of hematopoietic progenitor cells and disrupts macrophage development in vitro and in vivo. *Blood* **2004**, *104*, 1639–1647. [CrossRef] [PubMed]

10. Zhang, D.; Zhang, P.; Wang, N.; Hetherington, C.; Darlington, G.; Tenen, D. Absence of granulocyte colony-stimulating factor signaling and neutrophil development in CCAAT enhancer binding protein α-deficient mice. *Proc. Natl. Acad. Sci. USA* **1997**, *94*, 569–574. [CrossRef] [PubMed]

11. Marcinkowska, E.; Garay, E.; Gocek, E.; Chrobak, A.; Wang, X.; Studzinski, G. Regulation of C/EBPβ isoforms by MAPK pathways in HL60 cells induced to differentiate by 1,25-dihydroxyvitamin D_3. *Exp. Cell Res.* **2006**, *312*, 2054–2065. [CrossRef] [PubMed]

12. Pham, T.; Langmann, S.; Schwarzfischer, L.; El Chartouni, C.; Lichtinger, M.; Klug, M.; Krause, S.; Rehli, M. CCAAT enhancer-binding protein β regulates constitutive gene expression during late stages of monocyte to macrophage differentiation. *J. Biol. Chem.* **2007**, *282*, 21924–21933. [CrossRef] [PubMed]

13. Lekstrom-Himes, J. The role of C/EBPepsilon in the terminal stages of granulocyte differentiation. *Stem Cells* **2001**, *19*, 125–133. [CrossRef] [PubMed]

14. Tanaka, T.; Yoshida, N.; Kishimoto, T.; Akira, S. Defective adipocyte differentiation in mice lacking the C/EBPβ and/or C/EBPδ gene. *EMBO J.* **1997**, *16*, 7432–7443. [CrossRef] [PubMed]

15. Ford, A.; Bennett, C.; Healy, L.; Towatari, M.; Greaves, M.; Enver, T. Regulation of the myeloperoxidase enhancer binding proteins Pu1, C-EBP α, -β, and -δ during granulocyte-lineage specification. *Proc. Natl. Acad. Sci. USA* **1996**, *93*, 10838–10843. [CrossRef] [PubMed]

16. Rieger, M.A.; Hoppe, P.S.; Smejkal, B.M.; Eitelhuber, A.C.; Schroeder, T. Hematopoietic cytokines can instruct lineage choice. *Science* **2009**, *325*, 217–218. [CrossRef] [PubMed]

17. Chambon, P. A decade of molecular biology of retinoic acid receptors. *FASEB J.* **1996**, *10*, 940–954. [CrossRef] [PubMed]

18. Gritz, E.; Hirschi, K. Specification and function of hemogenic endothelium during embryogenesis. *Cell. Mol. Life Sci.* **2016**, *73*, 1547–1567. [CrossRef] [PubMed]

19. Cañete, A.; Cano, E.; Muñoz-Chápuli, R.; Carmona, R. Role of Vitamin A/Retinoic Acid in Regulation of Embryonic and Adult Hematopoiesis. *Nutrients* **2017**, *9*, 159. [CrossRef] [PubMed]

20. Hillestad, L. Acute promyelocytic leukemia. *Acta Med. Scand.* **1957**, *159*, 189–194. [CrossRef] [PubMed]

21. Adès, L.; Guerci, A.; Raffoux, E.; Sanz, M.; Chevallier, P.; Lapusan, S.; Recher, C.; Thomas, X.; Rayon, C.; Castaigne, S.; et al. Very long-term outcome of acute promyelocytic leukemia after treatment with all-trans retinoic acid and chemotherapy: the European APL Group experience. *Blood* **2010**, *115*, 1690–1696. [CrossRef] [PubMed]

22. Grande, A.; Montanari, M.; Tagliafico, E.; Manfredini, R.; Zanocco Marani, T.; Siena, M.; Tenedini, E.; Gallinelli, A.; Ferrari, S. Physiological levels of 1α, 25 dihydroxyvitamin D_3 induce the monocytic commitment of CD34+ hematopoietic progenitors. *J. Leukoc. Biol.* **2002**, *71*, 641–651. [PubMed]

23. Janik, S.; Nowak, U.; Łaszkiewicz, A.; Satyr, A.; Majkowski, M.; Marchwicka, A.; Śnieżewski, Ł.; Berkowska, K.; Gabryś, M.C.M.; Marcinkowska, E. Diverse Regulation of Vitamin D Receptor Gene Expression by 1,25-Dihydroxyvitamin D and ATRA in Murine and Human Blood Cells at Early Stages of Their Differentiation. *Int. J. Mol. Sci.* **2017**, *18*, 1323. [CrossRef] [PubMed]

24. Kastner, P.; Chan, S. Function of RARα during the maturation of neutrophils. *Oncogene* **2001**, *20*, 7178–7185. [CrossRef] [PubMed]

25. Yoshizawa, T.; Handa, Y.; Uematsu, Y.; Takeda, S.; Sekine, K.; Yoshihara, Y.; Kawakami, T.; Arioka, K.; Sato, H.; Uchiyama, Y.; et al. Mice lacking the vitamin D receptor exhibit impaired bone formation, uterine hypoplasia and growth retardation after weaning. *Nat. Genet.* **1997**, *16*, 391–396. [CrossRef] [PubMed]

26. Morosetti, R.; Park, D.; Chumakov, A.; Grillier, I.; Shiohara, M.; Gombart, A.; Nakamaki, T.; Weinberg, K.; Koeffler, H. A novel, myeloid transcription factor, C/EBPepsilon, is upregulated during granulocytic, but not monocytic, differentiation. *Blood* **1997**, *90*, 2591–2600. [PubMed]

27. Christakos, S.; Barletta, F.; Huening, M.; Dhawan, P.; Liu, Y.; Porta, A.; Peng, X. Vitamin D target proteins: function and regulation. *J. Cell. Biochem.* **2003**, *88*, 238–244. [CrossRef] [PubMed]

28. Corcoran, A.; Bermudez, M.; Seoane, S.; Perez-Fernandez, R.; Krupa, M.; Pietraszek, A.; Chodyński, M.; Kutner, A.; Brown, G.; Marcinkowska, E. Biological evaluation of new vitamin D₂ analogues. *J. Steroid Biochem. Mol. Biol.* **2016**, *164*, 66–71. [CrossRef] [PubMed]

29. Duprez, E.; Wagner, K.; Koch, H.; Tenen, D. C/EBPβ: A major PML-RARA-responsive gene in retinoic acid-induced differentiation of APL cells. *EMBO J.* **2003**, *22*, 5806–5816. [CrossRef] [PubMed]

30. Fujiki, A.; Imamura, T.; Sakamoto, K.; Kawashima, S.; Yoshida, H.; Hirashima, Y.; Miyachi, M.; Yagyu, S.; Nakatani, T.; Sugita, K.; et al. All-trans retinoic acid combined with 5-Aza-2′-deoxycitidine induces C/EBPα expression and growth inhibition in MLL-AF9-positive leukemic cells. *Biochem. Biophys. Res. Commun.* **2012**, *428*, 216–223. [CrossRef] [PubMed]

31. Marchwicka, A.; Cebrat, M.; Łaszkiewicz, A.; Śnieżewski, Ł.; Brown, G.; Marcinkowska, E. Regulation of vitamin D receptor expression by retinoic acid receptor α in acute myeloid leukemia cells. *J. Steroid Biochem. Mol. Biol.* **2016**, *159*, 121–130. [CrossRef] [PubMed]

32. Gocek, E.; Kielbinski, M.; Baurska, H.; Haus, O.; Kutner, A.; Marcinkowska, E. Different susceptibilities to 1,25-dihydroxyvitamin D₃-induced differentiation of AML cells carrying various mutations. *Leuk. Res.* **2010**, *34*, 649–657. [CrossRef] [PubMed]

33. Gocek, E.; Marchwicka, A.; Baurska, H.; Chrobak, A.; Marcinkowska, E. Opposite regulation of vitamin D receptor by ATRA in AML cells susceptible and resistant to vitamin D-induced differentiation. *J. Steroid Biochem. Mol. Biol.* **2012**, *132*, 220–226. [CrossRef] [PubMed]

34. Marchwicka, A.; Corcoran, A.; Berkowska, K.; Marcinkowska, E. Restored expression of vitamin D receptor and sensitivity to 1,25-dihydroxyvitamin D₃ in response to disrupted fusion *FOP2-FGFR1* gene in acute myeloid leukemia cells. *Cell Biosci.* **2016**, *6*, 7. [CrossRef] [PubMed]

35. Kahlen, J.; Carlberg, C. Identification of a vitamin D receptor homodimer-type response element in the rat calcitriol 24-hydroxylase gene promoter. *Biochem. Biophys. Res. Commun.* **1994**, *202*, 1366–1372. [CrossRef] [PubMed]

36. Dhawan, P.; Peng, X.; Sutton, A.; MacDonald, P.; Croniger, C.; Trautwein, C.; Centrella, M.; McCarthy, T.; Christakos, S. Functional cooperation between CCAAT/enhancer-binding proteins and the vitamin D receptor in regulation of 25-hydroxyvitamin D₃ 24-hydroxylase. *Mol. Cell. Biol.* **2005**, *25*, 472–487. [CrossRef] [PubMed]

37. Dhawan, P.; Wieder, R.; Christakos, S. CCAAT enhancer-binding protein α is a molecular target of 1,25-dihydroxyvitamin D3 in MCF-7 breast cancer cells. *J. Biol. Chem.* **2009**, *284*, 3086–3095. [CrossRef] [PubMed]

38. Brown, G.; Hughes, P.; Michell, R.; Rolink, A.; Ceredig, R. *Ordered Commitment of Hematopoietic Stem Cells to Lineage Options*; Nova Science Publishers Inc.: Hauppauge, NY, USA, 2008.

39. Scott, L.; Civin, C.; Rorth, P.; Friedman, A. A novel temporal expression pattern of three C/EBP family members in differentiating myelomonocytic cells. *Blood* **1992**, *80*, 1725–1735. [PubMed]

40. Taskesen, E.; Bullinger, L.; Corbacioglu, A.; Sanders, M.; Erpelinck, C.; Wouters, B.; van der Poel-van de Luytgaarde, S.; Damm, F.; Krauter, J.; Ganser, A.; et al. Prognostic impact, concurrent genetic mutations, and gene expression features of AML with CEBPA mutations in a cohort of 1182 cytogenetically normal AML patients: Further evidence for CEBPA double mutant AML as a distinctive disease entity. *Blood* **2011**, *117*, 2469–2475. [CrossRef] [PubMed]

41. Bullinger, L.; Döhner, K.; Döhner, H. Genomics of Acute Myeloid Leukemia Diagnosis and Pathways. *J. Clin. Oncol.* **2017**, *35*, 934–946. [CrossRef] [PubMed]

42. Pabst, T.; Mueller, B.; Harakawa, N.; Schoch, C.; Haferlach, T.; Behre, G.; Hiddemann, W.; Zhang, D.; Tenen, D. AML1-ETO downregulates the granulocytic differentiation factor C/EBPα in t(8;21) myeloid leukemia. *Nat. Med.* **2001**, *7*, 444–451. [CrossRef] [PubMed]

43. Cilloni, D.; Carturan, S.; Gottardi, E.; Messa, F.; Messa, E.; Fava, M.; Diverio, D.; Guerrasio, A.; Lo-Coco, F.; Saglio, G. Down-modulation of the C/EBPα transcription factor in core binding factor acute myeloid leukemias. *Blood* **2003**, *102*, 2705–2706. [CrossRef] [PubMed]

44. Balmer, J.; Blomhoff, R. A robust characterization of retinoic acid response elements based on a comparison of sites in three species. *J. Steroid Biochem. Mol. Biol.* **2005**, *96*, 347–354. [CrossRef] [PubMed]

45. Moutier, E.; Ye, T.; Choukrallah, M.; Urban, S.; Osz, J.; Chatagnon, A.; Delacroix, L.; Langer, D.; Rochel, N.; Moras, D.; et al. Retinoic acid receptors recognize the mouse genome through binding elements with diverse spacing and topology. *J. Biol. Chem.* **2012**, *287*, 26328–26341. [CrossRef] [PubMed]

46. Gombart, A.; Kwok, S.; Anderson, K.; Yamaguchi, Y.; Torbett, B.; Koeffler, H. Regulation of neutrophil and eosinophil secondary granule gene expression by transcription factors C/EBPepsilon and PU.1. *Blood* **2003**, *101*, 3265–3273. [CrossRef] [PubMed]

47. Baurska, H.; Kłopot, A.; Kiełbiński, M.; Chrobak, A.; Wijas, E.; Kutner, A.; Marcinkowska, E. Structure-function analysis of vitamin D_2 analogs as potential inducers of leukemia differentiation and inhibitors of prostate cancer proliferation. *J. Steroid Biochem. Mol. Biol.* **2011**, *126*, 46–54. [CrossRef] [PubMed]

48. Pfaffl, M. A new mathematical model for relative quantification in real-time RT-PCR. *Nucleic Acids Res.* **2001**, *29*, e45. [CrossRef] [PubMed]

49. Gocek, E.; Marchwicka, A.; Bujko, K.; Marcinkowska, E. NADPH-cytochrome p450 reductase is regulated by all-*trans* retinoic acid and by 1,25-dihydroxyvitamin D_3 in human acute myeloid leukemia cells. *PLoS ONE* **2014**, *9*, e91752. [CrossRef] [PubMed]

International Journal of
Molecular Sciences

MDPI

Article

Ferritin Heavy Subunit Silencing Blocks the Erythroid Commitment of K562 Cells via miR-150 up-Regulation and GATA-1 Repression

Fabiana Zolea [1,†], Anna Martina Battaglia [1,†], Emanuela Chiarella [2], Donatella Malanga [3], Carmela De Marco [3], Heather Mandy Bond [2], Giovanni Morrone [2], Francesco Costanzo [1] and Flavia Biamonte [1,*]

[1]   Research Center of Advanced Biochemistry and Molecular Biology, Department of Experimental and Clinical Medicine, University Magna Græcia, 88100 Catanzaro, Italy; fabiana.zolea@alice.it (F.Z.); annamartinabattaglia@gmail.com (A.M.B.); fsc@unicz.it (F.C.)
[2]   Laboratory of Molecular Haematopoiesis and Stem Cell Biology, Department of Experimental and Clinical Medicine, University Magna Græcia, 88100 Catanzaro, Italy; emanuelachiarella@libero.it (E.C.); bond@unicz.it (H.M.B.); morrone@unicz.it (G.M.)
[3]   Department of Experimental and Clinical Medicine, University of Catanzaro "Magna Graecia", 88100 Catanzaro, Italy; malanga@unicz.it (D.M.); cdemarco@unicz.it (C.D.M.)
*   Correspondence: flavia.biamonte.fb@gmail.com; Tel.: +39-0961-369-4105
†   These authors contributed equally to this work.

Received: 11 September 2017; Accepted: 12 October 2017; Published: 17 October 2017

Abstract: Erythroid differentiation is a complex and multistep process during which an adequate supply of iron for hemoglobinization is required. The role of ferritin heavy subunit, in this process, has been mainly attributed to its capacity to maintain iron in a non-toxic form. We propose a new role for ferritin heavy subunit (FHC) in controlling the erythroid commitment of K562 erythro-myeloid cells. FHC knockdown induces a change in the balance of GATA transcription factors and significantly reduces the expression of a repertoire of erythroid-specific genes, including α- and γ-globins, as well as CD71 and CD235a surface markers, in the absence of differentiation stimuli. These molecular changes are also reflected at the morphological level. Moreover, the ability of FHC-silenced K562 cells to respond to the erythroid-specific inducer hemin is almost completely abolished. Interestingly, we found that this new role for FHC is largely mediated via regulation of miR-150, one of the main microRNA implicated in the cell-fate choice of common erythroid/megakaryocytic progenitors. These findings shed further insight into the biological properties of FHCand delineate a role in erythroid differentiation where this protein does not act as a mere iron metabolism-related factor but also as a critical regulator of the expression of genes of central relevance for erythropoiesis.

Keywords: ferritin heavy subunit; differentiation; K562; miR-150; GATA-1

1. Introduction

Megakaryocytic and erythroid progenitors that give rise to platelets and red blood cells, respectively, derive from a megakaryocyte-erythroid common precursor (MEP) [1]. The MEPs differentiation fate is dynamically determined by the coordination of different molecular events, including the progressive loss of differentiation potential, the expression of lineage-specific markers, and the acquisition of specialized morphological features and functionalities [1,2]. During erythroid maturation, MEPs undergo a progressive decrease in cell size, nuclear condensation, and activation of the transcription of globin genes with the consequent accumulation of hemoglobin [3]. Immunophenotypic studies also highlighted the acquisition and/or the increase in the expression of surface markers such as CD71 (Transferrin receptor protein 1, TfR1) and CD235a (Glycophorin A), now considered erythroid-specific hallmarks [4,5].

The MEPs fate is orchestrated by thecoordinated action of specific transcription factors. The dynamic exchange of GATA-1 with GATA-2, the so-called "GATA factor switch", in which GATA-1 levels increase during the terminal erythroid maturation while GATA-2 is turned off, is one of the best-known examples [6,7].

This scenario is even more complex when taking into account the role played by microRNAs in the fine-tuning of haematopoiesis [8,9]. MicroRNAs are a class of small non-coding RNA whose activity as post-transcriptional inhibitor of gene expression belongs to the biological process, known as RNA interference [10]. Almost all of the stages of the haematopoietic lineage specification, from the maintenance of the haematopoietic stem cell pool to the generation of lineage-committed progenitors and of specific mature cells, are controlled by miRNAs [11]. Indeed, monocytopoiesis is regulated by miR-17/92 cluster while erythropoiesis is driven by miR-451, miR-16, and miR-144 and inhibited by miR-150, miR-155, miR-221, miR-222, and miR-223 [11]. MiR-125b supports myelopoiesis while B-cells maturation is promoted by the down-regulation of miR-34a [11]. Several reports indicate that miR-150 is involved in the control of multiple haematopoietic cell fates [12]. miR-150 is very highly expressed during advanced stages of both B and T cell maturation in bone marrow and thymus, respectively, and its premature expression leads to severe defects in B cell development through the down-regulation of target genes, such as *Myb* and *Foxp1* [13]. Within the myeloid lineage, a constant repression of miR-150 ensures the normal terminal erythroid development; on the contrary, its increased expression induces MEPs toward megakaryocytic maturation [14–16]. The role of miR-150 has been supported by several in vitro analyses: it has been shown that overexpression of miR-150 promotes the generation of colony-forming unit megakaryocyte (CFU-Mk), while its antagomiR-mediated suppression induces colony-forming unit erythrocyte (CFU-E) [17]; furthermore, forced expression of miR-150 significantly reduces hemin-dependent erythropoiesis, commitment to hemoglobinization and CD235a expression in the bipotent megakaryocyte/erythroid K562 human leukemia cells [18]. K562 cells can be terminally differentiated in vitro toward the erythroid and megakaryocytic lineages; thus, they are considered as a useful in vitro model for studying MEP commitment [1,2]. The molecular mechanisms underlying the effects of miR-150 on MEPs fate-decision are not fully elucidated. Different models have been proposed either associated with differentiation-related or proliferation-related pathways [15]. Moreover, gene expression profiling suggests that forced miR-150 expression in hemin-induced K562 cells suppress the activation of ErbB-MAPK-p38 and ErbB-PI3K-AKT pathways [18]. However, the upstream regulators of miR-150 have not yet been determined.

The MEPs function and fate are also affected by metabolic perturbations [19–21]. In particular, iron metabolism and erythropoiesis are intimately linked. An adequate supply of iron is indeed necessary to ensure sufficient hemoglobin synthesis and thus for the correct maturation of red blood cells [20,21]. However, an excessive amount of intracellular free iron may be harmful to the cells since it can trigger the generation of reactive oxygen species (ROS) through the Fenton reaction [22]. Ferritin, the main intracellular iron storage protein, tightly regulates iron levels by storing it in a nontoxic and bioavailable form for supply upon metabolic requirement of hemoglobinization [23]. Ferritin is a multimeric protein composed of a total of twenty-four subunits of two types, the ferritin heavy subunit(FHC, FTH) and the ferritin light subunit (FLC, FTL), assembled to form a shell that is able to sequester up to 4500 iron atoms [19,20]. FHC has a ferroxidase activity through which it converts Fe(II) to Fe(III) and protects cells against oxidative stress [24,25]. Indeed, we have recently demonstrated that FHC-silencing results in a significant increase in intracellular ROS in erythroleukemia K562 cells [25] as well as in other cell types [26]. At the same time, a growing body of experimental evidence has shed light on new and intriguing roles for FHC in the control of proliferation and migration of several cancer cell lines as well as in the regulation of many oncogenes and oncomiRNAs [24–27].

The role of FHC in the haematopoietic differentiation has been so far mainly explored in relation to its function in the iron intracellular metabolism. To date, the gene expression profiling after the hemin-mediated erythroid differentiation of K562 cells highlighted the occurrence of both transcriptional and translational up-regulation of the ferritin gene [23,28]. This results in an increase in ferritin synthesis that ultimately enhances the cellular capacity of iron storage for hemoglobin synthesis [23].

In this study, we investigated the role of FHC in K562 cells erythroid differentiation by exploring the effects of the perturbation of its intracellular amount on cell morphology, expression of representative genes and lineage-specific markers. Our results revealed that FHC knock-down induced a significant arrest in the erythroid commitment of K562 cells that was mostly mediated by the up-regulation of miR-150 and the parallel repression of GATA-1, and uncovers a new role of FHC in the lineage choice of the erythro-megakaryocytic K562 cells through the fine tuning of key regulatory molecules.

2. Results and Discussion

The K562 leukemia-derived cell line represents a useful in vitro model of MEP since they are situated at the common progenitor stage of erythroid and megakaryocytic lineages differentiation [1,2], and can be induced toward either of the above cell fates by a number of chemical agents, such as hemin and phorbol 12-myristate 13-acetate (PMA), respectively [29].

Ferritin is the main iron storage protein within the cell and is localized in cytoplasm, nucleus, and mitochondria [27]. The erythroid differentiation is accompanied by an enhanced expression of its heavy subunit (FHC), which has a ferroxidase activity, and this induction has been mainly attributed to the necessity for the cell to store in a non-toxic form the high amounts of iron required for optimal hemoglobinization [23,28]. On the other hand, in particular in K562 cells, FHC might play other roles besides the control of iron metabolism; it has been proposed that FHC, in its nuclear form, represses β-globin transcription [30], while its silencing modulates the expression of a repertoire of genes during hemin treatment [28]. Collectively, these data suggest that FHC might play multiple and key roles in erythropoiesis, however the underlying mechanisms are not fully elucidated.

2.1. Ferritin Heavy Subunit (FHC) Knockdown Negatively Regulates the Erythroid Fate of K562 Cells

We have previously found that FHC silencing strongly reduces the hemin-mediated induction of γ-globin synthesis in K562 cells [28]. Here, in order to more clearly define the role of the ferritin heavy subunit in K562 erythroid differentiation we extended the analysis to key molecules associated with erythroid development and functions, such as α-globin, transferrin receptor-1 (CD71), and the glycophorin A (CD235a) erythroid-lineage specific marker. Pools of stably FHC-silenced (K562^{shFHC}) and of control cells (K562^{shScr}), as previously described [24,28], were treated with 50μM hemin for 48 h. The FHC amounts, evaluated at both mRNA and protein levels, of hemin-treated and untreated K562^{shFHC} and K562^{shScr} cells are shown in Figure 1a,b, respectively. FHC was significantly downregulated in the K562^{shFHC} cells as compared to the control K562^{shScr} cells at both mRNA and protein levels; hemin treatment consistently increased FHC amounts both in the control K562$^{shScr(50\ \mu M\ hemin)}$ and in the K562$^{shFHC(50\ \mu M\ hemin)}$ cells when compared to their relative untreated cells (Figure 1a,b). As expected, hemin treatment strongly up-modulated the expression of α-*globin*, γ-*globin*, *CD71*, and *CD235a* mRNAs in K562^{shScr} control cells (Figure 1c). Notably, FHC-silencing was able, per se, to significantly reduce the expression of all four markers; moreover, FHC-silencing almost completely counteracted the effects of hemin treatment and effectively inhibited the induction of α-*globin*, γ-*globin* and *CD71* (Figure 1c). *CD235a* displayed a relatively different behaviour since its mRNA was the most sensitive to FHC knockdown and, in parallel, the only one still responsive to hemin treatment in FHC-silenced cells. We also performed fluorescence-activated cell sorting (FACS) analysis of CD235a in FHC-silenced and control cells, treated and untreated with hemin. FHC knockdown substantially reduced the CD235a positive cells as shown by the substantially lower values of both the mean fluorescence intensity of CD235a and the percentage of CD235a-positive cells in Figure 1d (FLH-2 mean: 79 vs. 26.7; % CD235a$^+$: 85.4 vs. 27.3); after hemin treatment K562^{shScr} became nearly completely positive (FLH-2 mean: 97.1; % CD235a$^+$: 94.3) while the CD235a$^+$ K562^{shFHC} appeared almost completely unaffected (FLH-2 mean: 23.8; % CD235a$^+$: 31.4). Along with the modifications of marker expression, microscopic analyses highlighted considerable alterations in the morphology of K562^{shFHC} cells as assessed by May-Grünwald-Giemsa staining, which included features typically not associated to an erythroid phenotype such as increased size, vacuole-rich cytoplasm, and polylobulated nuclei (Figure 1e). After 48 h of culture in the presence of hemin, K562^{shScr} cells showed a globular shape morphology with round nuclei eccentrically located and the cell pellet developed a distinctive red color; conversely, no morphological changes have been observed in K562^{shFHC} cells and the cell pellets became barely pink upon hemin treatment.

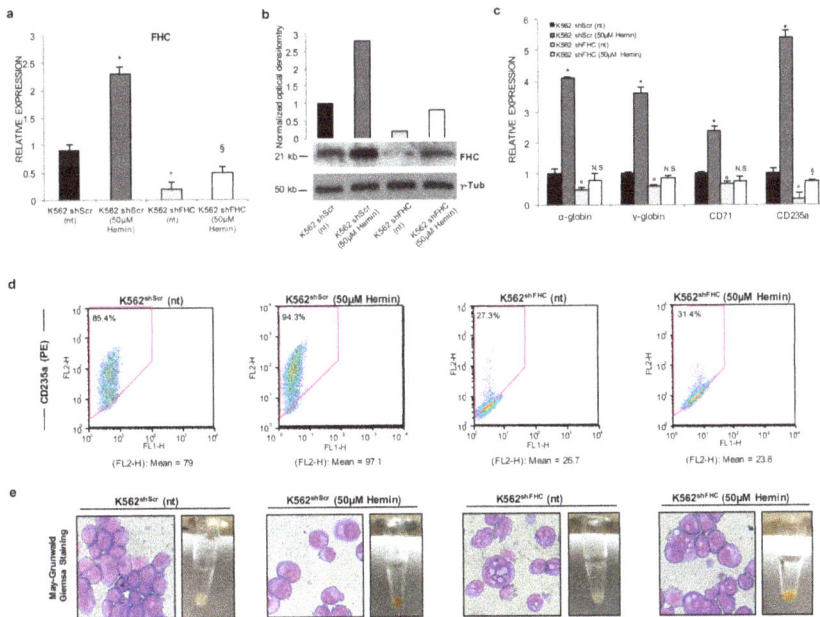

Figure 1. FHC silencing inhibits hemin-induced erythroid differentiation of K562 cells. (**a**) qRT-PCR and (**b**) Western Blot analyses of FHC mRNA and protein levels, respectively, in the control K562^{shScr} and FHC-silenced K562^{shFHC} cells untreated (nt) and treated with 50 μM hemin for 48 h; (**c**) Relative expression of α-*globin*, γ-*globin*, *CD71* and *CD235a* of K562^{shScr} and K562^{shFHC} cells untreated or treated with hemin. All data represent mean ± SD (*n* = 3). * *p* < 0.05 compared with K562^{shScr} cells, ° *p* < 0.05 compared with K562^{shScr} cells, § *p* < 0.05 compared with K562^{shFHC} cells, N.S.: not significant; (**d**) Flow cytometry analysis of CD235a-positive cells in K562^{shScr} and K562^{shFHC} cells untreated or treated with hemin. Data are reported both as mean fluorescence intensity of CD235a (FL2-H) and as percentage (%) of CD235$^+$ cells; (**e**) May-Grünwald-Giemsa staining of K562^{shScr} and K562^{shFHC} cell untreated or treated with hemin.

To rule out possible off-target effects of the shRNA, we investigated the expression of α-*globin*, γ-*globin*, *CD71* and *CD235a* in K562 cells where FHC expression was transiently silenced by using a specific FHC-siRNA. Figure 2a shows a representative Western Blot of FHC protein levels at 24 h, 48 h and 72 h after the siRNA transfection while in Figure 2b H-ferritin mRNA levels are represented as the mean of three independent experiments. Consistently with the findings obtained in the stably FHC-silenced cells, the qRT-PCR analyses, shown in Figure 2c–f, highlighted a reduction of the erythroid-specific markers upon transient FHC silencing at each time point with a statistical significance at 48 h and 72 h ($p < 0.05$). Also in these set of experiments *CD235a* showed the highest responsiveness to FHC knockdown. CD235a and CD71 were further investigated, 72 h upon FHC transient silencing, at the cell surface level: Figure 2g–h report representative FACS analyses that highlight markedly reduced levels of both markers in K562^{siFHC} cells compared to the control ones (CD235a, FL2-H mean: 46 vs. 67; % CD235a$^+$: 22 vs. 86.7) (CD71, FL2-H mean: 53.3 vs. 121, CD71$^+$: 83.4 vs. 87.2).

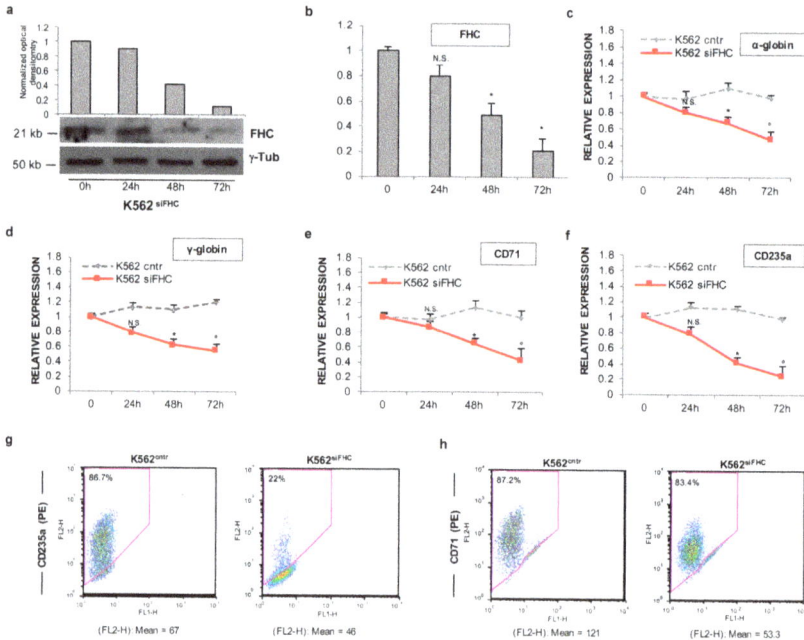

Figure 2. Transient FHC silencing down-regulates erythroid markers expression in K562 cells. FHC amounts of K562 cells at 0, 24, 48 and 72 h after a transient transfection of a specific FHC siRNA at protein (**a**) and mRNA (**b**) levels; Data represent mean ± SD ($n = 3$). * $p < 0.05$ compared with 0 h, N.S.: not significant. The relative expression of *α-globin* (**c**); *γ-globin* (**d**); *CD71* (**e**) and *CD235a* (**f**) in K562^{siFHC} cells at each time point was measured using qRT-PCR. Data represent mean ± SD ($n = 3$). * $p < 0.05$ compared with K562^{cntr} at 48 h; ° $p < 0.05$ as compared with K562^{cntr} at 72 h; N.S.: not significant. Flow cytometry analysis of CD235a- (**g**) and CD71- (**h**) positive cells in K562^{cntr} and K562^{siFHC} cells at 72 h upon FHC siRNA transfection. Data are reported both as mean fluorescence intensity of CD235a and CD71 (FL2-H) and as percentage (%) of CD235$^+$ and CD71$^+$ cells.

Taken together, these results indicate that ferritin heavy subunit knockdown is able to strongly inhibit the erythroid transcriptional program in K562 cells, and to almost completely abrogate the transcriptional induction of these genes in response to hemin treatment, thus suggesting that appropriate intracellular amounts of FHC are needed for the correct expression of these key genes.

We recently found that FHC knockdown restrains K562 cell proliferation [24]. Because the proliferation rate of a given cell population can be correlated to its differentiation state, here, by using direct cell counting assays, we re-confirmed the decline of K562^{shFHC} proliferative rate in comparison with control cells at 24 h (37×10^4 vs. 48×10^4) and 48 h (51×10^4 vs. 64×10^4) (Figure 3a). Using flow cytometry and Annexin V/7-AAD double staining, we assayed apoptosis of K562^{shScr} and K562^{shFHC} cells and found that FHC silencing slightly increased late apoptosis (Annexin V$^+$/7-AAD$^+$, 6.2% vs. 13.7%, $p < 0.05$), but not early apoptosis (Annexin V$^+$/7-AAD$^-$, 2.8% vs. 3.8%) (Figure 3b).

The analysis of cell cycle, using PI staining and flow cytometric detection, further confirmed the presence of a higher percentage of K562^{shFHC} cells in the G0/G1 phase as compared to control K562^{shScr} cells (33.6% vs. 19.9%, $p < 0.05$) (Figure 3c). The moderate extent of the FHC silencing-induced G0/G1 cell cycle arrest, in comparison with the strong interference with the erythroid commitment, makes it difficult to correlate these two phenomena. Proliferation and differentiation can either be coupled or inversely correlated in function on the differentiation stage [31], thus further studies are needed to deeply investigate this point. Finally, through clonogenic assays we found that K562^{shFHC} cells displayed an approximately 2-fold decrease of clonogenic potential as compared to the control cells, as shown by representative microscopy images and quantified in the histogram ($p < 0.05$) (Figure 3d).

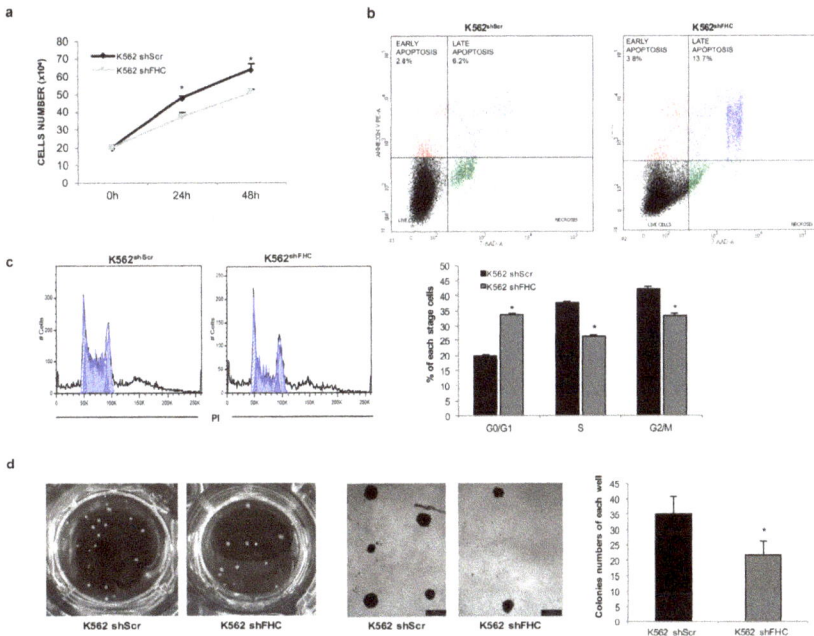

Figure 3. FHC silencing suppresses the proliferative rate and the clonogenic potential of K562 cells. (**a**) Direct cell counting of K562^{shScr} and K562^{shFHC} at 0, 24 and 48 h; (**b**) Apoptosis analysis in K562^{shScr} and K562^{shFHC} using Annexin V/7-AAD double staining. The reported plots are representative of two independent experiments; (**c**) Representative flow cytometry plots of cell cycle analysis in K562^{shScr} and K562^{shFHC} with statistics on the right side. All data represent mean $\pm$ SD ($n = 3$). * $p < 0.05$ compared with K562^{shScr} cells; (**d**) Representative microscopy images and statistics of colony formation assay in K562^{shScr} and K562^{shFHC}; colonies were observed and counted in each well of 12-well plate. Scale bar: 200 μm. All data represent mean $\pm$ SD ($n = 3$). * $p < 0.05$ compared with K562^{shScr} cells.

2.2. FHC Effects on K562 Erythroid Fate Are Mediated by miR-150

Transcription factors play a pivotal role in the cell differentiation fate since they are responsible for the expression of lineage-specific genes and for the parallel repression of stemness-related markers [7,32]. In erythropoiesis, the precise balance of GATAs transcription factors is one of the best-known examples [7]. GATA-2 overexpression, in the presence of a reduction of GATA-1, drives megakaryocytic differentiation at the expense of the erythroid one; on the contrary, terminal erythroid differentiation is associated with GATA-1 increase, that in turn, down-regulates GATA-2 expression [6,7].

However, transcription factors alone cannot account for every aspect of haematopoietic differentiation. Post-transcriptional regulatory mechanisms, such as those played by microRNAs, ensure a finer tuning and a more rapid response to stimuli [8,9]. Many recent studies have shown that a number of miRNAs are involved in the fine regulation of the haematopoietic stem cells differentiation, such as miR-15/16, miR-222, miR-150, miR-451, miR-210, and let-7d [8,9]. miR-150, in particular, has been implicated in the control of lineage choice in MEPs fate decision: high levels of miR-150 drive MEPs toward megakaryocytic maturation, whereas low levels lead to erythroid commitment [12,18].

Recently, we have demonstrated that FHC controls the expression of a set of miRNAs in a variety of cancer cell types, including K562 cells [24] and that, in SKOV-3 ovarian carcinoma cells, FHC promotes the expansion of a subset of cells with stem-like features through the modulation of miR-150 [27].

In order to establish whether modulation of miR-150 expression in response to FHC knockdown may contribute to the phenotypes observed in K562 and described above, we analyzed the steady state amounts of *GATA-1* and *GATA-2* mRNAs and those of miR-150 in both transiently and stably FHC-silenced cells. As shown in Figure 4a,b, FHC silencing down-regulates *GATA-1* and, at the same time, strongly induces *GATA-2* expression. As highlighted in Figure 4c,d, this is accompanied by a significant increase in miR-150 levels upon both transient and stable FHC silencing. Notably, both *GATAs* switch and miR-150 up-regulation are consistent with the FHC-mediated arrest of K562 erythroid differentiation.

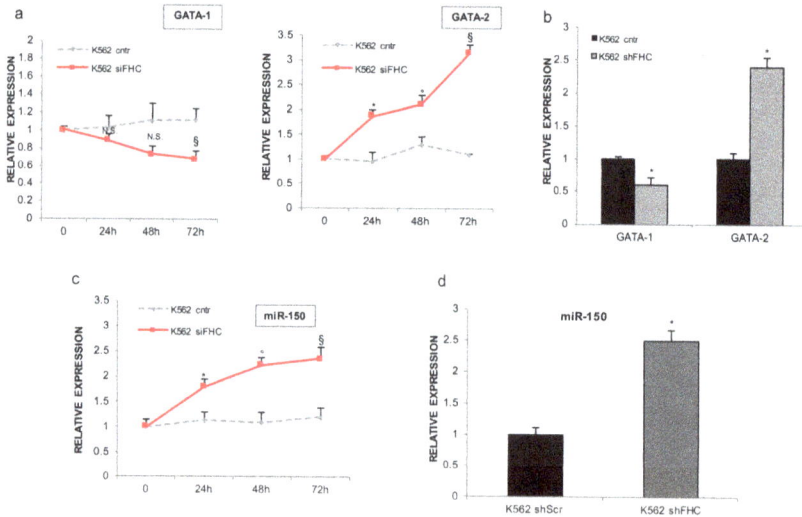

Figure 4. FHC silencing reduces GATA-1 and induces miR-150 up-regulation. (a) qRT-PCR analysis of *GATA-1* and *GATA-2* genes in transiently FHC-silenced K562[siFHC] cells at 0 h, 24 h, 48 h, and 72 h, compared to control K562[cntr] cells. All data represent mean ± SD ($n = 3$). * $p < 0.05$ compared with K562[cntr] at 24 h; ° $p < 0.05$ compared with K562[cntr] at 48 h; § $p < 0.05$ as compared with K562[cntr] at 72 h. N.S. not significant; (b) qRT-PCR analysis of *GATA-1* and *GATA-2* genes in stably FHC-silenced K562[shFHC] cells compared to control K562[shScr] cells. All data represent mean ± SD ($n = 3$). * $p < 0.05$ compared with K562[shScr] cells; (c) TaqMan analysis of miR-150 in transiently FHC-silenced K562[siFHC] cells at 0 h, 24 h, 48 h, and 72 h, compared to control K562[cntr] cells. All data represent mean ± SD ($n = 3$). * $p < 0.05$ compared with K562[cntr] at 24 h; ° $p < 0.05$ compared with K562[cntr] at 48 h; § $p < 0.05$ compared with K562[cntr] at 72 h; (d) TaqMan analysis of miR-150 in stably FHC-silenced K562[shFHC] cells compared to control K562[shScr] cells. All data represent mean ± SD ($n = 3$). * $p < 0.05$ compared with K562[shScr] cells.

To shed further light on the role of miR-150 in the cascade of molecular events induced by FHC silencing, we either reconstituted FHC or inhibited miR-150 inK562^{shFHC} cells. FHC reconstitution, whose efficiency is shown in Figure 5a, determined a significant reduction of miR-150, thus confirming the existence of an inverse correlation between FHC and miR-150 steady state amounts (Figure 5b). The results of a triplicate set of experiments, graphically represented in Figure 5c, indicate that both FHC reconstitution and miR-150 inhibition were able to revert the *GATAs* switch. This is the first report showing that GATAs transcription factors are potential downstream molecules of miR-150. Since the bioinformatic online prediction tool FINDTAR3 did not underline any complementary regions between *GATAs* and miR-150, we propose the existence of an indirect relationship between these two molecules, mediated by an intermediate still to be determined. The down-modulation of *α-globin*, *γ-globin* and *CD-71* were efficiently counteracted by both FHC reconstitution and miR-150 inhibition (Figure 5d). It has been already reported that the expression of both *globins* and *CD71* is transcriptionally regulated by GATA1 [33,34]. Taken all together our data suggest a model in which FHC amounts modulates, through miR-150, *GATA1* that in turn regulates globins and *CD71*. *CD235a* appears to be not included in this pathways since its amounts strongly responded to the FHC reconstitution, increasing by approximately 5-fold, but were unaffected by miR-150 inhibition.

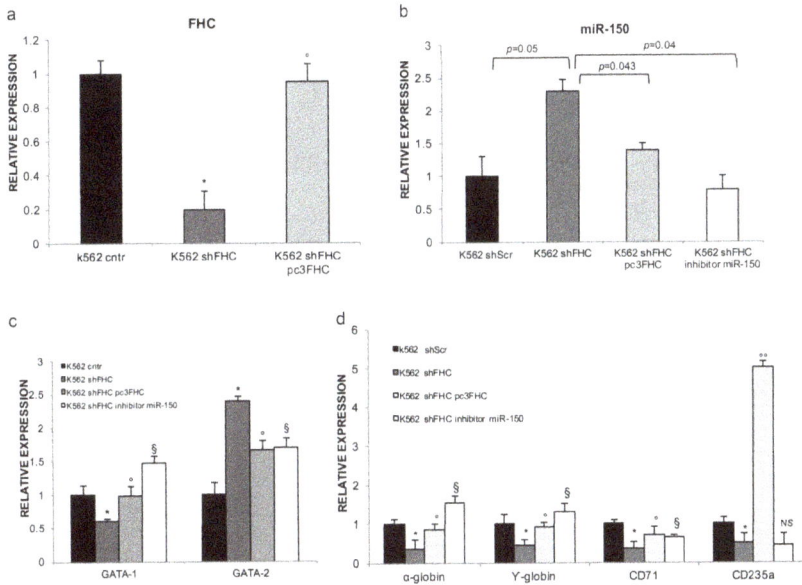

Figure 5. miR-150 mediates the control exerted by FHC on *GATAs*, *α-globin*, *γ-globin* and *CD71* expression. (**a**) pc3FHC expression vector almost completely restore FHC amounts in FHC-silenced K562 cells (K562$^{shFHC/pc3FHC}$). Data represent mean ± SD of 3 qRT-PCR analysis. * $p < 0.05$ compared with K562^{shScr}; ° $p < 0.05$ compared with K562^{shFHC}; (**b**) TaqMan analysis of miR-150 levels after FHC reconstitution and miR-150 inhibition highlight a significant down-regulation in K562$^{shFHC/pc3FHC}$ and K562$^{shFHC/miR-150\ inhibitor}$ compared to K562^{shFHC} cells; (**c**) qRT-PCR analysis of *GATA-1* and *GATA-2* genes in K562$^{shFHC/pc3FHC}$ and K562$^{shFHC/miR-150\ inhibitor}$ as compared to K562^{shFHC} cells. Data represent mean ± SD ($n = 3$). * $p < 0.05$ compared with K562^{shScr} at; ° $p < 0.05$ compared with K562^{shFHC}; § $p < 0.05$ compared with K562^{shFHC}; (**d**) qRT-PCR analysis of *α-globin*, *γ-globin*, *CD71* and *CD235a* genes in K562$^{shFHC/pc3FHC}$ and K562$^{shFHC/miR-150\ inhibitor}$ compared to K562^{shFHC} cells. Data represent mean ± SD ($n = 3$). * $p < 0.05$ compared with K562^{shScr} at; ° $p < 0.05$ when compared with K562^{shFHC}; °° $p < 0.001$ compared with K562^{shFHC}; § $p < 0.05$ compared with K562^{shFHC}; N.S. not significant.

Notably, in FHC-silenced K562 cells, the up-regulation of miR-150 is unable to promote megakaryocytic differentiation since the analysis of gene and surface expression of CD41, CD61, and CD110 MK-specific markers did not highlight any modification in their steady state amount (Figure S1). Megakaryocytic differentiation is driven not only by miR-150 but also by other microRNAs such as miR-34a, miR-146a, miR-27a, and miR-28 [35]. The natural conclusion of our observations is that, at least in FHC-silenced K562 cells, miR-150 is a central hub in the molecular phenomena resulting in the arrest of the erythroid commitment while the megakaryocytic differentiation requires other so far uncovered regulatory molecules.

The mechanisms by which FHC inhibits miR-150 expression require further and more targeted analyses. To the best of our knowledge, in K562 cells, the nuclear form of H ferritin has been suggested as specific repressor of the β-globin promoter [30] but no further evidences have emerged in this direction. On the other hand, it has been recently shown that the transcript of FTH1P3, one intronless member of the FHC multigene family, acts as molecular sponge of miR-224-5p in oral squamous cell carcinoma cells [36]. Interestingly, as reported in Figure S2, the FINDTAR3 bioinformatic prediction software highlights the existence of two regions of complementarity between FHC mRNA and miR-150 (FHC mRNA NM_002032 positions: 38–73; 165–188). This observation constitutes the molecular ground for future studies aimed at analyzing a possible role of FHC mRNA as competitive endogenous RNA (ceRNA).

In conclusion, we demonstrate that a costant expression of FHC is essential for K562 erythroid differentiation since either its stable or its transient silencing is associated with an arrest toward this commitment. Furthermore, our data delineate a new regulatory axis through which FHC controls erythroid differentiation and lend strong support to the notion that, in addition to its role in iron metabolism, FHC can act as a relevant regulator of gene expression through the modulation of miR-150.

3. Materials and Methods

3.1. Cell Culture

K562 cells were obtained from the American Type Culture Collection (ATCC, CCL-243; Manassas, VA, USA). K562 cells were grown in RPMI 1640 media (Sigma Aldrich, St. Louis, MO, USA) supplemented with 10% FBS (Thermo Fisher Scientific, Waltham, MA, USA) at 37 °C in 5% CO_2. Hemin (50 µM) (Sigma Aldrich) was used to induce erythroid differentiation of K562 cells.

3.2. K562 Cells Transfection and Transduction

FHC silencing of K562 cells was performed using two different methods: (i) stable transduction of a lentiviral DNA containing an shRNA targeting the 196–210 region of the FHC mRNA (sh29432) (K562[shFHC]) or a control shRNA without significant homology to known human mRNAs (K562[shScr]) and (ii) transient transfection of a pre-cast siRNAspecific for FHC (K562[siFHC]) or a negative control siRNA (K562[cntr]) (Thermo Fisher Scientific). Stable transduction was achieved, as previously reported [19,23]. Transient transfections were performed using the Amaxa Nucleofactor kit (Lonza, Basel, Switzerland) [37]. FHC stable knock-down was verified by Western Blot analysis and RT-qPCR while FHC transient knockdown was checked at 24, 48 and 72 h by RT-qPCR analysis. FHC-reconstitution in K562[shFHC] cells was performed using the expression vector containing the full length of human FHC cDNA (pc3FHC) (K562[shFHCpc3FHC]).

K562[shFHC] cells were also transiently transfected with a final concentration of 100 nM of a specific miR-150 inhibitor (K562[shFHCmiR−150 inhibitor]) obtained from Thermo Fisher Scientific, using the Amaxa Nucleofactor kit (Lonza). 48 h after transfection, miR-150 levels were measured by TaqMan microRNA Assay (Thermo Fisher Scientific).

3.3. RNA Extraction and Quantitative Real-Time PCR (qRT-PCR)

Total RNA was extracted from K562 cells with the TRizol RNA isolation method (Thermo Fisher Scientific), according to the manufacturer's instructions. The purity and the integrity of each RNA sample were checked, as previously reported [26]. Then, 1 µg of RNA from each sample was reverse transcribed by using High Capacity cDNA Reverse Transcription Kit (Thermo Fisher Scientific). Gene expression analysis was performed by using SYBR®Green qPCR Master Mix (Thermo Fisher Scientific) [24]. Primers to detect *FHC*, *CD235a*, *CD71*, *α-globin*, *γ-globin*, *GATA-1* and *GATA-2* are reported in Table 1. Relative quantification between samples and control transcript levels was performed by using the comparative $2^{-\Delta\Delta Ct}$ method [26]. Each sample was normalized to its glyceraldehyde 3-phosphate dehydrogenase (*GAPDH*) content.

Table 1. Primers used in qRT-PCR analysis.

Gene	Forward Primer	Reverse Primer
GAPDH	5′-TGA TGA CAT CAA GAA GGT GGT GAA G-3′	5′-TCC TTG GAG GCC ATG TGG GCC AT-3′
FHC	5′-CAT CAA CCG CCA GAT CAA C-3′	5′-GAT GGC TTT CAC CTG CTC AT-3′
α-globin	5′-GTG GAC GAC ATG CCC AAC-3′	5′-TAT TTG GAG GTC AGC ACG GT-3′
γ-globin	5′-CAG AAA TAC ACA TAC ACA CTT CC-3′	5′-GAG AGA TCA CAC ATG ATT TTC TT-3′
CD71	5′-ACT GGT CCA TGC TAA TTT TGG T-3′	5′-AGT TCT GCG TTA ACA ATG GGA-3′
CD235a	5′-GAG AAA GGG TAC AAC TTG CC-3′	5′-CAT TGA TCA CTT GTC TCT GG-3′
GATA-1	5′-GAT GAA TGG GCA GAA CAG GC-3′	5′-TAG CTT GTA GTA GAG GCC GC-3′
GATA-2	5′-GAA CCG ACC ACT CAT CAA GC-3′	5′-GCA GCT TGT AGT AGA GGC CA-3′

3.4. TaqMan miR-150 Analysis

Specific cDNA synthesis for miR-150, was performed using TaqMan® MicroRNA Reverse Transcription Kit (Thermo Fisher Scientific) containing microRNA-specific RT primers and Taqman miRNA assay. To measure miRNAs expression levels, 1.33 µL of each cDNA was added to the specific TaqMan microRNA Assay (20×) and TaqMan 2× Universal PCR Master MiX (Thermo Fisher Scientific). The amplification conditions for miRNA expression profile were the following: 10 min at 95 °C, 40 cycles at 95 °C for 15 s, and 60 °C for 60 s. The experiments were performed in duplicate and the analysis was performed using the $2^{-\Delta\Delta Ct}$ formula using snRNA U6 as housekeeping microRNA [24].

3.5. Western Blot Analysis

Whole-cell lysis, protein extraction and Western Blot analyses of cultured K562[shScr], K562[shFHC], K562[cntr], and K562[siFHC] cells were performed, as previously reported [25,26]. For FHC protein quantification, the incubation of the anti-rabbit polyclonal primary anti-FHC (H-53) (1:200; sc-25617, Santa Cruz Biotechnology, Dallas, TX, USA) followed by the incubation with the HRP-conjugated secondary antibody (1:3000 Cell Signaling) was carried out. The goat polyclonal anti-γ-Tubulin antibody (C-20) (1:3000; sc-7396, Santa Cruz Biotechnology) was used as loading control. The immunoreactive bands were visualized with the ECL Western blotting detection system (Santa Cruz Biotechnology) and the bands intensity was quantified by using ImageJ (National Institutes of Health, Bethesda, MD, USA).

3.6. Flow Cytometry Analyses

The expression of erythroid-specific cell-surface markers was determined by direct immunofluorescence using the following conjugated antibodies: PE-CD235a and PE-CD71 (Miltenyi Biotec, Bergisch Gladbach, Germany). Briefly, 2×10^5 cells were suspended in 100 µL PBS supplemented and then stained with fluorochrome-conjugated antibodies for 30 min on ice. Cells were washed twice with PBS before analysis. Flow cytometry analyses were carried out 1 h after staining. The analysis was performed using FACScan flow cytometry (Becton Dickinson, Franklin Lakes, NJ, USA) and the data files were analyzed by FlowJo software (FlowJo v8.8.6, Becton Dickinson, Franklin Lakes, NJ, USA).

Int. J. Mol. Sci. **2017**, *18*, 2167

For cell cycle analysis, 5×10^5 cells were washed in PBS twice, fixed by adding ethanol 100%, in a drop wise manner, and then was stored at $-20\ ^\circ$C overnight. Cells were then rinsed twice to remove ethanol and incubated with PI solution for 1 h, in the dark, at room temperature prior to FACS analysis.

Apoptosis analysis was carried out using the Annexin V-PE Apoptosis Detection Kit (Becton Dickinson, Franklin Lakes, NJ, USA). Cells were washed twice with cold PBS and then resuspended in $1\times$ Binding Buffer at a concentration of 1×10^6 cells/mL. Then, 5 µL of PE Annexin V and 5 µL 7-AAD were added to the cells incubated for 15 min at room temperature in the dark. Finally, 400 µL of $1\times$ Binding Buffer was added to each sample before flow cytometry analysis. The analysis was performed in duplicate; here, a representative plot has been reported.

3.7. Direct Cell Counting

Briefly, 20×10^4 cells/well were seeded during the exponential phase of proliferating cells. Cell pellets, obtained by centrifugation at 1000 rpm $\times$ 5 min, were washed with fresh PBS and then resuspended in 5 mL by vigorously pipetting to disperse any clumps. The cell count was performed by mixing 50 µL of sample with 50 µL of 0.4% trypan blue solution, afterwards the mixture was loaded into the Bürker chamber. Each cell-count was performed in triplicate by using a $10\times$ objective according to the standard methods.

3.8. Haematopoietic Colony Formation Assay

For colony formation assays, K562 cells (5×10^2/mL) were plated in tissue culture 12 well plates in MethoCult H4100 medium (StemCell Technologies, Vancouver, BC, Canada) consisting of RPMI 1640 supplemented with 1% methylcellulose and 10% FBS. After 14 days, plates were scored for colony forming units (CFUs) using an inverted microscope (Leica, Wetzlar, Germany).

3.9. Cell Morphology Assay

For evaluating cell morphology, cells were cytocentrifuged onto glass slides, fixed in methanol and stained with May Grunwald-Giemsa (Thermo Fisher Scientific) and photographed with $40\times$ magnification with a digital camera Leica DFC420 C and Leica Application Suite Software (v1.9.0, Leica).

3.10. Statistical Analysis

Statistical significance of data was assessed by Student's *t*-test. *p* values < 0.05 were considered significant.

3.11. Bioinformatic Analysis

We used the publicly available software programs FINDTAR3 to predict the existence of complementary regions between miRNAs and mRNAs based on seed-pairing, free energy of miRNA:mRNA duplex, and proper dynamic programming score.

Supplementary Materials: The following are available online at www.mdpi.com/1422-0067/18/10/2167/s1.

Acknowledgments: This work was supported by funds from Associazione Italiana per la Ricerca sul Cancro (AIRC) and PON03PE_00009_2 (ICaRe: Infrastruttura Calabrese per la Medicina Rigenerativa). We thank Caterina Alessi (Department of Experimental and Clinical Medicine, University of Catanzaro "Magna Graecia", 88100 Catanzaro, Italy) for editorial assistance.

Author Contributions: Flavia Biamonte, Giovanni Morrone and Francesco Costanzo conceived and designed the experiments; Flavia Biamonte, Fabiana Zolea, Anna Martina Battaglia and Emanuela Chiarella performed the experiments; Flavia Biamonte, Fabiana Zolea, Emanuela Chiarella and Heather Mandy Bond analyzed the data; Emanuela Chiarella, Heather Mandy Bond, Donatella Malanga and Carmela De Marco contributed reagents/materials/analysis tools; Flavia Biamonte, Heather Mandy Bond, Giovanni Morrone and Francesco Costanzo wrote the paper.

Conflicts of Interest: The authors declare no conflict of interest.

References

1. Seita, J.; Weissman, I.L. Hematopoietic stem cell: Self-renewal versus differentiation. *Wiley Interdiscip. Rev. Syst. Biol. Med.* **2010**, *2*, 640–653. [CrossRef] [PubMed]
2. Weissman, I.L.; Anderson, D.J. Stem and progenitor cells: Origins, phenotypes, lineage commitments, and transdifferentiations. *Annu. Rev. Cell Dev. Biol.* **2001**, *17*, 387–403. [CrossRef] [PubMed]
3. Stamatoyannopoulos, G. Control of globin gene expression during development and erythroid differentiation. *Exp. Hematol.* **2005**, *33*, 259–271. [CrossRef] [PubMed]
4. Fajtova, M.; Kovarikova, A.; Svec, P.; Kankuri, E.; Sedlak, J. Immunophenotypic profile of nucleated erythroid progenitors during maturation in regenerating bone marrow. *Leuk. Lymphoma* **2013**, *54*, 2523–2530. [CrossRef] [PubMed]
5. Hu, J.; Liu, J.; Xue, F.; Halverson, G.; Reid, M.; Guo, A.; Chen, L.; Raza, A.; Galili, N.; Jaffray, J.; et al. Isolation and functional characterization of human erythroblasts at distinct stages: Implications for understanding of normal and disordered erythropoiesis in vivo. *Blood* **2013**, *121*, 3246–3253. [CrossRef] [PubMed]
6. Suzuki, M.; Kobayashi-Osaki, M.; Tsutsumi, S.; Pan, X.; Ohmori, S.; Takai, J.; Moriguchi, T.; Ohneda, O.; Ohneda, K.; Shimizu, R.; et al. GATA factor switching from GATA2 to GATA1 contributes to erythroid differentiation. *Genes Cells* **2013**, *18*, 921–933. [CrossRef] [PubMed]
7. Krumsiek, J.; Marr, C.; Schroeder, T.; Theis, F.J. Hierarchical differentiation of myeloid progenitors is encoded in the transcription factor network. *PLoS ONE* **2011**, *6*, e22649. [CrossRef] [PubMed]
8. Teruel-Montoya, R.; Kong, X.; Abraham, S.; Ma, L.; Kunapuli, S.P.; Holinstat, M.; Shaw, C.A.; McKenzie, S.E.; Edelstein, L.C.; Bray, P.F. MicroRNA expression differences in human hematopoietic cell lineages enable regulated transgene expression. *PLoS ONE* **2014**, *9*, e102259. [CrossRef] [PubMed]
9. Lazare, S.S.; Wojtowicz, E.E.; Bystrykh, L.V.; de Haan, G. MicroRNAs in hematopoiesis. *Exp. Cell Res* **2014**, *329*, 234–238. [CrossRef] [PubMed]
10. Agrawal, N.; Dasaradhi, P.V.; Mohmmed, A.; Malhotra, P.; Bhatnagar, R.K.; Mukherjee, S.K. RNA interference: Biology, Mechanisms, and Applications. *Microbiol. Mol. Biol.* **2003**, *67*, 657–685. [CrossRef]
11. Bissels, U.; Bosio, A.; Wagner, W. MicroRNAs are shaping the hematopoietic landscape. *Haematologica* **2012**, *97*, 160–167. [CrossRef] [PubMed]
12. He, Y.; Jiang, X.; Chen, J. The role of miR-150 in normal and malignant hematopoiesis. *Oncogene* **2014**, *33*, 3887–3893. [CrossRef] [PubMed]
13. Zhou, B.; Wang, S.; Mayr, C.; Bartel, D.P.; Lodish, H.F. miR-150, a microRNA expressed in mature B and T cells, blocks early B cell development when expressed prematurely. *Proc. Natl. Acad. Sci. USA* **2007**, *104*, 7080–7085. [CrossRef] [PubMed]
14. Doré, L.C.; Crispino, J.D. Transcription factor networks in erythroid cell and megakaryocyte development. *Blood* **2011**, *118*, 231–239. [CrossRef] [PubMed]
15. García, P.; Frampton, J. Hematopoietic lineage commitment: miRNAs add specificity to a widely expressed transcription factor. *Dev. Cell* **2008**, *14*, 815–816. [CrossRef] [PubMed]
16. Lu, J.; Guo, S.; Ebert, B.L.; Zhang, H.; Peng, X.; Bosco, J.; Pretz, J.; Schlanger, R.; Wang, J.Y.; Mak, R.H.; et al. MicroRNA-mediated control of cell fate in megakaryocyte-erythrocyte progenitors. *Dev. Cell* **2008**, *14*, 843–853. [CrossRef] [PubMed]
17. Zhang, L.; Sankaran, V.G.; Lodish, H.F. MicroRNAs in erythroid and megakaryocytic differentiation and megakaryocyte–erythroid progenitor lineage commitment. *Leukemia* **2012**, *26*, 2310–2316. [CrossRef] [PubMed]
18. Sun, Z.; Wang, Y.; Han, X.; Zhao, X.; Peng, Y.; Li, Y.; Peng, M.; Song, J.; Wu, K.; Sun, S.; et al. miR-150 inhibits terminal erythroid proliferation and differentiation. *Oncotarget* **2015**, *6*, 43033–43047. [CrossRef] [PubMed]
19. Li, H.; Ginzburg, Y.Z. Crosstalk between Iron Metabolism and Erythropoiesis. *Adv. Hematol.* **2010**, 605435. [CrossRef] [PubMed]
20. Goodnough, L.T.; Skikne, B.; Brugnara, C. Erythropoietin, iron, and erythropoiesis. *Blood* **2000**, *96*, 823–833. [PubMed]
21. Eliades, A.; Matsuura, S.; Ravid, K. Oxidases and reactive oxygen species during hematopoiesis: A focus on megakaryocytes. *J. Cell. Physiol.* **2012**, *227*, 3355–3362. [CrossRef] [PubMed]

22. Zheng, Q.Q.; Zhao, Y.S.; Guo, J.; Zhao, S.D.; Song, L.X.; Fei, C.M.; Zhang, Z.; Li, X.; Chang, C.K. Iron overload promotes erythroid apoptosis through regulating HIF-1a/ROS signaling pathway in patients with myelodysplastic syndrome. *Leuk. Res.* **2017**, *58*, 55–62. [CrossRef] [PubMed]

23. Iwasaki, K.; MacKenzie, E.L.; Hailemariam, K.; Sakamoto, K.; Tsuji, Y. Hemin-Mediated Regulation of an Antioxidant-Responsive Element of the Human Ferritin H Gene and Role of Ref-1 during Erythroid Differentiation of K562 Cells. *Mol. Cell. Biol.* **2006**, *26*, 2845–2856. [CrossRef] [PubMed]

24. Biamonte, F.; Zolea, F.; Bisognin, A.; Di Sanzo, M.; Saccoman, C.; Scumaci, D.; Aversa, I.; Panebianco, M.; Faniello, M.C.; Bortoluzzi, S.; et al. H-ferritin-regulated microRNAs modulate gene expression in K562 cells. *PLoS ONE* **2015**, *10*, e0122105. [CrossRef] [PubMed]

25. Zolea, F.; Biamonte, F.; Candeloro, P.; Di Sanzo, M.; Cozzi, A.; Di Vito, A.; Quaresima, B.; Lobello, N.; Trecroci, F.; Di Fabrizio, E.; et al. H ferritin silencing induces protein misfolding in K562 cells: A Raman analysis. *Free Radic. Biol. Med.* **2015**, *89*, 614–623. [CrossRef] [PubMed]

26. Zolea, F.; Biamonte, F.; Battaglia, A.M.; Faniello, M.C.; Cuda, G.; Costanzo, F. Caffeine Positively Modulates Ferritin Heavy Chain Expression in H460 Cells: Effects on Cell Proliferation. *PLoS ONE* **2016**, *11*, e0163078. [CrossRef] [PubMed]

27. Lobello, N.; Biamonte, F.; Pisanu, M.E.; Faniello, M.C.; Jakopin, Ž.; Chiarella, E.; Giovannone, E.D.; Mancini, R.; Ciliberto, G.; Cuda, G.; et al. Ferritin heavy chain is a negative regulator of ovarian cancer stem cell expansion and epithelial to mesenchymal transition. *Oncotarget* **2016**, *7*, 62019–62033. [CrossRef] [PubMed]

28. Misaggi, R.; Di Sanzo, M.; Cosentino, C.; Bond, H.M.; Scumaci, D.; Romeo, F.; Stellato, C.; Giurato, G.; Weisz, A.; Quaresima, B.; et al. Identification of H ferritin-dependent and independent genes in K562 differentiating cells by targeted gene silencing and expression profiling. *Gene* **2014**, *535*, 327–335. [CrossRef] [PubMed]

29. Huo, X.F.; Yu, J.; Peng, H.; Du, Z.W.; Liu, X.L.; Ma, Y.N.; Zhang, X.; Zhang, Y.; Zhao, H.L.; Zhang, J.W. Differential expression changes in K562 cells during the hemin-induced erythroid differentiation and the phorbolmyristate acetate (PMA)-induced megakaryocytic differentiation. *Mol. Cell. Biochem.* **2006**, *292*, 155–167. [CrossRef] [PubMed]

30. Broyles, R.H.; Belegu, V.; DeWitt, C.R.; Shah, S.N.; Stewart, C.A.; Pye, Q.N.; Floyd, R.A. Specific repression of β-globin promoter activity by nuclear ferritin. *Proc. Natl. Acad. Sci. USA* **2001**, *98*, 9145–9150. [CrossRef] [PubMed]

31. Wang, F.; Yu, J.; Yang, G.H.; Wang, X.S.; Zhang, J.W. Regulation of erythroid differentiation by miR-376a and its targets. *Cell Res.* **2011**, *21*, 1196–1209. [CrossRef] [PubMed]

32. Nakajima, H. Role of transcription factors in differentiation and reprogramming of hematopoietic cells. *Keio J. Med.* **2011**, *60*, 47–55. [CrossRef] [PubMed]

33. Welch, J.J.; Watts, J.A.; Vakoc, C.R.; Yao, Y.; Wang, H.; Hardison, R.C.; Blobel, G.A.; Chodosh, L.A.; Weiss, M.J. Global regulation of erythroid gene expression by transcription factor GATA-1. *Blood* **2004**, *104*, 3136–3147. [CrossRef] [PubMed]

34. Whitelaw, E.; Tsai, S.F.; Hogben, P.; Orkin, S.H. Regulated expression of globin chains and the erythroid transcription factor GATA-1 during erythropoiesis in the developing mouse. *Mol. Cell. Biol.* **1990**, *10*, 6596–6606. [CrossRef] [PubMed]

35. Edelstein, L.C.; McKenzie, S.E.; Shaw, C.; Holinstat, M.A.; Kunapuli, S.P.; Bray, P.F. MicroRNAs in platelet production and activation. *J. Thromb. Haemost.* **2013**, *1*, 340–350. [CrossRef] [PubMed]

36. Zhang, C.Z. Long non-coding RNA FTH1P3 facilitates oral squamous cell carcinoma progression by acting as a molecular sponge of miR-224–5p to modulate fizzled 5 expression. *Gene* **2017**, *607*, 47–55. [CrossRef] [PubMed]

37. Chiarella, E.; Carrà, G.; Scicchitano, S.; Codispoti, B.; Mega, T.; Lupia, M.; Pelaggi, D.; Marafioti, M.G.; Aloisio, A.; Giordano, M.; et al. UMG Lenti: Novel lentiviral vectors for efficient transgene- and reporter gene expression in human early hematopoietic progenitors. *PLoS ONE* **2014**, *12*, e114795. [CrossRef] [PubMed]

International Journal of
Molecular Sciences

MDPI

Review

Molecular Regulation of Differentiation in Early B-Lymphocyte Development

Mikael Sigvardsson [1,2]

[1]   Division of Molecular Hematology, Lund Stem Cell Center, Department of Laboratory Medicine,
     Lund University, 22184 Lund, Sweden; mikael.sigvardsson@med.lu.se; Tel.: +46-708-320-120
[2]   Department of Clinical and Experimental Medicine, Linköping University, SE-581 85 Linköping, Sweden

Received: 30 May 2018; Accepted: 28 June 2018; Published: 30 June 2018

Abstract: B-lymphocyte differentiation is one of the best understood developmental pathways in the hematopoietic system. Our understanding of the developmental trajectories linking the multipotent hematopoietic stem cell to the mature functional B-lymphocyte is extensive as a result of efforts to identify and prospectively isolate progenitors at defined maturation stages. The identification of defined progenitor compartments has been instrumental for the resolution of the molecular features that defines given developmental stages as well as for our understanding of the mechanisms that drive the progressive maturation process. Over the last years it has become increasingly clear that the regulatory networks that control normal B-cell differentiation are targeted by mutations in human B-lineage malignancies. This generates a most interesting link between development and disease that can be explored to improve diagnosis and treatment protocols in lymphoid malignancies. The aim of this review is to provide an overview of our current understanding of molecular regulation in normal and malignant B-cell development.

Keywords: B-lymphocyte; development; transcription factors; lymphoid leukemia

---

1. Introduction

The generation of B-lymphocytes in the bone marrow (BM) is a highly complex process guiding multipotent hematopoietic stem cells to become immunoglobulin-expressing B-cells. The differentiation process depends on the orchestrated activities of transcription factor (TF) networks and extracellular signals acting in conjunction to drive the expansion and maturation of progenitor populations. The process is complicated by the fact that cells must undergo *Immunoglobulin* (Ig) gene recombination and both positive and negative selection events to ensure proper functionality (reviewed in [1]). Even though much of our understanding of this developmental pathway is based on mouse models, there exist several similarities between mouse and human B-cell differentiation [2–4]. Furthermore, it is now evident that the same mechanisms that control normal B-lymphoid development in mice and humans are targeted in B-lymphoid malignancies (reviewed in [5]). The aim of this review is to provide an overview of our knowledge about developmental trajectories and regulatory networks in normal early B-lymphocyte development and their potential involvement in malignant transformation.

2. Resolving Developmental Trajectories in B-Cell Development

In order to understand the process controlling the generation of highly specified blood cells, it is of critical importance to identify and prospectively isolate cells at defined maturation stages. B-lymphocyte development has been suggested to proceed from the hematopoietic stem cell, through the lymphoid primed multipotent progenitor (LMPP) [6] stage, to generate a lymphoid-restricted common lymphoid progenitor (CLP) [7]. CLPs have the capacity to generate B-lineage-restricted B220+ Fraction A compartment [8], proceeding in differentiation to generate CD19+ cells.

While the progenitor cells within the classical CLP compartment retain lymphoid linage potentials and display a reduced capacity to generate myeloid cells [7], the inclusion of additional surface markers in the staining protocols has revealed a molecular and functional heterogeneity within this population. Surface expression of Integrin α(2)β(7) (LPAM1) or CXCR6 identifies a subpopulation of cells with reduced B but preserved NK/T lineage potential [9], and BST2 expression identifies a dendritic cell population [10]. It is further possible to isolate a B220+ population with preserved combined B and T-lineage potential within the classical CLP compartment [11,12]. Hence, it has become increasingly clear that the CLP compartment is highly heterogeneous and likely harbors a variety of more or less lineage-restricted progenitors.

One of the earliest markers associated with B-cell progenitors is B220, a heavily glycosylated splice form of the CD45 protein (CD45R) (reviewed in [13]). Expression of B220 in combination with other surface markers, such as CD43 (S7), CD24 (HSA), BP1, CD19, KIT (CD117), CD93 (AA4.1) [8,14–16], and CD25 [17,18], can be used to identify specific subpopulations of B-cell progenitors. Combined with functional and molecular analysis this has allowed for the establishment of a developmental hierarchy instrumental for our understanding of B-cell development (Figure 1). However, while a substantial fraction of the CD19⁻ B-cell progenitors express B220, functional analysis fails to link B220 expression exclusively to B-lineage-committed progenitors. Rather, a fraction of the B220+ cells retain T-cell [11,12,15], NK [19], and even myeloid potential [20,21].

These findings could be seen as evidence that early B-cell development does not follow one distinct path but rather proceeds through multiple pathways whereby lineage potentials are lost in a more or less stochastic manner (Figure 1). This model for lymphocyte development is supported by the finding that early thymic progenitors display combined T-macrophage potential but most have a limited ability to generate B-lineage cells [22]. Furthermore, the fetal liver contains cells with combined B-macrophage or T-macrophage potential [23]. Additional complexity in developmental trajectories in the fetal liver comes with the identification of B/T and B/NK bi-potent progenitors [9,24]. Hence, the difficulty of identifying CD19⁻ B-lineage committed progenitors could be a consequence of non-linear developmental paths not subject to the restrictions predicted from a hematopoietic tree (Figure 1).

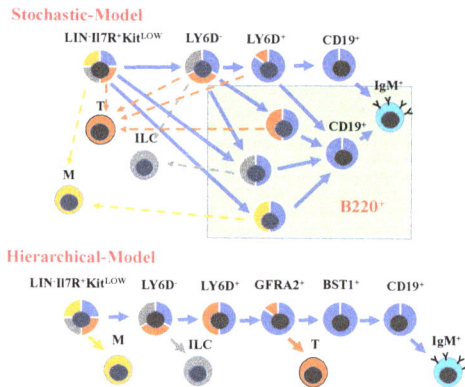

Figure 1. Developmental trajectories in B-cell development. Schematic drawing displaying two models for the developmental trajectories in B-cell development. Yellow indicates myeloid potential (M), gray indicates potential to generate innate lymphoid cells (ILC), orange indicates T lineage potential (T), and blue indicates B-cell potential. The arrows indicate potential developmental trajectories for the defined lineages. The green square indicates B220+ populations.

While conventional surface marker expression did not allow for the prospective isolation of committed CD19⁻ B-cell progenitors, expression of a reporter gene under the control of the Igll1 (Lambda 5) promoter [25] allowed for the identification of B-lymphoid- restricted progenitors within the classical CLP compartment [26]. Despite the fact that the *Igll1* gene, encoding one of the surrogate light chains [27], is not crucial for the earliest stages of B-cell development [28,29], the gene is transcribed in primitive progenitors serving as a marker for lineage commitment [30,31]. Continued analysis of the cellular heterogeneity within the CLP compartment identified the surface marker Ly6D as being expressed in a subpopulation of cells [32]. Transplantation experiments revealed that the Ly6D⁺ "CLP" population was largely restricted to the production of B-lineage cells; subsequently, these progenitors were denoted B-lymphocyte progenitors BLPs [32]. While the BLP population displayed a minimal T-linage potential in vivo, in vitro differentiation analysis suggested that a fraction of the cells retained the ability to generate T-lineage cells in response to a strong Notch signal [12]. This indicates further heterogeneity within the BLP compartment. This heterogeneity is largely resolved using the expression of the surface markers GNDF family receptor α2 (GFRA2) and bone marrow stroma cell antigen 1 (BST1), to prospectively isolate BLP1, BLP2, and BLP3 cells displaying a progressive degree of commitment to B-cell development [33] (Figure 1). Even though the progenitor populations defined by GFRA2 and BST1 displayed variable levels of B220, their gene expression patterns as well as functional analysis suggest that they represent highly similar or even identical developmental stages. These findings would be in line with observations made using multiparameter molecular analysis of human B-lineage development suggesting a hierarchical model based on progressive and ordered loss of lineage potentials (Figure 1) [34].

While CD19 expression marks stably committed B-lineage progenitors, this population is complex and can be further subdivided into defined developmental stages. The most immature cells express KIT, CD127 (IL7Ra), CD49E, CD11A, CD54, and CD43 [17,18,35,36], while CD25 (IL2Rα) is restricted to cells with a functional pre-BCR [17]. The expression of CD2 resembles that of CD25 because it is restricted to cells with cytoplasmic immunoglobulin heavy (IgH) chains [37]. Despite that surface antigens are most useful in the identification of progenitor stages in early B-cell development, a degree of heterogeneity can be observed even within a defined developmental stage [18]. This argues for the use of multiple markers for the identification of any given progenitor population. In all, the development of protocols for the prospective isolation of defined progenitor compartments provides a detailed map of the developmental trajectories in B-cell development generally compatible with the hematopoietic tree.

3. Transcription Factor Networks Regulating Early B-Cell Development

With our increasing understanding of the developmental trajectories in early B-cell development, it has been possible to explore molecular interplay at defined developmental stages. While lineage-specific gene expression is often associated with functional lineage commitment, it has been reported that the expression of lineage-associated genes can be detected already in multipotent progenitors [38,39]. These early transcriptional programs are suggested to be associated with defined cell fates [40], indicating the existence of functional lineage priming in non-committed progenitors (Figure 2). Lymphoid lineage priming is mainly observed as activation of genes expressed in both B- and T-cells, including *Rag1* and *Dntt* [39,41]. The activation of these genes is dependent on the TFs TCF3 [42–44], IKZF1 (IKAROS) [45], SPI1 (PU.1) [46,47], and MYB [48] acting in a concerted manner to promote the development of lymphoid progenitors. While SPI1 and IKZF1 create a regulatory loop controlling lymphoid versus myeloid cell fate [49], TCF3 initiates the B-lineage-restricted transcriptional program by activation of the *FoxO1* gene [50] (Figure 2). FOXO1 acts in a feed-forward loop with the TF EBF1 [51] to activate the transcription of B-lineage genes during B-cell specification [52–54]. The ability of EBF1 to coordinate the activation of transcriptionally inactive genes in epigenetically silenced chromatin [55,56] is likely a consequence of EBF1 associating with chromatin remodeling complexes, thereby directly impacting the structure and the epigenetic landscape [57,58]. Despite the LY6D⁺ "CLP" compartment being intact or even increased in the absence of EBF1 [59,60], the GFRA2⁺ compartment is dramatically decreased [33], the transcription of B-lineage-restricted genes is lost [51,52], and the cells are not properly lineage-restricted [60]. This highlights the essential role of EBF1 in B-lineage specification.

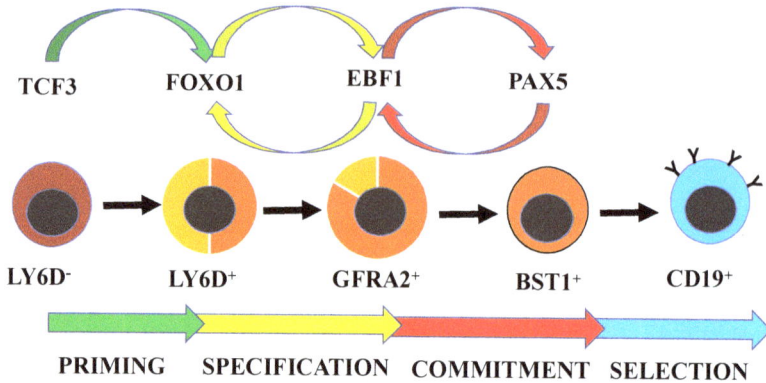

Figure 2. Schematic drawing of the transcription factor networks involved in priming (green arrows), specification (yellow arrows), commitment (red arrows), and selection (blue arrows) in early B-cell development. Red indicates B/T and NK cell potential, orange indicates B-cell potential, and yellow indicates residual T-cell potential. Blue indicates IgM$^+$ cells.

Even though EBF1 has the ability to repress genes associated with alternative cell fates [61,62] and the loss of EBF1 in B-lineage cells results in plasticity [63], stable lineage commitment depends on the TF PAX5. Despite a large portion of the B-lineage-restricted transcriptional program being activated in the absence of PAX5 [64,65], the progenitor cells are not stably committed to B-lineage cell fate [65–69]. In vitro differentiation or transplantation of PAX5-deficient pro-B cells, as well as deletion of the *Pax5* gene in B-lineage cells, results in the formation of both myeloid and T-lineage cells in vivo and in vitro [65–71]. Lineage restriction is likely achieved through direct repression of target genes such as *Colony-stimulating factor receptor 1* (*Csf1r* or *c-fms*) gene [72] and *Notch1* [73]. Even though this suggests that lineage specification can be separated from commitment [65], the finding that PAX5 is a direct EBF1 target [74] links these two processes in normal development. PAX5 and EBF1 target control elements in the *Ebf1* gene [75,76], creating a second regulatory loop and resulting in functional lineage commitment (Figure 2). The importance of the reciprocal regulation and collaboration between EBF1 and PAX5 is highlighted by the finding that normal as well as malignant pro-B cells carrying trans heterozygote mutations in the *Ebf1* and *Pax5* genes display lineage plasticity [77,78].

In addition to their function in stable lineage commitment, EBF1 and PAX5 are critical regulators of genes encoding proteins forming the pre-B cell receptor (pre-BCR) [54,79–82]. This receptor is formed as a newly generated IGH chain complex with surrogate light chains IGLL1 (λ5) and VPREB as well as the signal transduction proteins CD79α and CD79β (reviewed in [1,83]). Combined signaling through the pre-BCR and the IL7 receptor stimulates a proliferative burst [84] and causes a reduction in RAG protein levels [85]. While the proliferative burst is of great importance for the expansion and overall production of B-lineage cells, progressive development and Ig light chain rearrangement depend on the cells exiting the cell cycle [86,87]. This maturation step is suggested to depend on a pre-BCR-mediated activation of a regulatory network involving interferon regulatory factor (IRF4) and PAX5 [88]. PAX5 targets and activates the *Ikzf3* gene [89], encoding a TF suggested to collaborate with IKZF1 to repress genes encoding surrogate light chain components [90]. This results in a reduction of pre-BCR levels on the surface of the pro-B cell, reducing the proliferative signal. IRF4 has a somewhat different role in this process since it collaborates with the transcription factor FOXO1, stabilized by the pre-BCR signal [88], to drive differentiation and reactivate *Rag* gene expression critical for recombination of the Ig light chain genes [91–93]. Furthermore, IRF4 increases the expression of the CXCL12-responsive chemokine receptor CXCR4 [87]. This has been suggested to stimulate migration of the pre-B cells to a micromolecular niche with low levels of Il7 to further reduce the proliferative signal [87]. Hence, B-cell development is driven by an intricate interplay between stage-specific regulatory TF networks that orchestrate the differentiation process.

4. Transcription Factor Networks Link Development to B-Lymphoid Malignancies

While these TF networks clearly play crucial roles in normal B-cell development, it is becoming increasingly clear that they are closely connected to malignant transformation (reviewed in [5]). This is because genetic alterations in the *PAX5, EBF1,* or *IKZF1* genes are observed in a majority of B-ALL patients [94–96]. Even though there are reports of translocations of TF coding genes resulting in deregulated expression [97,98] or formation of fusion proteins [99–103], the most common genetic alterations result in partial inactivation of one or several TFs [94–96]. Even though inherited point mutations in PAX5 have been reported to result in increased leukemia incidence [104], the reduced functional TF activity more commonly depends on the inactivation of the TF genes via somatic heterozygote mutations and deletions [94–96]. The importance of TF dose for normal blood cell development in mice is well established because heterozygote inactivation of the *Myb* [48], *Spi1* [105], *Bcl11a* [106], *Ikzf1* [107], or *Ebf1* [108] gene results in disturbances in B-cell differentiation. Despite the fact that the heterozygote inactivation of the *Pax5* gene in mice does not appear to result in any dramatic developmental block [109,110], it has been suggested to result in alterations of cellular metabolism that may promote transformation [111]. Furthermore, *PAX5* deletions are often found in combination with complex karyotypes and other genetic aberrations, including recurrent translocations like t(12;21)(p13;q22) (ETV6-RUNX1) or t(1;19)(q23;p13) (TCF3-PBX1) [94,96]. *PAX5* mutations are also found in combination with genetic alterations in other TFs such as *IKZF1* and *EBF1* [94,96], likely augmenting the effect of a reduced PAX5 dose. Interestingly, the phenotypic changes in B-cell development in mice carrying heterozygote inactivation of a TF gene are often exacerbated upon combined targeting of two TFs. Combined heterozygote inactivation of *Ebf1/Tcf3* [108], *Ebf1/Runx1* [112], or *Ebf1/Pax5* [113] results in more dramatic phenotypes than what is observed in the single mutants. This highlights the importance of functional and correctly balanced TF networks in B-cell development.

Even though leukemia is generally considered to be confined to one defined hematopoietic lineage, about 7% of patients display a more complex disease [114,115]. These malignancies are denoted as acute leukemia of ambiguous lineage (ALAL). ALAL can be manifested either as bi-lineal leukemia, involving several lineages, or bi-phenotypic disease, with expansion of cells displaying combined expression of normally lineage-restricted surface markers [114–116]. It has been reported that the level of PAX5 regulates the formation of bi-phenotypic leukemia [117] and that B-ALL cells carrying mutations in *Pax5* can be converted into other lineages with preserved malignant features [78,118]. Furthermore, in MYC-induced lymphoma, oscillations in EBF1 and PAX5 levels result in lineage plasticity [119]. Additionally, dramatic phenotypic changes with preserved cytogenetic features have occasionally been reported from patients experiencing relapse of disease [120], further challenging the idea that leukemia is restricted to a given lineage of cells. While this has been considered uncommon in clinical practice [114,115], novel treatment protocols may reveal a higher degree of complexity in B-ALL. Recently it was reported that 13 out of 20 patients relapsing after treatment of B-ALL with genetically manipulated T-lymphocytes (chimeric antigen receptor (CAR-T) cells) targeting CD19 developed CD19 negative leukemia even at the clonal level [121,122]. This makes lineage plasticity a central mechanism for resistance development upon targeted treatment of leukemia. Hence, disruptions in transcription factor networks in leukemia may impact not only the transformation process per se but may also underlie the development of resistance to lineage-targeted therapies.

5. Integration of External Signals and Transcription Factor Networks in Early B-Cell Development

Even though intrinsic cell events such as regulatory loops created by TF networks are of critical importance for B-cell differentiation, the normal development and expansion of progenitors depend on extracellular signals in the microenvironment. While some of these signals are shared with other hematopoietic progenitors, others are restricted to lymphoid progenitor compartments. Among the former are Kit ligand (Steel factor, Stem Cell Factor SCF), acting via the receptor tyrosine kinase cKIT [123]. This receptor is expressed in a variety of hematopoietic progenitor cells including the multipotent hematopoietic stem cells [124], myeloid progenitors [125,126] as well as the CLPs [7]. Subsequently, disruption of this signaling pathway results in defective formation of multiple hematopoietic lineages [127–129]. The expression of cKIT is rather restricted in B-lymphoid progenitors, but a substantial fraction of the pro-B cell compartment retains this surface receptor as well as an ability to respond to the cytokine [130]. CXCL12 is another broadly acting cytokine involved in the homing of cells to specific niches in the BM [131,132]. The protein acts via its surface receptor CXCR4 and both the ligand and the receptor are crucial for normal homing of hematopoietic progenitor cells [131,132], including B-cell progenitors [133]. The expression of CXCL12 is restricted to specific subpopulations of stroma cells in the BM [131], contributing to the ability of this chemokine to act as an organizer of the BM microenvironment.

B-cell development is also influenced by cytokines with more restricted activity. These include FLT3 ligand (FL) acting through the FLT3 receptor expressed in the earliest lineage-restricted progenitor cells. Both the LMPP [6] and the CLPs [134] express FLT3; however, upon progression of B-cell development, the expression is downregulated as a consequence of the *Flt3* gene being repressed by PAX5 [135]. Ectopic expression of the FLT3 ligand causes alterations in blood cell development [136] and disruption of this signaling pathway results in reductions in LMPPs and CLPs [137]. However, this occurs without dramatic changes to the peripheral CD19$^+$ B-cell compartments [137]. The phenotype is exacerbated when combined with inactivation of Il7 signaling since this results in a complete block in BM B-cell development [138]. In line with the idea that the Il7R is expressed on CLPs with all lymphoid lineage potentials [7], lineage tracing analysis suggests that all B- and T-lymphoid cells as well as a substantial portion of the NK cells in the adult mouse have a history of Il7 expression [139]. The IL7 receptor is expressed on B-cell progenitors and a deficiency in either the receptor [140] or the ligand [141] results in a dramatic impairment in B-cell development already in the B-cell-restricted CD19$^-$ compartments in the mouse BM [60,142]. The Il7 receptor α (IL7Rα) chain is also a component of the receptor for thymic stromal lymphopoietin (TSLP), a cytokine acting via the Il7Rα and a specific TSLP-receptor [143]. IL7 and TSLP appear to be functionally redundant since ectopic expression of the latter largely rescues B-cell development in IL7-deficient mice [144].

Many cytokines would appear to be permissive, stimulating proliferation and reducing apoptosis in the B-cell progenitors rather than driving development to a specific lineage in an instructive manner. However, while T-cell development can be largely rescued by overexpression of BCL2 in mice deficient in IL7 signaling, this does not fully rescue B-cell development [145,146]. This suggests partially distinct functions for IL-7 signaling in the formation of different lymphoid lineages. One potential explanation for this could be that STAT5 activation, resulting from IL7R signaling [147], induces *Ebf1* transcription, potentially driving the progenitor towards B-cell fate [148]. Furthermore, ectopic expression of EBF1 partially rescues B-cell development in mice lacking IL7 [148] or the BTB/POZ domain transcription factor ZBTB17 (MIZ1), crucial for functional IL7 signaling [149]. While this would suggest that IL7 has unique functions in the induction of the genetic program in early B-cell progenitors, the finding that ectopic expression of FL can rescue B-cell development independently of IL7 argues for a more permissive function [150]. Furthermore, the developmental block imposed by conditional deletion of STAT5, a key mediator of Il7 signaling [147], can be partially rescued by ectopic expression of anti-apoptotic proteins [151]. In all, it would appear that the IL7 signaling pathway has both permissive and instructive components, as supported by the finding that deletion of the pro-apoptotic protein BIM rescues survival but not differentiation of B-cell progenitors in Il7-deficient mice [152].

Despite the function of Il7 in human B-cell development being somewhat disputed [153], inactivating mutations in the common gamma chain results in severe combined immunodeficiency in humans [154] and activating mutations in the IL7 signaling pathway are commonly detected in human malignancies [155]. Interestingly, heterozygote deletion of *Pax5* or *Ebf1*, in combination with transgenic expression of a constitutive active STAT5, causes a synergistic increase in the formation of B-lineage leukemia [156]. Hence, the interplay between TF networks and extracellular signals is critical for normal B-cell development and disturbances may result in impaired immune response or lymphoid malignancies.

6. Concluding Remarks

While the detailed understanding of maturation pathways is often considered a subject mainly relevant to developmental biology, our increased understanding of molecular events involved in malignant transformation highlights the relevance of cell differentiation in malignant transformation. Understanding developmental trajectories can be important for diagnosis since, even though leukemia is caused by expansion of progenitor B-cells, the heterogeneous expression of surface IG suggests that leukemia can reside in both the pro- and pre-B cell compartments [114,115]. The use of more advanced FACS staining protocols may resolve an even higher heterogeneity and possibly better classify leukemia in both the CD19-positive and CD19-negative (ALAL) groups. Furthermore, it is becoming increasingly clear that the regulatory networks that drive normal development are targeted in the transformation process. This is knowledge that can be explored to identify novel diagnostic and therapeutic approaches. Additionally, the understanding of the molecular regulation of lineage stability can be used to predict the risk of relapse through lineage conversion in association with targeted therapies. Hence, it can be predicted that basic developmental biology will become of increasing importance for the improvement of modern cancer care in the near future.

Funding: This work was supported by grants from the Swedish Cancer Society, the Swedish Childhood Cancer Foundation, the Swedish Research Council including the Stem Therapy and BioCare programs, Knut and Alice Wallenberg's Foundation, and a donation from Henry Hallberg.

Acknowledgments: I wish to acknowledge my colleagues at Lund and Linköping Universities for stimulating discussions.

Conflicts of Interest: The author declares no conflict of interest.

References

1. Melchers, F. Checkpoints that control B cell development. *J. Clin. Investig.* **2015**, *125*, 2203–2210. [CrossRef] [PubMed]
2. Ghia, P.; ten Boekel, E.; Rolink, A.G.; Melchers, F. B-cell development: A comparison between mouse and man. *Immunol. Today* **1998**, *19*, 480–485. [CrossRef]
3. Ghia, P.; ten Boekel, E.; Sanz, E.; de la Hera, A.; Rolink, A.; Melchers, F. Ordering of human bone marrow B lymphocyte precursors by single-cell polymerase chain reaction analyses of the rearrangement status of the immunoglobulin H and L chain gene loci. *J. Exp. Med.* **1996**, *184*, 2217–2229. [CrossRef] [PubMed]
4. Ghia, P.; Gratwohl, A.; Signer, E.; Winkler, T.H.; Melchers, F.; Rolink, A.G. Immature B cells from human and mouse bone marrow can change their surface light chain expression. *Eur. J. Immunol.* **1995**, *25*, 3108–3114. [CrossRef] [PubMed]
5. Somasundaram, R.; Prasad, M.A.; Ungerback, J.; Sigvardsson, M. Transcription factor networks in B-cell differentiation link development to acute lymphoid leukemia. *Blood* **2015**, *126*, 144–152. [CrossRef] [PubMed]
6. Adolfsson, J.; Mansson, R.; Buza-Vidas, N.; Hultquist, A.; Liuba, K.; Jensen, C.T.; Bryder, D.; Yang, L.; Borge, O.J.; Thoren, L.A.; et al. Identification of flt3+ lympho-myeloid stem cells lacking erythro-megakaryocytic potential a revised road map for adult blood lineage commitment. *Cell* **2005**, *121*, 295–306. [CrossRef] [PubMed]
7. Kondo, M.; Weissman, I.L.; Akashi, K. Identification of clonogenic common lymphoid progenitors in mouse bone marrow. *Cell* **1997**, *91*, 661–672. [CrossRef]
8. Hardy, R.R.; Carmack, C.E.; Shinton, S.A.; Kemp, J.D.; Hayakawa, K. Resolution and characterization of pro-B and pre-pro-B cell stages in normal mouse bone marrow. *J. Exp. Med.* **1991**, *173*, 1213–1225. [CrossRef] [PubMed]

9. Possot, C.; Schmutz, S.; Chea, S.; Boucontet, L.; Louise, A.; Cumano, A.; Golub, R. Notch signaling is necessary for adult, but not fetal, development of RORγt(+) innate lymphoid cells. *Nat. Immunol.* **2011**, *12*, 949–958. [CrossRef] [PubMed]

10. Medina, K.L.; Tangen, S.N.; Seaburg, L.M.; Thapa, P.; Gwin, K.A.; Shapiro, V.S. Separation of plasmacytoid dendritic cells from B-cell-biased lymphoid progenitor (BLP) and Pre-pro B cells using PDCA-1. *PLoS ONE* **2013**, *8*, e78408. [CrossRef] [PubMed]

11. Martin, C.H.; Aifantis, I.; Scimone, M.L.; von Andrian, U.H.; Reizis, B.; von Boehmer, H.; Gounari, F. Efficient thymic immigration of B220+ lymphoid-restricted bone marrow cells with T precursor potential. *Nat. Immunol.* **2003**, *4*, 866–873. [CrossRef] [PubMed]

12. Mansson, R.; Zandi, S.; Welinder, E.; Tsapogas, P.; Sakaguchi, N.; Bryder, D.; Sigvardsson, M. Single-cell analysis of the common lymphoid progenitor compartment reveals functional and molecular heterogeneity. *Blood* **2010**, *115*, 2601–2609. [CrossRef] [PubMed]

13. Rolink, A.; Melchers, F. B lymphopoiesis in the mouse. *Adv. Immunol.* **1993**, *53*, 123–156. [PubMed]

14. Hardy, R.R.; Hayakawa, K. B-lineage differentiation stages resolved by multiparameter flow cytometry. *Ann. N. Y. Acad. Sci.* **1995**, *764*, 19–24. [CrossRef] [PubMed]

15. Rumfelt, L.L.; Zhou, Y.; Rowley, B.M.; Shinton, S.A.; Hardy, R.R. Lineage specification and plasticity in CD19− early B cell precursors. *J. Exp. Med.* **2006**, *203*, 675–687. [CrossRef] [PubMed]

16. Li, Y.S.; Wasserman, R.; Hayakawa, K.; Hardy, R.R. Identification of the earliest B lineage stage in mouse bone marrow. *Immunity* **1996**, *5*, 527–535. [CrossRef]

17. Rolink, A.; Grawunder, U.; Winkler, T.H.; Karasuyama, H.; Melchers, F. IL-2 receptor α chain (CD25, TAC) expression defines a crucial stage in pre-B cell development. *Int. Immunol.* **1994**, *6*, 1257–1264. [CrossRef] [PubMed]

18. Jensen, C.T.; Lang, S.; Somasundaram, R.; Soneji, S.; Sigvardsson, M. Identification of Stage-Specific Surface Markers in Early B Cell Development Provides Novel Tools for Identification of Progenitor Populations. *J. Immunol.* **2016**, *197*, 1937–1944. [CrossRef] [PubMed]

19. Rolink, A.; ten Boekel, E.; Melchers, F.; Fearon, D.T.; Krop, I.; Andersson, J. A subpopulation of B220+ cells in murine bone marrow does not express CD19 and contains natural killer cell progenitors. *J. Exp. Med.* **1996**, *183*, 187–194. [CrossRef] [PubMed]

20. Balciunaite, G.; Ceredig, R.; Massa, S.; Rolink, A.G. A B220+ CD117+ CD19± hematopoietic progenitor with potent lymphoid and myeloid developmental potential. *Eur. J. Immunol.* **2005**, *35*, 2019–2030. [CrossRef] [PubMed]

21. Alberti-Servera, L.; von Muenchow, L.; Tsapogas, P.; Capoferri, G.; Eschbach, K.; Beisel, C.; Ceredig, R.; Ivanek, R.; Rolink, A. Single-cell RNA sequencing reveals developmental heterogeneity among early lymphoid progenitors. *EMBO J.* **2017**, *36*, 3619–3633. [CrossRef] [PubMed]

22. Bell, J.J.; Bhandoola, A. The earliest thymic progenitors for T cells possess myeloid lineage potential. *Nature* **2008**, *452*, 764–767. [CrossRef] [PubMed]

23. Katsura, Y. Redefinition of lymphoid progenitors. *Nat. Rev. Immunol.* **2002**, *2*, 127–132. [CrossRef] [PubMed]

24. Pereira de Sousa, A.; Berthault, C.; Granato, A.; Dias, S.; Ramond, C.; Kee, B.L.; Cumano, A.; Vieira, P. Inhibitors of DNA binding proteins restrict T cell potential by repressing Notch1 expression in Flt3-negative common lymphoid progenitors. *J. Immunol.* **2012**, *189*, 3822–3830. [CrossRef] [PubMed]

25. Martensson, I.L.; Melchers, F.; Winkler, T.H. A transgenic marker for mouse B lymphoid precursors. *J. Exp. Med.* **1997**, *185*, 653–661. [CrossRef] [PubMed]

26. Mansson, R.; Zandi, S.; Anderson, K.; Martensson, I.L.; Jacobsen, S.E.; Bryder, D.; Sigvardsson, M. B-lineage commitment prior to surface expression of B220 and CD19 on hematopoietic progenitor cells. *Blood* **2008**, *112*, 1048–1055. [CrossRef] [PubMed]

27. Karasuyama, H.; Kudo, A.; Melchers, F. The proteins encoded by the VpreB and l5 pre-B cell-specific genes can associate with each other and with the m heavy chain. *J. Exp. Med.* **1990**, *172*, 969–972. [CrossRef] [PubMed]

28. Rolink, A.; Karasuyama, H.; Grawunder, U.; Haasner, D.; Kudo, A.; Melchers, F. B cell development in mice with a defective lambda 5 gene. *Eur. J. Immunol.* **1993**, *23*, 1284–1288. [CrossRef] [PubMed]

29. Kitamura, D.; Kudo, A.; Schaal, S.; Muller, W.; Melchers, F.; Rajewsky, K. A critical role of lambda 5 protein in B cell development. *Cell* **1992**, *69*, 823–831. [CrossRef]

30. Kudo, A.; Thalmann, P.; Sakaguchi, N.; Davidson, W.F.; Pierce, J.H.; Kearney, J.F.; Reth, M.; Rolink, A.; Melchers, F. The expression of the mouse VpreB/lambda 5 locus in transformed cell lines and tumors of the B lineage differentiation pathway. *Int. Immunol.* **1992**, *4*, 831–840. [CrossRef] [PubMed]

31. Karasuyama, H.; Rolink, A.; Shinkai, Y.; Young, F.; Alt, F.W.; Melchers, F. The expression of Vpre-B/lambda 5 surrogate light chain in early bone marrow precursor B cells of normal and B cell-deficient mutant mice. *Cell* **1994**, *77*, 133–143. [CrossRef]

32. Inlay, M.A.; Bhattacharya, D.; Debashis, S.; Serwold, T.; Seita, J.; Karsunky, H.; Plevritis, S.K.; Dill, D.L.; Weissman, I.L. Ly6d marks the earliest stage of B-cell specification and identifies the branchpoint between B-cell and T-cell development. *Genes Dev.* **2009**, *23*, 2376–2381. [CrossRef] [PubMed]

33. Jensen, C.T.; Ahsberg, J.; Sommarin, M.N.E.; Strid, T.; Somasundaram, R.; Okuyama, K.; Ungerback, J.; Kupari, J.; Airaksinen, M.S.; Lang, S.; et al. Dissection of progenitor compartments resolves developmental trajectories in B-lymphopoiesis. *J. Exp. Med.* **2018**. [CrossRef] [PubMed]

34. Bendall, S.C.; Davis, K.L.; Amir el, A.D.; Tadmor, M.D.; Simonds, E.F.; Chen, T.J.; Shenfeld, D.K.; Nolan, G.P.; Pe'er, D. Single-cell trajectory detection uncovers progression and regulatory coordination in human B cell development. *Cell* **2014**, *157*, 714–725. [CrossRef] [PubMed]

35. Chen, J.; Ma, A.; Young, F.; Alt, F.W. IL-2 receptor α chain expression during early B lymphocyte differentiation. *Int. Immunol.* **1994**, *6*, 1265–1268. [CrossRef] [PubMed]

36. Hardy, R.R.; Kincade, P.W.; Dorshkind, K. The protean nature of cells in the B lymphocyte lineage. *Immunity* **2007**, *26*, 703–714. [CrossRef] [PubMed]

37. Milne, C.D.; Fleming, H.E.; Paige, C.J. IL-7 does not prevent pro-B/pre-B cell maturation to the immature/sIgM+ stage. *Eur. J. Immunol.* **2004**, *34*, 2647–2655. [CrossRef] [PubMed]

38. Hu, M.; Krause, D.; Greaves, M.; Sharkis, S.; Dexter, M.; Heyworth, C.; Enver, T. Multilineage gene expression precedes commitment in the hemopoietic system. *Genes Dev.* **1997**, *11*, 774–785. [CrossRef] [PubMed]

39. Mansson, R.; Hultquist, A.; Luc, S.; Yang, L.; Anderson, K.; Kharazi, S.; Al-Hashmi, S.; Liuba, K.; Thoren, L.; Adolfsson, J.; et al. Molecular evidence for hierarchical transcriptional lineage priming in fetal and adult stem cells and multipotent progenitors. *Immunity* **2007**, *26*, 407–419. [CrossRef] [PubMed]

40. Arinobo, Y.; Mizuno, S.; Chong, Y.; Shigematsu, H.; Iino, T.; Iwasaki, H.; Graf, T.; Mayfield, R.; Chan, S.; Kastner, P.; et al. Reciprocal activation of Gata-1 and PU.1 marks initial specification of hematopoietic stem cells into myeloerythroid and myelolymphoid lineages. *Cell Stem Cell* **2007**, *1*, 416–427. [CrossRef] [PubMed]

41. Igarashi, H.; Gregory, S.C.; Yokota, T.; Sakaguchi, N.; Kincade, P.W. Transcription from the RAG1 locus marks the earliest lymphocyte progenitors in bone marrow. *Immunity* **2002**, *17*, 117–130. [CrossRef]

42. Semerad, C.L.; Mercer, E.M.; Inlay, M.A.; Weissman, I.L.; Murre, C. E2A proteins maintain the hematopoietic stem cell pool and promote the maturation of myelolymphoid and myeloerythroid progenitors. *Proc. Natl. Acad. Sci. USA* **2009**, *106*, 1930–1935. [CrossRef] [PubMed]

43. Dias, S.; Mansson, R.; Gurbuxani, S.; Sigvardsson, M.; Kee, B.L. E2A proteins promote development of lymphoid-primed multipotent progenitors. *Immunity* **2008**, *29*, 217–227. [CrossRef] [PubMed]

44. Miyai, T.; Takano, J.; Endo, T.A.; Kawakami, E.; Agata, Y.; Motomura, Y.; Kubo, M.; Kashima, Y.; Suzuki, Y.; Kawamoto, H.; et al. Three-step transcriptional priming that drives the commitment of multipotent progenitors toward B cells. *Genes Dev.* **2018**, *32*, 112–126. [CrossRef] [PubMed]

45. Yoshida, T.; Ng, S.Y.; Zuniga-Pflucker, J.C.; Georgopoulos, K. Early hematopoietic lineage restrictions directed by Ikaros. *Nat. Immunol.* **2006**, *7*, 382–391. [CrossRef] [PubMed]

46. Carotta, S.; Dakic, A.; D'Amico, A.; Pang, S.H.; Greig, K.T.; Nutt, S.L.; Wu, L. The transcription factor PU.1 controls dendritic cell development and Flt3 cytokine receptor expression in a dose-dependent manner. *Immunity* **2010**, *32*, 628–641. [CrossRef] [PubMed]

47. DeKoter, R.P.; Lee, H.J.; Singh, H. PU.1 regulates expression of the interleukin-7 receptor in lymphoid progenitors. *Immunity* **2002**, *16*, 297–309. [CrossRef]

48. Greig, K.T.; de Graaf, C.A.; Murphy, J.M.; Carpinelli, M.R.; Pang, S.H.; Frampton, J.; Kile, B.T.; Hilton, D.J.; Nutt, S.L. Critical roles for c-Myb in lymphoid priming and early B-cell development. *Blood* **2010**, *115*, 2796–2805. [CrossRef] [PubMed]

49. Spooner, C.J.; Cheng, J.X.; Pujadas, E.; Laslo, P.; Singh, H. A recurrent network involving the transcription factors PU.1 and Gfi1 orchestrates innate and adaptive immune cell fates. *Immunity* **2009**, *31*, 576–586. [CrossRef] [PubMed]

50. Welinder, E.; Mansson, R.; Mercer, E.M.; Bryder, D.; Sigvardsson, M.; Murre, C. The transcription factors E2A and HEB act in concert to induce the expression of FOXO1 in the common lymphoid progenitor. *Proc. Natl. Acad. Sci. USA* **2011**, *108*, 17402–17407. [CrossRef] [PubMed]

51. Mansson, R.; Welinder, E.; Ahsberg, J.; Lin, Y.C.; Benner, C.; Glass, C.K.; Lucas, J.S.; Sigvardsson, M.; Murre, C. Positive intergenic feedback circuitry, involving EBF1 and FOXO1, orchestrates B-cell fate. *Proc. Natl. Acad. Sci. USA* **2012**, *109*, 21028–21033. [CrossRef] [PubMed]

52. Zandi, S.; Mansson, R.; Tsapogas, P.; Zetterblad, J.; Bryder, D.; Sigvardsson, M. EBF1 is essential for B-lineage priming and establishment of a transcription factor network in common lymphoid progenitors. *J. Immunol.* **2008**, *181*, 3364–3372. [CrossRef] [PubMed]

53. Lin, H.; Grosschedl, R. Failure of B-cell differentiation in mice lacking the transcription factor EBF. *Nature* **1995**, *376*, 263–267. [CrossRef] [PubMed]

54. Lin, Y.C.; Jhunjhunwala, S.; Benner, C.; Heinz, S.; Welinder, E.; Mansson, R.; Sigvardsson, M.; Hagman, J.; Espinoza, C.A.; Dutkowski, J.; et al. A global network of transcription factors, involving E2A, EBF1 and Foxo1, that orchestrates B cell fate. *Nat. Immunol.* **2010**, *11*, 635–643. [CrossRef] [PubMed]

55. Maier, H.; Ostraat, R.; Gao, H.; Fields, S.; Shinton, S.A.; Medina, K.L.; Ikawa, T.; Murre, C.; Singh, H.; Hardy, R.R.; et al. Early B cell factor cooperates with Runx1 and mediates epigenetic changes associated with mb-1 transcription. *Nat. Immunol.* **2004**, *5*, 1069–1077. [CrossRef] [PubMed]

56. Li, R.; Cauchy, P.; Ramamoorthy, S.; Boller, S.; Chavez, L.; Grosschedl, R. Dynamic EBF1 occupancy directs sequential epigenetic and transcriptional events in B-cell programming. *Genes Dev.* **2018**, *32*, 96–111. [CrossRef] [PubMed]

57. Gao, H.; Lukin, K.; Ramirez, J.; Fields, S.; Lopez, D.; Hagman, J. Opposing effects of SWI/SNF and Mi-2/NuRD chromatin remodeling complexes on epigenetic reprogramming by EBF and Pax5. *Proc. Natl. Acad. Sci. USA* **2009**, *106*, 11258–22363. [CrossRef] [PubMed]

58. Yang, C.Y.; Ramamoorthy, S.; Boller, S.; Rosenbaum, M.; Gil, A.R.; Mittler, G.; Imai, Y.; Kuba, K.; Grosschedl, R. Interaction of CCR4-NOT with EBF1 regulates gene-specific transcription and mRNA stability in B lymphopoiesis. *Genes Dev.* **2016**, *30*, 2310–2324. [CrossRef] [PubMed]

59. Gyory, I.; Boller, S.; Nechanitzky, R.; Mandel, E.; Pott, S.; Liu, E.; Grosschedl, R. Transcription factor Ebf1 regulates differentiation stage-specific signaling, proliferation, and survival of B cells. *Genes Dev.* **2012**, *25*, 668–682. [CrossRef] [PubMed]

60. Tsapogas, P.; Zandi, S.; Ahsberg, J.; Zetterblad, J.; Welinder, E.; Jonsson, J.I.; Mansson, R.; Qian, H.; Sigvardsson, M. IL-7 mediates Ebf-1-dependent lineage restriction in early lymphoid progenitors. *Blood* **2011**, *118*, 1283–1290. [CrossRef] [PubMed]

61. Thal, M.A.; Carvalho, T.L.; He, T.; Kim, H.G.; Gao, H.; Hagman, J.; Klug, C.A. Ebf1-mediated down-regulation of Id2 and Id3 is essential for specification of the B cell lineage. *Proc. Natl. Acad. Sci. USA* **2009**, *106*, 552–557. [CrossRef] [PubMed]

62. Pongubala, J.M.; Northrup, D.L.; Lancki, D.W.; Medina, K.L.; Treiber, T.; Bertolino, E.; Thomas, M.; Grosschedl, R.; Allman, D.; Singh, H. Transcription factor EBF restricts alternative lineage options and promotes B cell fate commitment independently of Pax5. *Nat. Immunol.* **2008**, *9*, 203–215. [CrossRef] [PubMed]

63. Nechanitzky, R.; Akbas, D.; Scherer, S.; Gyory, I.; Hoyler, T.; Ramamoorthy, S.; Diefenbach, A.; Grosschedl, R. Transcription factor EBF1 is essential for the maintenance of B cell identity and prevention of alternative fates in committed cells. *Nat. Immunol.* **2013**, *14*, 867–875. [CrossRef] [PubMed]

64. Nutt, S.L.; Urbanek, P.; Rolink, A.; Busslinger, M. Essential functions of Pax5 (BSAP) in pro-B cell development: Difference between fetal and adult B lymphopoiesis and reduced V-to-DJ recombination at the IgH locus. *Genes Dev.* **1997**, *11*, 476–491. [CrossRef] [PubMed]

65. Zandi, S.; Ahsberg, J.; Tsapogas, P.; Stjernberg, J.; Qian, H.; Sigvardsson, M. Single-cell analysis of early B-lymphocyte development suggests independent regulation of lineage specification and commitment in vivo. *Proc. Natl. Acad. Sci. USA* **2012**, *109*, 15871–15876. [CrossRef] [PubMed]

66. Mikkola, I.; Heavey, B.; Horcher, M.; Busslinger, M. Reversion of B cell commitment upon loss of Pax5 expression. *Science* **2002**, *297*, 110–113. [CrossRef] [PubMed]

67. Nutt, S.L.; Heavey, B.; Rolink, A.G.; Busslinger, M. Commitment to the B-lymphoid lineage depends on the transcription factor Pax5. *Nature* **1999**, *401*, 556–562. [CrossRef] [PubMed]

68. Rolink, A.G.; Nutt, S.L.; Melchers, F.; Busslinger, M. Long-term in vivo reconstitution of T-cell development by Pax5- deficient B-cell progenitors. *Nature* **1999**, *401*, 603–606. [CrossRef] [PubMed]

69. Hoflinger, S.; Kesavan, K.; Fuxa, M.; Hutter, C.; Heavey, B.; Radtke, F.; Busslinger, M. Analysis of Notch1 function by in vitro T cell differentiation of Pax5 mutant lymphoid progenitors. *J. Immunol.* **2004**, *173*, 3935–3944. [CrossRef] [PubMed]

70. Rolink, A.G.; Schaniel, C.; Bruno, L.; Melchers, F. In vitro and in vivo plasticity of Pax5-deficient pre-B I cells. *Immunol. Lett.* **2002**, *82*, 35–40. [CrossRef]

71. Schaniel, C.; Bruno, L.; Melchers, F.; Rolink, A.G. Multiple hematopoietic cell lineages develop in vivo from transplanted Pax5-deficient pre-B I-cell clones. *Blood* **2002**, *99*, 472–478. [CrossRef] [PubMed]

72. Tagoh, H.; Ingram, R.; Wilson, N.; Salvagiotto, G.; Warren, A.J.; Clarke, D.; Busslinger, M.; Bonifer, C. The mechanism of repression of the myeloid-specific c-fms gene by Pax5 during B lineage restriction. *EMBO J.* **2006**, *25*, 1070–1080. [CrossRef] [PubMed]

73. Souabni, A.; Cobaleda, C.; Schebesta, M.; Busslinger, M. Pax5 promotes B lymphopoiesis and blocks T cell development by repressing Notch1. *Immunity* **2002**, *17*, 781–793. [CrossRef]

74. Decker, T.; Pasca di Magliano, M.; McManus, S.; Sun, Q.; Bonifer, C.; Tagoh, H.; Busslinger, M. Stepwise activation of enhancer and promoter regions of the B cell commitment gene *Pax5* in early lymphopoiesis. *Immunity* **2009**, *30*, 508–520. [CrossRef] [PubMed]

75. Roessler, S.; Gyory, I.; Imhof, S.; Spivakov, M.; Williams, R.R.; Busslinger, M.; Fisher, A.G.; Grosschedl, R. Distinct promoters mediate the regulation of *Ebf1* gene expression by interleukin-7 and Pax5. *Mol. Cell. Biol.* **2007**, *27*, 579–594. [CrossRef] [PubMed]

76. Smith, E.M.; Gisler, R.; Sigvardsson, M. Cloning and Characterization of a Promoter Flanking the Early B Cell Factor (EBF) Gene Indicates Roles for E-Proteins and Autoregulation in the Control of EBF Expression. *J. Immunol.* **2002**, *169*, 261–270. [CrossRef] [PubMed]

77. Ungerback, J.; Ahsberg, J.; Strid, T.; Somasundaram, R.; Sigvardsson, M. Combined heterozygous loss of Ebf1 and Pax5 allows for T-lineage conversion of B cell progenitors. *J. Exp. Med.* **2015**, *212*, 1109–1123. [CrossRef] [PubMed]

78. Somasundaram, R.; Ahsberg, J.; Okuyama, K.; Ungerback, J.; Lilljebjorn, H.; Fioretos, T.; Strid, T.; Sigvardsson, M. Clonal conversion of B lymphoid leukemia reveals cross-lineage transfer of malignant states. *Genes Dev.* **2016**, *30*, 2486–2499. [CrossRef] [PubMed]

79. Treiber, T.; Mandel, E.M.; Pott, S.; Gyory, I.; Firner, S.; Liu, E.T.; Grosschedl, R. Early B Cell Factor 1 Regulates B Cell Gene Networks by Activation, Repression, and Transcription-Independent Poising of Chromatin. *Immunity* **2010**, *32*, 714–725. [CrossRef] [PubMed]

80. Revilla, I.D.R.; Bilic, I.; Vilagos, B.; Tagoh, H.; Ebert, A.; Tamir, I.M.; Smeenk, L.; Trupke, J.; Sommer, A.; Jaritz, M.; et al. The B-cell identity factor Pax5 regulates distinct transcriptional programmes in early and late B lymphopoiesis. *EMBO J.* **2012**, *31*, 3130–3146. [CrossRef] [PubMed]

81. Nutt, S.L.; Thevenin, C.; Busslinger, M. Essential functions of Pax-5 (BSAP) in pro-B cell development. *Immunobiology* **1997**, *198*, 227–235. [CrossRef]

82. Nutt, S.L.; Morrison, A.M.; Dorfler, P.; Rolink, A.; Busslinger, M. Identification of BSAP (Pax-5) target genes in early B-cell development by loss- and gain-of-function experiments. *EMBO J.* **1998**, *17*, 2319–2333. [CrossRef] [PubMed]

83. Martensson, I.L.; Rolink, A.; Melchers, F.; Mundt, C.; Licence, S.; Shimizu, T. The pre-B cell receptor and its role in proliferation and Ig heavy chain allelic exclusion. In *Seminars in Immunology*; Academic Press: Cambridge, MA, USA, 2002; Volume 14, pp. 335–342.

84. Erlandsson, L.; Licence, S.; Gaspal, F.; Lane, P.; Corcoran, A.E.; Martensson, I.L. Both the pre-BCR and the IL-7Ralpha are essential for expansion at the pre-BII cell stage in vivo. *Eur. J. Immunol.* **2005**, *35*, 1969–1976. [CrossRef] [PubMed]

85. Grawunder, U.; Leu, T.M.J.; Scatz, D.G.; Werner, A.; Rolink, A.G.; Melchers, F.; Winkler, T.H. Down-regulation of *RAG1* and *RAG2* gene expression in preB cells after functional immunoglobulin heavy chain rearrangement. *Immunity* **1995**, *3*, 601–608. [CrossRef]

86. Mandal, M.; Powers, S.E.; Ochiai, K.; Georgopoulos, K.; Kee, B.L.; Singh, H.; Clark, M.R. Ras orchestrates exit from the cell cycle and light-chain recombination during early B cell development. *Nat. Immunol.* **2009**, *10*, 1110–1117. [CrossRef] [PubMed]

87. Johnson, K.; Hashimshony, T.; Sawai, C.M.; Pongubala, J.M.; Skok, J.A.; Aifantis, I.; Singh, H. Regulation of immunoglobulin light-chain recombination by the transcription factor IRF-4 and the attenuation of interleukin-7 signaling. *Immunity* **2008**, *28*, 335–345. [CrossRef] [PubMed]

88. Ochiai, K.; Maienschein-Cline, M.; Mandal, M.; Triggs, J.R.; Bertolino, E.; Sciammas, R.; Dinner, A.R.; Clark, M.R.; Singh, H. A self-reinforcing regulatory network triggered by limiting IL-7 activates pre-BCR signaling and differentiation. *Nat. Immunol.* **2012**, *13*, 300–307. [CrossRef] [PubMed]

89. Pridans, C.; Holmes, M.L.; Polli, M.; Wettenhall, J.M.; Dakic, A.; Corcoran, L.M.; Smyth, G.K.; Nutt, S.L. Identification of Pax5 target genes in early B cell differentiation. *J. Immunol.* **2008**, *180*, 1719–1728. [CrossRef] [PubMed]

90. Thompson, E.C.; Cobb, B.S.; Sabbattini, P.; Meixlsperger, S.; Parelho, V.; Liberg, D.; Taylor, B.; Dillon, N.; Georgopoulos, K.; Jumaa, H.; et al. Ikaros DNA-binding proteins as integral components of B cell developmental-stage-specific regulatory circuits. *Immunity* **2007**, *26*, 335–344. [CrossRef] [PubMed]

91. Lu, R.; Medina, K.; Lancki, D.; Singh, H. IRF-4,8 orchestrate the pre-B-to-B transition in lymphocyte development. *Genes Dev.* **2003**, *17*, 1703–1708. [CrossRef] [PubMed]

92. Amin, R.H.; Schlissel, M.S. Foxo1 directly regulates the transcription of recombination-activating genes during B cell development. *Nat. Immunol.* **2008**, *9*, 613–622. [CrossRef] [PubMed]

93. Dengler, H.S.; Baracho, G.V.; Omori, S.A.; Bruckner, S.; Arden, K.C.; Castrillon, D.H.; DePinho, R.A.; Rickert, R.C. Distinct functions for the transcription factor Foxo1 at various stages of B cell differentiation. *Nat. Immunol.* **2008**, *9*, 1388–1398. [CrossRef] [PubMed]

94. Mullighan, C.G.; Goorha, S.; Radtke, I.; Miller, C.B.; Coustan-Smith, E.; Dalton, J.D.; Girtman, K.; Mathew, S.; Ma, J.; Pounds, S.B.; et al. Genome-wide analysis of genetic alterations in acute lymphoblastic leukaemia. *Nature* **2007**, *446*, 758–764. [CrossRef] [PubMed]

95. Mullighan, C.G.; Miller, C.B.; Radtke, I.; Phillips, L.A.; Dalton, J.; Ma, J.; White, D.; Hughes, T.P.; le Beau, M.M.; Pui, C.H.; et al. BCR-ABL1 lymphoblastic leukaemia is characterized by the deletion of Ikaros. *Nature* **2008**, *453*, 110–114. [CrossRef] [PubMed]

96. Kuiper, R.P.; Schoenmakers, E.F.; van Reijmersdal, S.V.; Hehir-Kwa, J.Y.; van Kessel, A.G.; van Leeuwen, F.N.; Hoogerbrugge, P.M. High-resolution genomic profiling of childhood ALL reveals novel recurrent genetic lesions affecting pathways involved in lymphocyte differentiation and cell cycle progression. *Leukemia* **2007**, *21*, 1258–1266. [CrossRef] [PubMed]

97. Iida, S.; Rao, P.H.; Nallasivam, P.; Hibshoosh, H.; Butler, M.; Louie, D.C.; Dyomin, V.; Ohno, H.; Chaganti, R.S.; Dalla-Favera, R. The t(9;14)(p13;q32) chromosomal translocation associated with lymphoplasmacytoid lymphoma involves the *PAX-5* gene. *Blood* **1996**, *88*, 4110–4117. [PubMed]

98. Bouamar, H.; Abbas, S.; Lin, A.P.; Wang, L.; Jiang, D.; Holder, K.N.; Kinney, M.C.; Hunicke-Smith, S.; Aguiar, R.C. A capture-sequencing strategy identifies IRF8, EBF1, and APRIL as novel IGH fusion partners in B-cell lymphoma. *Blood* **2013**, *122*, 726–733. [CrossRef] [PubMed]

99. Roberts, K.G.; Morin, R.D.; Zhang, J.; Hirst, M.; Zhao, Y.; Su, X.; Chen, S.C.; Payne-Turner, D.; Churchman, M.L.; Harvey, R.C.; et al. Genetic alterations activating kinase and cytokine receptor signaling in high-risk acute lymphoblastic leukemia. *Cancer Cell* **2012**, *22*, 153–166. [CrossRef] [PubMed]

100. Coyaud, E.; Struski, S.; Prade, N.; Familiades, J.; Eichner, R.; Quelen, C.; Bousquet, M.; Mugneret, F.; Talmant, P.; Pages, M.P.; et al. Wide diversity of PAX5 alterations in B-ALL: A Groupe Francophone de Cytogenetique Hematologique study. *Blood* **2010**, *115*, 3089–3097. [CrossRef] [PubMed]

101. Kamps, M.P.; Murre, C.; Sun, X.H.; Baltimore, D. A new homeobox gene contributes the DNA binding domain of the t(1;19) translocation protein in pre-B ALL. *Cell* **1990**, *60*, 547–555. [CrossRef]

102. Inaba, T.; Roberts, W.M.; Shapiro, L.H.; Jolly, K.W.; Raimondi, S.C.; Smith, S.D.; Look, A.T. Fusion of the leucine zipper gene *HLF* to the *E2A* gene in human acute B-lineage leukemia. *Science* **1992**, *257*, 531–534. [CrossRef] [PubMed]

103. Fazio, G.; Palmi, C.; Rolink, A.; Biondi, A.; Cazzaniga, G. PAX5/TEL acts as a transcriptional repressor causing down-modulation of CD19, enhances migration to CXCL12, and confers survival advantage in pre-BI cells. *Cancer Res.* **2008**, *68*, 181–189. [CrossRef] [PubMed]

104. Shah, S.; Schrader, K.A.; Waanders, E.; Timms, A.E.; Vijai, J.; Miething, C.; Wechsler, J.; Yang, J.; Hayes, J.; Klein, R.J.; et al. A recurrent germline PAX5 mutation confers susceptibility to pre-B cell acute lymphoblastic leukemia. *Nat. Genet.* **2013**, *45*, 1226–1231. [CrossRef] [PubMed]

105. DeKoter, R.P.; Singh, H. Regulation of B lymphocyte and macrophage development by graded expression of PU.1. *Science* **2000**, *288*, 1439–1441. [CrossRef] [PubMed]

106. Yu, Y.; Wang, J.; Khaled, W.; Burke, S.; Li, P.; Chen, X.; Yang, W.; Jenkins, N.A.; Copeland, N.G.; Zhang, S.; et al. Bcl11a is essential for lymphoid development and negatively regulates p53. *J. Exp. Med.* **2012**, *209*, 2467–2483. [CrossRef] [PubMed]

107. Ferreiros-Vidal, I.; Carroll, T.; Taylor, B.; Terry, A.; Liang, Z.; Bruno, L.; Dharmalingam, G.; Khadayate, S.; Cobb, B.S.; Smale, S.T.; et al. Genome-wide identification of Ikaros targets elucidates its contribution to mouse B-cell lineage specification and pre-B-cell differentiation. *Blood* **2013**, *121*, 1769–1782. [CrossRef] [PubMed]

108. O'Riordan, M.; Grosschedl, R. Coordinate regulation of B cell differentiation by the transcription factors EBF and E2A. *Immunity* **1999**, *11*, 21–31. [CrossRef]

109. Urbánek, P.; Wang, Z.-Q.; Fetka, I.; Wagner, E.F.; Busslinger, M. Complete block of early B cell differentiation and altered patterning of the posterior midbrain in mice lacking Pax5/BSAP. *Cell* **1994**, *79*, 901–912. [CrossRef]

110. Ahsberg, J.; Ungerback, J.; Strid, T.; Welinder, E.; Stjernberg, J.; Larsson, M.; Qian, H.; Sigvardsson, M. Early B-cell Factor 1 regulates the expansion of B-cell progenitors in a dose dependent manner. *J. Biol. Chem.* **2013**, *288*, 33449–33461. [CrossRef] [PubMed]

111. Chan, L.N.; Chen, Z.; Braas, D.; Lee, J.W.; Xiao, G.; Geng, H.; Cosgun, K.N.; Hurtz, C.; Shojaee, S.; Cazzaniga, V.; et al. Metabolic gatekeeper function of B-lymphoid transcription factors. *Nature* **2017**, *542*, 479–483. [CrossRef] [PubMed]

112. Lukin, K.; Fields, S.; Lopez, D.; Cherrier, M.; Ternyak, K.; Ramirez, J.; Feeney, A.J.; Hagman, J. Compound haploinsufficiencies of *Ebf1* and *Runx1* genes impede B cell lineage progression. *Proc. Natl. Acad. Sci. USA* **2010**, *107*, 7869–7874. [CrossRef] [PubMed]

113. Prasad, M.A.; Ungerback, J.; Ahsberg, J.; Somasundaram, R.; Strid, T.; Larsson, M.; Mansson, R.; de Paepe, A.; Lilljebjorn, H.; Fioretos, T.; et al. Ebf1 heterozygosity results in increased DNA damage in pro-B cells and their synergistic transformation by Pax5 haploinsufficiency. *Blood* **2015**, *125*, 4052–4059. [CrossRef] [PubMed]

114. Jennings, C.D.; Foon, K.A. Recent advances in flow cytometry: Application to the diagnosis of hematologic malignancy. *Blood* **1997**, *90*, 2863–2892. [PubMed]

115. Craig, F.E.; Foon, K.A. Flow cytometric immunophenotyping for hematologic neoplasms. *Blood* **2008**, *111*, 3941–3967. [CrossRef] [PubMed]

116. Manola, K.N. Cytogenetic abnormalities in acute leukaemia of ambiguous lineage: An overview. *Br. J. Haematol.* **2013**, *163*, 24–39. [CrossRef] [PubMed]

117. Simmons, S.; Knoll, M.; Drewell, C.; Wolf, I.; Mollenkopf, H.J.; Bouquet, C.; Melchers, F. Biphenotypic B-lymphoid/myeloid cells expressing low levels of Pax5: Potential targets of BAL development. *Blood* **2012**, *120*, 3688–3698. [CrossRef] [PubMed]

118. Jacoby, E.; Nguyen, S.M.; Fountaine, T.J.; Welp, K.; Gryder, B.; Qin, H.; Yang, Y.; Chien, C.D.; Seif, A.E.; Lei, H.; et al. CD19 CAR immune pressure induces B-precursor acute lymphoblastic leukaemia lineage switch exposing inherent leukaemic plasticity. *Nat. Commun.* **2016**, *7*, 12320. [CrossRef] [PubMed]

119. Yu, D.; Allman, D.; Goldschmidt, M.H.; Atchison, M.L.; Monroe, J.G.; Thomas-Tikhonenko, A. Oscillation between B-lymphoid and myeloid lineages in Myc-induced hematopoietic tumors following spontaneous silencing/reactivation of the EBF/Pax5 pathway. *Blood* **2003**, *101*, 1950–1955. [CrossRef] [PubMed]

120. Dorantes-Acosta, E.; Pelayo, R. Lineage switching in acute leukemias: A consequence of stem cell plasticity? *Bone Marrow Res.* **2012**, *2012*, 406796. [CrossRef] [PubMed]

121. Park, J.H.; Geyer, M.B.; Brentjens, R.J. CD19-targeted CAR T-cell therapeutics for hematologic malignancies: Interpreting clinical outcomes to date. *Blood* **2016**, *127*, 3312–3320. [CrossRef] [PubMed]

122. Gardner, R.; Wu, D.; Cherian, S.; Fang, M.; Hanafi, L.A.; Finney, O.; Smithers, H.; Jensen, M.C.; Riddell, S.R.; Maloney, D.G.; et al. Acquisition of a CD19-negative myeloid phenotype allows immune escape of MLL-rearranged B-ALL from CD19 CAR-T-cell therapy. *Blood* **2016**, *127*, 2406–2410. [CrossRef] [PubMed]

123. Huang, E.; Nocka, K.; Beier, D.R.; Chu, T.Y.; Buck, J.; Lahm, H.W.; Wellner, D.; Leder, P.; Besmer, P. The hematopoietic growth factor KL is encoded by the Sl locus and is the ligand of the c-kit receptor, the gene product of the W locus. *Cell* **1990**, *63*, 225–233. [CrossRef]

124. Ikuta, K.; Weissman, I.L. Evidence that hematopoietic stem cells express mouse c-kit but do not depend on steel factor for their generation. *Proc. Natl. Acad. Sci. USA* **1992**, *89*, 1502–1506. [CrossRef] [PubMed]

125. Akashi, K.; Traver, D.; Miyamoto, T.; Weissman, I.L. A clonogenic common myeloid progenitor that gives rise to all myeloid lineages. *Nature* **2000**, *404*, 193–197. [CrossRef] [PubMed]
126. Pronk, C.; Rossi, D.; Månsson, R.; Attema, J.; Norddahl, G.; Chan, C.; Sigvardsson, M.; Weissman, I.; Bryder, D. Elucidation of the phenotype, functional, and molecular topography of a myeloerythroid progenitor cell hierarchy. *Cell Stem Cell* **2007**, *1*, 428–442. [CrossRef] [PubMed]
127. Waskow, C.; Paul, S.; Haller, C.; Gassmann, M.; Rodewald, H.R. Viable c-Kit(W/W) mutants reveal pivotal role for c-kit in the maintenance of lymphopoiesis. *Immunity* **2002**, *17*, 277–288. [CrossRef]
128. Miller, C.L.; Rebel, V.I.; Lemieux, M.E.; Helgason, C.D.; Lansdorp, P.M.; Eaves, C.J. Studies of W mutant mice provide evidence for alternate mechanisms capable of activating hematopoietic stem cells. *Exp. Hematol.* **1996**, *24*, 185–194. [PubMed]
129. Sharma, Y.; Astle, C.M.; Harrison, D.E. Heterozygous kit mutants with little or no apparent anemia exhibit large defects in overall hematopoietic stem cell function. *Exp. Hematol.* **2007**, *35*, 214–220. [CrossRef] [PubMed]
130. Rolink, A.; Streb, M.; Nishikawa, S.; Melchers, F. The c-kit-encoded tyrosine kinase regulates the proliferation of early pre-B cells. *Eur. J. Immunol.* **1991**, *21*, 2609–2612. [CrossRef] [PubMed]
131. Sugiyama, T.; Kohara, H.; Noda, M.; Nagasawa, T. Maintenance of the hematopoietic stem cell pool by CXCL12-CXCR4 chemokine signaling in bone marrow stromal cell niches. *Immunity* **2006**, *25*, 977–988. [CrossRef] [PubMed]
132. Greenbaum, A.; Hsu, Y.M.; Day, R.B.; Schuettpelz, L.G.; Christopher, M.J.; Borgerding, J.N.; Nagasawa, T.; Link, D.C. CXCL12 in early mesenchymal progenitors is required for haematopoietic stem-cell maintenance. *Nature* **2013**, *495*, 227–230. [CrossRef] [PubMed]
133. D'Apuzzo, M.; Rolink, A.; Loetscher, M.; Hoxie, J.A.; Clark-Lewis, I.; Melchers, F.; Baggiolini, M.; Moser, B. The chemokine SDF-1, stromal cell-derived factor 1, attracts early stage B cell precursors via the chemokine receptor CXCR4. *Eur. J. Immunol.* **1997**, *27*, 1788–1793. [CrossRef] [PubMed]
134. Karsunky, H.; Inlay, M.A.; Serwold, T.; Bhattacharya, D.; Weissman, I.L. Flk2+ common lymphoid progenitors possess equivalent differentiation potential for the B and T lineages. *Blood* **2008**, *111*, 5562–5570. [CrossRef] [PubMed]
135. Holmes, M.L.; Carotta, S.; Corcoran, L.M.; Nutt, S.L. Repression of Flt3 by Pax5 is crucial for B-cell lineage commitment. *Genes Dev.* **2006**, *20*, 933–938. [CrossRef] [PubMed]
136. Tsapogas, P.; Swee, L.K.; Nusser, A.; Nuber, N.; Kreuzaler, M.; Capoferri, G.; Rolink, H.; Ceredig, R.; Rolink, A. In vivo evidence for an instructive role of fms-like tyrosine kinase-3 (FLT3) ligand in hematopoietic development. *Haematologica* **2014**, *99*, 638–646. [CrossRef] [PubMed]
137. Sitnicka, E.; Bryder, D.; Theilgaard-Monch, K.; Buza-Vidas, N.; Adolfsson, J.; Jacobsen, S.E. Key role of flt3 ligand in regulation of the common lymphoid progenitor but not in maintenance of the hematopoietic stem cell pool. *Immunity* **2002**, *17*, 463–472. [CrossRef]
138. Sitnicka, E.; Brakebusch, C.; Martensson, I.L.; Svensson, M.; Agace, W.W.; Sigvardsson, M.; Buza-Vidas, N.; Bryder, D.; Cilio, C.M.; Ahlenius, H.; et al. Complementary signaling through flt3 and interleukin-7 receptor alpha is indispensable for fetal and adult B cell genesis. *J. Exp. Med.* **2003**, *198*, 1495–1506. [CrossRef] [PubMed]
139. Schlenner, S.M.; Madan, V.; Busch, K.; Tietz, A.; Laufle, C.; Costa, C.; Blum, C.; Fehling, H.J.; Rodewald, H.R. Fate mapping reveals separate origins of T cells and myeloid lineages in the thymus. *Immunity* **2010**, *32*, 426–436. [CrossRef] [PubMed]
140. Peschon, J.J.; Morrissey, P.J.; Grabstein, K.H.; Ramsdell, F.J.; Maraskovsky, E.; Gliniak, B.C.; Park, L.S.; Ziegler, S.F.; Williams, D.E.; Ware, C.B.; et al. Early lymphocyte expansion is severely impaired in interleukin 7 receptor-deficient mice. *J. Exp. Med.* **1994**, *180*, 1955–1960. [CrossRef] [PubMed]
141. Von Freeden-Jeffry, U.; Vieira, P.; Lucian, L.A.; McNeil, T.; Burdach, S.E.; Murray, R. Lymphopenia in interleukin (IL)-7 gene-deleted mice identifies IL-7 as a nonredundant cytokine. *J. Exp. Med.* **1995**, *181*, 1519–1526. [CrossRef] [PubMed]
142. Miller, J.P.; Izon, D.; DeMuth, W.; Gerstein, R.; Bhandoola, A.; Allman, D. The earliest step in B lineage differentiation from common lymphoid progenitors is critically dependent upon interleukin 7. *J. Exp. Med.* **2002**, *196*, 705–711. [CrossRef] [PubMed]

143. Park, L.S.; Martin, U.; Garka, K.; Gliniak, B.; Di Santo, J.P.; Muller, W.; Largaespada, D.A.; Copeland, N.G.; Jenkins, N.A.; Farr, A.G.; et al. Cloning of the murine thymic stromal lymphopoietin (TSLP) receptor: Formation of a functional heteromeric complex requires interleukin 7 receptor. *J. Exp. Med.* **2000**, *192*, 659–670. [CrossRef] [PubMed]

144. Chappaz, S.; Flueck, L.; Farr, A.G.; Rolink, A.G.; Finke, D. Increased TSLP availability restores T- and B-cell compartments in adult IL-7 deficient mice. *Blood* **2007**, *110*, 3862–3870. [CrossRef] [PubMed]

145. Kondo, M.; Akashi, K.; Domen, J.; Sugamura, K.; Weissman, I.L. Bcl-2 rescues T lymphopoiesis, but not B or NK cell development, in common gamma chain-deficient mice. *Immunity* **1997**, *7*, 155–162. [CrossRef]

146. Akashi, K.; Kondo, M.; von Freeden-Jeffry, U.; Murray, R.; Weissman, I.L. Bcl-2 rescues T lymphopoiesis in interleukin-7 receptor-deficient mice. *Cell* **1997**, *89*, 1033–1041. [CrossRef]

147. Van der Plas, D.C.; Smiers, F.; Pouwels, K.; Hoefsloot, L.H.; Lowenberg, B.; Touw, I.P. Interleukin-7 signaling in human B cell precursor acute lymphoblastic leukemia cells and murine BAF3 cells involves activation of STAT1 and STAT5 mediated via the interleukin-7 receptor alpha chain. *Leukemia* **1996**, *10*, 1317–1325. [PubMed]

148. Dias, S.; Silva, H., Jr.; Cumano, A.; Vieira, P. Interleukin-7 is necessary to maintain the B cell potential in common lymphoid progenitors. *J. Exp. Med.* **2005**, *201*, 971–979. [CrossRef] [PubMed]

149. Kosan, C.; Saba, I.; Godmann, M.; Herold, S.; Herkert, B.; Eilers, M.; Moroy, T. Transcription factor miz-1 is required to regulate interleukin-7 receptor signaling at early commitment stages of B cell differentiation. *Immunity* **2010**, *33*, 917–928. [CrossRef] [PubMed]

150. Von Muenchow, L.; Alberti-Servera, L.; Klein, F.; Capoferri, G.; Finke, D.; Ceredig, R.; Rolink, A.; Tsapogas, P. Permissive roles of cytokines interleukin-7 and Flt3 ligand in mouse B-cell lineage commitment. *Proc. Natl. Acad. Sci. USA* **2016**, *113*, E8122–E8130. [CrossRef] [PubMed]

151. Malin, S.; McManus, S.; Cobaleda, C.; Novatchkova, M.; Delogu, A.; Bouillet, P.; Strasser, A.; Busslinger, M. Role of STAT5 in controlling cell survival and immunoglobulin gene recombination during pro-B cell development. *Nat. Immunol.* **2010**, *11*, 171–179. [CrossRef] [PubMed]

152. Oliver, P.M.; Wang, M.; Zhu, Y.; White, J.; Kappler, J.; Marrack, P. Loss of Bim allows precursor B cell survival but not precursor B cell differentiation in the absence of interleukin 7. *J. Exp. Med.* **2004**, *200*, 1179–1187. [CrossRef] [PubMed]

153. Pribyl, J.; LeBien, T.W. Interleukin 7 independent development of human B cells. *Proc. Natl. Acad. Sci. USA* **1996**, *93*, 10348–10353.

154. Puel, A.; Ziegler, S.F.; Buckley, R.H.; Leonard, W.J. Defective *IL7R* expression in T⁻B⁺NK⁺ severe combined immunodeficiency. *Nat. Genet.* **1998**, *20*, 394–397. [CrossRef] [PubMed]

155. Shochat, C.; Tal, N.; Bandapalli, O.R.; Palmi, C.; Ganmore, I.; te Kronnie, G.; Cario, G.; Cazzaniga, G.; Kulozik, A.E.; Stanulla, M.; et al. Gain-of-function mutations in interleukin-7 receptor-α (IL7R) in childhood acute lymphoblastic leukemias. *J. Exp. Med.* **2011**, *208*, 901–908. [CrossRef] [PubMed]

156. Heltemes-Harris, L.M.; Willette, M.J.; Ramsey, L.B.; Qiu, Y.H.; Neeley, E.S.; Zhang, N.; Thomas, D.A.; Koeuth, T.; Baechler, E.C.; Kornblau, S.M.; et al. Ebf1 or Pax5 haploinsufficiency synergizes with STAT5 activation to initiate acute lymphoblastic leukemia. *J. Exp. Med.* **2011**, *208*, 1135–1149. [CrossRef] [PubMed]

International Journal of
Molecular Sciences

MDPI

Review

Murine Bone Marrow Niches from Hematopoietic Stem Cells to B Cells

Michel Aurrand-Lions *,† and Stéphane J. C. Mancini *,†

Aix Marseille University, CNRS, INSERM, Institut Paoli-Calmettes, CRCM, 13009 Marseille, France
* Correspondence: michel.aurrand-lions@inserm.fr (M.A.-L.); stephane.mancini@inserm.fr (S.J.C.M.);
  Tel.: +33-486-977-291 (S.J.C.M.)
† These authors contributed equally.

Received: 20 July 2018; Accepted: 8 August 2018; Published: 10 August 2018

Abstract: After birth, the development of hematopoietic cells occurs in the bone marrow. Hematopoietic differentiation is finely tuned by cell-intrinsic mechanisms and lineage-specific transcription factors. However, it is now clear that the bone marrow microenvironment plays an essential role in the maintenance of hematopoietic stem cells (HSC) and their differentiation into more mature lineages. Mesenchymal and endothelial cells contribute to a protective microenvironment called hematopoietic niches that secrete specific factors and establish a direct contact with developing hematopoietic cells. A number of recent studies have addressed in mouse models the specific molecular events that are involved in the cellular crosstalk between hematopoietic subsets and their niches. This has led to the concept that hematopoietic differentiation and commitment towards a given hematopoietic pathway is a dynamic process controlled at least partially by the bone marrow microenvironment. In this review, we discuss the evolving view of murine hematopoietic–stromal cell crosstalk that is involved in HSC maintenance and commitment towards B cell differentiation.

Keywords: early hematopoiesis; B lymphopoiesis; bone marrow niches; stromal cells

1. Introduction on Early Hematopoiesis and B Lymphopoiesis

In mammals, adult hematopoiesis occurs in the bone marrow (BM). Deciphering the different stages of hematopoiesis and the developmental cues driving stem cell commitment towards a particular lineage is essential in regenerative medicine and in the development of treatments for hematopoietic diseases. Lymphoid development from hematopoietic stem cells (HSC) has been extensively dissected in mouse models through genetic ablation of key genes and phenotypic characterization of cell subsets at different maturation steps.

HSC have the lifelong capacity to self-renew and to give rise to all hematopoietic lineages. The existence of HSC was first demonstrated by Till, McCulloch, and colleagues, who reported that the bone marrow contains cells having the capacity to reconstitute lethally irradiated mice and form myelo-erythroid colonies in the spleen (Colony-Forming Unite Spleen, CFU-S) [1,2]. Later, Weissman and colleagues strongly contributed to the phenotypic identification of hematopoietic cells enriched for HSC in mouse. The multipotent and self-renewal potentials were first shown to be properties of a subset of cells lacking markers of committed hematopoietic lineages—so-called lineage-negative (Lin$^-$)—and expressing Sca1 and low levels of Thy1.1 [3]. It was then demonstrated that the Lin$^-$CD117$^+$Sca1$^+$ (LSK) fraction retained the multipotent potential and could be further fractioned into long-term (LT) and short-term (ST) repopulating subsets [4,5]. The acquisition of CD34 expression by murine HSC marks the transition from LT- to ST-HSC [6]. Later on, the introduction of differentially expressed markers such as CD135 (Flk2/Flt3) and signaling lymphocytic activation molecule (SLAM) family proteins led to the identification of different multipotent progenitor (MPP) subsets: the LSK CD150$^+$CD48$^-$CD34$^-$ subset LT-HSC on top of the hierarchy and more differentiated

MPPs with a full spectrum of lineage reconstitution (ST-HSC/MPP1) or with a biased (but not definitive) engagement towards particular lineages (MPP2 to MPP4; Figure 1) [6–12]. Finally, advances in single-cell technology demonstrated that the phenotypic boundaries between subsets are not strict, but rather represent a continuum of progenitor states acquiring lineage restrictions progressively. This is usually illustrated as a landscape of hills and valleys branching from LT-HSC up to the different committed lineages [13–18].

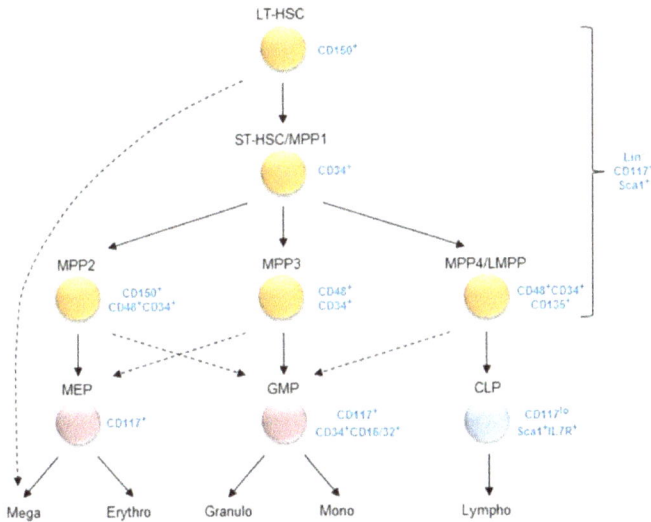

Figure 1. Murine early hematopoiesis from long-term hematopoietic stem cells towards the different lineages. Plain and dashed arrows show the main and the alternative branch points, respectively. Hematopoietic progenitors do not express markers of the different hematopoietic lineages and are therefore said to be lineage-negative (Lin⁻). Markers used to characterize the different hematopoietic progenitors are indicated. LT-HSC: long-term hematopoietic stem cell (HSC); ST-HSC: short-term HSC; MPP: multipotent progenitor; LMPP: lymphoid-biased multipotent progenitor; MEP: megakaryocyte-erythroid progenitor; GMP: granulocyte-monocyte progenitor; CLP: common lymphoid progenitor; Mega: megakaryocyte; Erythro: erythrocyte; Granulo: granulocyte; Mono: monocyte; Lympho: lymphocyte; Lin: lineage markers.

Progression from MPP4/LMPP to the common lymphoid progenitor (CLP) subset marks the entry into the lymphoid lineage and is characterized by interleukin-7 receptor (IL7R) upregulation [9,19]. While CLPs (CD117^{lo}Sca1$^+$IL7R$^+$) have the capacity to differentiate into all lymphoid subsets, the upregulation of Ly6D marks the engagement into the B cell lineage [20]. The low natural killer and T cell potentials retained by the Ly6D$^+$ CLPs (called B lymphoid progenitors, BLP) are lost upon entry in the earliest pre-pro-B cell stage, as indicated by B220 expression. The following B cell differentiation steps have a specific and crucial role in the acquisition of a non-autoreactive B cell receptor (BCR, or immunoglobulin, Ig) repertoire which is essential for efficient adaptive immune responses. More than two decades ago, both Hardy and Rolink established independent combinations of markers, known as the Philadelphia and Basel nomenclatures, respectively, which are still standards for the phenotypic characterization of the different BM B cell subsets in mice (Figure 2A) [21,22]. Both nomenclatures use B220 and CD19 as lineage markers as well as the surface expression of BCR as a marker of maturity. Hardy and colleagues established their classification based on CD43, CD24, and BP1 expression, while Rolink and colleagues used CD43, CD117, and CD25 [23,24]. Importantly, recent advances in multi-parameter flow cytometry now give the opportunity to consider both strategies together and improve the resolution of each subset (Figure 2B–D). These results clearly show that both phenotypic strategies of B cell maturation intermediates are valid and overlapping.

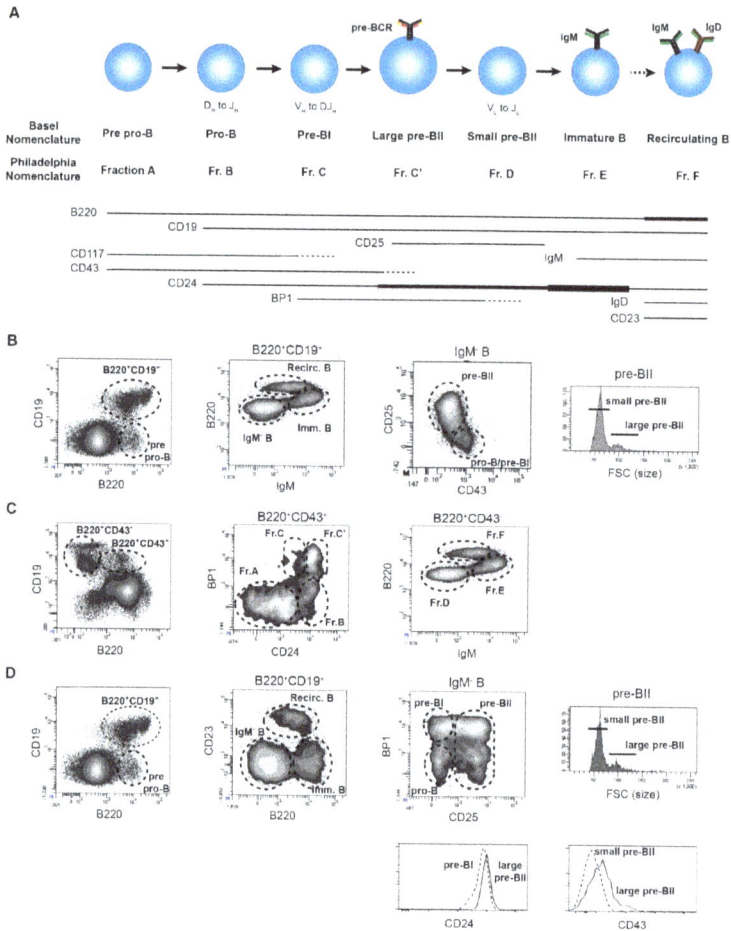

Figure 2. Murine bone marrow (BM) B cell differentiation. (**A**) Schematic representation of the different B cell differentiation stages with their denomination according to the Basel and Philadelphia nomenclatures. The pattern of expression of the main markers used to characterize each subset is shown with a line. The thickness of the line is representative of the level of expression. The dotted lines indicate subsets where expression is progressively lost. VDJ$_H$ and VJ$_L$ rearrangements are indicated; (**B**) BM B lymphopoiesis analysis by flow cytometry according to the Basel nomenclature; (**C**) BM B lymphopoiesis analysis by flow cytometry according to the Philadelphia nomenclature. The gating strategy and the main subsets are indicated in the panels; (**D**) BM B lymphopoiesis analysis by flow cytometry taking into consideration both nomenclatures. This strategy improves the resolution of the subsets, particularly of the pro-B, the pre-BI and the large pre-BII fractions (Fr. B to Fr. C'). Indeed, while the separation between pro-B and pre-BI cells (Fr. B and C) is not possible with the Basel nomenclature and the distinction between pre-BI and large pre-BII (Fr. C and C') with CD24 is dependent on mouse strains, the simultaneous use of BP1 and CD25 allows a clear definition of the three subsets.

B cell specification initiates at the pre-pro-B stage and is driven by the E2A transcription factor [25]. However, the definitive commitment to the B cell lineage occurs at the pro-B cell stage and is controlled by Pax5. Indeed, loss of Pax5 results in B cell differentiation arrest at the pro-B cell stage, and *Pax5*$^{-/-}$ pro-B cells acquire the capacity to differentiate into other lymphoid and myeloid lineages [26–28]. The rearrangements of genes encoding the Ig heavy chain (IgH) are initiated at the pro-B cell stage between the D_H and J_H segments, while the complete VDJ_H recombination takes place at the pre-BI stage. If a functional Igµ protein is generated from the recombination product, it is associated with the surrogate light chain (SLC), composed of the invariant λ5 and VpreB proteins, and with the Igα/Igβ signaling complex to form the pre-BCR [29–32]. Pre-BCR expression by large pre-BII cells induces proliferation and differentiation towards the small pre-BII stage, where pre-BCR expression is downmodulated, and Ig light chain (IgL) gene rearrangements are initiated. Upon expression of a functional IgL chain, the BCR is formed by association with the Igµ chain at the immature B cell stage. Finally, immature B cells expressing an autoreactive BCR receive a 'no-go' signal and have the possibility of reinitiating rearrangements through receptor editing [33], while immature B cells expressing a non-autoreactive BCR leave the BM to complete their maturation in the periphery.

Self-renewal, differentiation, and commitment events occurring during hematopoietic progenitor development as well as B lymphopoiesis are driven by complex intrinsic regulatory networks described in details elsewhere [25,34,35]. However, these networks are triggered and regulated by extrinsic signals delivered by cells of the BM microenvironment through the secretion of growth factors and direct cellular interactions. These supportive regions, called niches, are composed of mesenchymal, endothelial, and hematopoietic cells. Advances in the analysis of stromal cell niches in vivo have been possible thanks to the development of reporter and tissue-specific deletion systems in mouse models. In this review, we will thus focus on the known molecular mechanisms by which murine mesenchymal and endothelial cells control HSC behavior and B cell commitment.

2. The HSC Niche

2.1. Pioneer Views on the HSC Niche

The influence of cells from the BM microenvironment on the long-term growth of HSC was first demonstrated using in vitro cultures in which adherent cells were present and composed of macrophages, adipocytes, endothelial cells, and fibroblasts [36,37]. The analysis of the relationship between hematopoietic cells and their microenvironment in vivo has long been limited to the spatial localization of the different lineages in the BM conduit. The microscopic analysis of bone transversal sections following injection of tritiated thymidine suggested a centripetal differentiation of hematopoietic precursors from endosteal regions towards the center of the BM [38]. Furthermore, by measuring the progenitor self-renewal property using CFU-S assays, it was shown that the number of colonies formed was higher by transplanting cells from the sub-endosteal region rather than from the central marrow (reviewed in [39]). These results suggested that cells of the microenvironment located close to the bone border were involved in HSC maintenance.

The development of mouse models allowing tissue-specific deletion of genes or expression of fluorescent proteins as well as the technological progress in confocal microscopy have been crucial for the dissection of BM organization. The existence of an osteoblastic niche for HSC was first proposed. In a first model, the number of HSC could be significantly increased by manipulating osteoblast (OB) number through the injection of parathyroid hormone (PTH) or by specifically over-expressing the PTH receptor using the OB-specific *Col1a1* promoter [40]. Similarly, increasing spindle-shaped OB numbers through conditional deletion of the *Bmpr1a* gene in *Mx-Cre/Bmpr1a*$^{lox/lox}$ mice induced a parallel increase in HSC [41]. Altogether, these results suggested the importance of an osteoblastic niche in the maintenance of HSC.

2.2. The Peri-Sinusoidal HSC Niche

The existence of an osteoblastic niche was then challenged by the development of a mouse model in which green fluorescent protein (GFP) was knocked-in downstream of the C–X–C Motif Chemokine Ligand 12 (CXCL12) promoter that allowed visualization of CXCL12-producing cells [42,43]. CXCL12 was first described as a growth factor for early B cells, but deletion of CXCR4, its receptor, also induced major hematopoietic defects [44,45]. Deletion of *Cxcr4* in adult HSC using *Mx-Cre/Cxcr4*[lox/−] mice injected with poly(I:C) induced a strong decrease in HSC number, suggesting a crucial role of CXCL12 in HSC maintenance [43]. However, expansion of the HSC population was observed when *ROSA-Cre ERT2* and Tamoxifen were used to delete *Cxcr4*, suggesting that HSC maintenance and expansion are tightly dependent on the chemokine context of the bone marrow [46]. CXCL12 was further shown to be a chemoattractant for human and mouse hematopoietic progenitors and to allow their retention in the BM [47–49]. In vivo tracing of CXCL12-expressing cells using *Cxcl12-GFP* or *Cxcl12-dsRed* knock-in mice showed that GFP was strongly expressed by reticular cells (called CAR cells for CXCL12-abundant reticular cells), which were scattered throughout the BM and in contact with the vasculature. In contrast, the expression of *Cxcl12* by BM endothelial cells (BMEC) and OB was 100 and 1000 times lower, respectively [42,43,50]. Accordingly, CD150[+]CD48[−] HSC were essentially localized in peri-sinusoidal regions and in contact with CAR cells [7,43]. Specific deletion of *Cxcl12* in peri-sinusoidal stromal (PSS) cells, but not in OBs, led to an increase in circulating HSC (Figure 3). Furthermore, specific deletion in BMEC induced a decrease in HSC frequency but no loss of retention, indicating that CXCL12 plays a differential role in BMEC and PSS cells by allowing HSC maintenance and retention, respectively [50–52]. Stem cell factor (SCF), the ligand of the receptor tyrosine kinase c-kit, was also shown to be implicated in stem cell maintenance [53]. The use of *SCF-GFP* knock-in and *SCF-GFP/CXCL12-dsRed* double knock-in mice showed that SCF is expressed by BMEC and co-expressed with CXCL12 by PSS cells [50,54]. Specific deletion of *Kitl* encoding SCF in PSS cells decreased HSC maintenance and retention. In contrast, deletion in BMEC or OB resulted, respectively, in decreased HSC maintenance or in an absence of phenotype (Figure 3 [54]). Altogether, these results cast doubt over the existence of an osteoblastic niche and demonstrate the importance of perivascular niches, and more particularly of PSS cells, for HSC maintenance and retention.

In light of the recent knowledge accumulated on mesenchymal cell niches and development, it seems likely that the parallel increase in OB and HSC numbers is only correlative and that HSC are instead regulated by an osteoblastic progenitor. Indeed, in vitro differentiation assays have shown that CAR and PSS cells have the capacity to differentiate into osteoblasts or adipocytes [55,56]. Furthermore, PTH/PTHR signaling, which was shown to increase OB number, is able to directly stimulate PSS cell number and to favor differentiation into OB [40,56]. Finally, inducible and non-inducible lineage-tracing mouse models confirmed that PSS cells contain progenitors of osteoblasts in adult BM [57,58].

2.3. The Endosteal/Peri-Arteriolar Niche

Despite the clear involvement of peri-sinusoidal niches in HSC maintenance, some results still argue in favor of a function for the endosteal niche. Indeed, because of bone remodeling activity, the local concentration of calcium at the endosteal surface is high. Interestingly, HSC, which express the calcium-sensing receptor (CaSR), were found to be strongly decreased in *Casr*[−/−] mice. In addition, transplanted *Casr*[−/−] HSC, unlike wild-type (WT) HSC, failed to localize close to the endothelium [59]. The adhesion to extra-cellular matrix proteins as well as the migration towards CXCL12 of HSC treated in vitro with a CaSR agonist were increased. Homing and engraftment into the BM of in vitro treated HSC were also improved [60]. Finally, the frequency of quiescent HSC was reported to be higher close to the bone surface in endosteal and trabecular regions [61], indicating that these regions may indeed play a role in HSC maintenance.

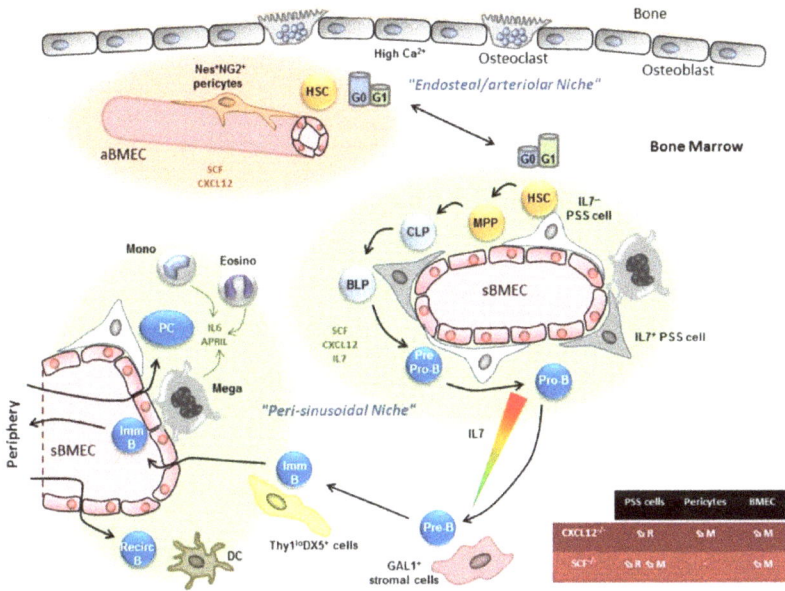

Figure 3. Bone marrow niches for hematopoietic stem cells and B cells. HSC are located in both endosteal/arteriolar and in peri-sinusoidal regions which express high levels of CXCL12 and stem cell factor (SCF). Quiescent HSC are enriched in the endosteal/arteriolar niche. Differentiation of MPP up to the pro-B cell stage takes place in the peri-sinusoidal niche, where the level of CXCL12 and IL7 are high. Pre-B cell then relocalize close to GAL1-expressing stromal cells located away from the sinusoids. At the next immature B cell stage, cells expressing an auto-reactive B cell receptor (BCR) are retained in the BM in order to initiate receptor editing, while non-autoreactive cells leave the BM to finish their maturation in the periphery. Mature/recirculating B cells and plasma cells follow CXCL12 gradients to home to the BM. Recirculating B cell survival relies on dendritic cells. PC survival relies on the secretion of IL6 and A proliferation-inducing ligand (APRIL) by monocytes, eosinophil, and megakaryocytes. The colored triangle represents the gradient of IL7 expression from high (red) to low (green). The table in the bottom right summarizes the influence of CXCL12 and SCF specific deletion in PSS cells, pericytes, or BMEC on HSC retention (R) and maintenance (M). MPP; multipotent progenitor; CLP: common lymphoid progenitor; BLP: B lymphoid progenitor; Imm. B: immature B cell; Recirc. B: recirculating B cell; PC: plasma cell; Mono: monocyte; Eosino: eosinophil; Mega: megakaryocyte; DC: dendritic cell; aBMEC: arteriolar bone marrow endothelial cell; sBMEC: sinusoidal BMEC; PSS cell: peri-sinusoidal stromal cell.

Importantly, endosteal and trabecular regions are enriched in arteries, while sinusoidal venous structures are more central [62,63]. By taking into consideration the large extent of the sinusoidal network in the BM compared to arteriolar structures, Frenette and colleagues showed that the proportion of HSC in the vicinity of arterioles was highly significant [62]. Furthermore, quiescent HSC identified by their capacity to retain EdU (5-ethynyl-2′-deoxyuridine) labelling in the long term, by their low levels of reactive oxygen species (ROS) or by their high expression of hypoxia-inducible transcription factor 1 (HIF-1α) [6,64,65], were found to be frequently located close to arteriolar cells [62,66,67]. Finally, low vessel permeability ensures HSC quiescence and retention in the BM [67]. Indeed, sinusoidal BMEC (sBMEC), but not arteriolar BMEC (aBMEC), were shown to be permeable and to be the site of trans-endothelial trafficking of HSPC. Notably, when the integrity of aBMEC was affected, HSC and HSPC numbers were decreased, while HSPC trafficking was increased, confirming the importance of arteriolar niches for HSC quiescence. Also, NG2$^+$Nestin$^+$ pericytes associated with arterioles were shown to express HSC niche genes, including CXCL12 [51,62]. These cells were mainly located at the metaphysis and adjacent to cortical bone at the diaphysis [67]. When pericytes were specifically depleted upon injection of tamoxifen and diphtheria toxin in *NG2-CreERT/ROSA26iDTR* mice (iDTR: inducible Diphteria Toxin Receptor), HSC were decreased and less quiescent and relocalized away from arteries [62]. A similar effect was observed upon the specific deletion of CXCL12, but not of SCF, from pericytes [51].

Altogether, these results suggest that cells of the endosteal/peri-arteriolar region control HSC maintenance and quiescence, while cells present in the peri-sinusoidal region control HSC maintenance and retention (Figure 3). CXCL12 expressed by PSS cells, pericytes, and BMEC as well as SCF expressed by PSS cells and BMEC play a crucial role in these functions, but other factors have also been identified. Furthermore, the complexity of the HSC niche is not limited to mesenchymal cells and BMEC, as the involvement of hematopoietic cells (megakaryocytes and macrophages) and Schwann cells have also been demonstrated. The function of these cells in the maintenance of HSC homeostasis has been extensively discussed elsewhere [68,69].

3. Niches for Lymphoid Progenitors

The localization of HSPCs was first performed by tracking injected stained cells using bi-photon live imaging [70]. While HSC were found close to the bone surface, MPPs and more differentiated progenitors were found further away, but their association to particular regions was not studied. Because of the strong influence of CXCL12 on HSC maintenance, it has been difficult to assess its specific role on the development of more committed progenitors. In contrast, the function of IL7 in the differentiation of BLPs, but not of CLPs, has been clearly established using *Il7$^{-/-}$* mice [71]. Contradictory results have been obtained when *Cxcl12* and *Il7* were specifically deleted in OBs. Indeed, CLP number decreased upon *Cxcl12* deletion using the *Col2.3-Cre* but not the *Bglap-Cre* systems, although both promoters were supposed to be expressed in OBs [50,52]. Conversely, CLPs were decreased upon *Il7* deletion using the *Bglap-Cre*, but not the *Col2.3-Cre*, system [72,73]. Of note, both promoters drive partial Cre expression in perivascular stromal cells of the BM [54,74]. Therefore, it can be speculated that CLPs are controlled by stromal cells located away from the endosteum.

We previously demonstrated using the *Il7-Cre/Rosa-eYFP* mouse model and by qPCR on sorted cells, that PSS cells expressing CXCL12 were the main source of IL7 [75]. A more recent study confirmed our results by suggesting the existence of IL7$^-$ and IL7$^+$ PSS cells [73]. Importantly, BMEC, but not OBs, express low levels of IL7, confirming that the phenotype observed upon *Il7* deletion using *Bglap-Cre* or *Col2.3-Cre* could be attributed to Cre expression in perivascular stromal cells. Specific deletion of *Cxcr4* in MPPs or CLPs showed the crucial role played by CXCL12 on MPP development and on the BLP fraction of CLP [73]. Furthermore, CXCL12 was shown to position BLPs close to IL7$^+$ PSS cells in order to get an efficient stimulation through the IL7R signaling pathway. Specific deletion of *Il7* in PSS cells using the *LepR-Cre* system induced a specific decrease in BLPs, confirming the phenotype observed in *Il7$^{-/-}$* mice [71,73]. Finally, HSC and MPPs were found indifferently in contact with both IL7$^-$ and IL7$^+$ PSS cells [73]. However, it cannot be excluded that these subsets may have differential functions. Indeed, while *Cxcl12* specific deletion in all PSS cells using the *LepR-Cre* system induced a loss of retention of HSC, specific deletion in IL7$^+$ PSS cells using *Il7-Cre* mice had only a marginal effect on HSC and MPP numbers and no effect on HSC retention [50,51,73]. Furthermore, specific deletion of *Kitl* (SCF) in all PSS cells induced a decrease in HSC maintenance and retention, but only a decrease in HSC maintenance when deleted in IL7$^+$ PSS cells. Therefore, the specific function of the different PSS subsets remains to be addressed.

Taking into consideration the most recent advances in niche biology, HSC, MPP, and CLP subsets are most likely preferentially located in peri-sinusoidal regions (Figure 3). These different data are in agreement with the hemosphere model proposed by Adams and colleagues, in which clonal hematopoiesis was observed in micro-anatomical structures composed of stromal and endothelial cells [76].

4. B Cell Niches

4.1. Pioneer Views on B Cell Niches

Early on, Whitlock and Witte demonstrated that adherent BM cells sustain long-term B cell cultures [77]. By removing glucocorticoids, known to impair lymphocyte development, from Dexter type cultures, differentiation of B cells, including IgM$^+$ immature B cells, could be observed for more than 10 weeks, and the recovered cells could be clonally expanded. Stromal cell lines were then derived from these cultures, leading to the identification of the main factors involved in early B cell development, namely, CXCL12 and IL7 [44,78,79]. The generation of cell lines with distinct phenotypic characteristics and supportive functions suggested the existence of heterogeneity between stromal cells. However, this heterogeneity may have been related to stromal cell instability, as modifications in surface marker expression could be observed upon cloning [80]. These in vitro co-culture systems were further developed by Rolink and Melchers to decipher B cell differentiation mechanisms and the influence of extrinsic cues [81,82]. They demonstrated that stromal cells and IL7 were able to maintain long-term proliferation of pro-B/pre-BI, while removal of IL7 induced their differentiation into immature B cells [81,83]. Importantly, IL7-dependent pro-B/pre-BI cell growth was inhibited by a blocking anti c-kit/CD117 antibody, indicating that c-kit, the receptor for SCF, was an important co-factor for IL7-dependent early B cell proliferation [84].

Contacts between B cells and stromal reticular cells were first observed in situ by electron microscopy on BM histological sections [85]. In this study, B cell development was proposed to be centripetal, with early B cells being mainly located close to the endosteum and in intermediate zones, while IgM$^+$ B cells were positioned in the vicinity of the central sinus. Furthermore, lymphoid cells, unlike myeloid cells, were closely associated to β1 integrin-expressing reticular cells [86]. Finally, β1 integrin expression revealed heterogeneity between stromal cells, with β1-integrin$^+$ cells mainly localized in the peripheral region of the BM.

4.2. Early B Cell Niches

The evidence for the existence of stromal cell niches for early B cell development became clear with the identification of CXCL12 and IL7 as crucial growth factors produced by the BM microenvironment. B lymphopoiesis is strongly affected in the fetal liver and in the BM of *Cxcl12*- and *Cxcr4*-deficient mice [45,87,88]. Nevertheless, because of the crucial role played by the CXCL12/CXCR4 axis in HSC and BLP [43,73,89], it is difficult to evaluate their influence on early B cell development in non-conditional KO mice. An increase in pro-B cells was however observed in the fetal blood of *Cxcr4*$^{-/-}$ mice as well as in the blood of adult WT mice reconstituted with *Cxcr4*$^{-/-}$ fetal liver cells, indicating a role for the CXCR4/CXCL12 axis in early B cell retention [48]. This effect was further confirmed in B cell-specific *CD19-Cre/CXCR4*$^{lox/lox}$ knock-out mice [90]. Although CXCR4-deficient IgM$^-$ B cells retain the capacity to proliferate in the presence of IL7 in in vitro co-cultures and to differentiate into IgM$^+$ B cells in the absence of IL7, a role for CXCL12 on early B cell development cannot be excluded, since differentiation towards the immature B cell stage is impaired in the absence of CXCR4 in vivo [48,90]. Furthermore, CXCL12, together with IL7 or SCF, stimulates the proliferation of pre-B cell clones and the survival of B cell progenitors in vitro, suggesting that CXCL12 may act synergistically with other factors in the earliest steps of B cell differentiation [44,89]. Finally, CXCL12 has been shown to induce α4β1 integrin-dependent adhesion of pro-B and pre-B cells to VCAM-1 through the activation of focal adhesion kinase (FAK) [91]. Accordingly, B cell development is affected in the BM of α4 integrin- and FAK-deficient mice [92,93], and pro-B cell egress in the periphery of FAK$^{-/-}$ mice.

As said earlier, IL7 plays an important role in pre-pro-B and pro-B cell proliferation [78,81]. Interestingly, pro-B cells need high concentrations of IL7 to proliferate, while a decrease in IL7 concentration favors differentiation toward the pre-BI stage, in which recombination between V and DJ gene segments and then intracellular expression of the Igμ chain are induced [94]. IL7 has also been implicated in B cell differentiation and survival [95,96]. As a consequence, deletion of the *Il7* and *Il7r* genes leads to a severe block at the earliest stages of B cell development [97,98]. Finally, IL7 was further shown to control B cell potential already from CLPs by regulating EBF1 and Pax5 expression [71].

The identification of the niches for early B cells is still controversial. Pre-pro-B and pro-B cells were first proposed to be associated with distinct niches, expressing, respectively, CXCL12 or IL7 [42]. However, we and others have shown that both factors are co-expressed by PSS cells [73,75]. The discrepancies between the studies may be due to the low level of IL7 expression in vivo and thus to the difficulty to detect positive cells using reporter systems or antibodies [99,100]. Therefore, one could not exclude that IL7 expression in CXCL12-expressing cells was missed in the early study by Tokoyoda and collaborators because of the lack of antibody staining sensitivity.

Other studies have proposed that OBs may play a role in early B cell development. Acute depletion of osteoblastic cells using mice expressing the herpes virus thymidine kinase under the control of the Col1a1 promoter (*Col2.3Δ-TK* Tg mice) induced a decrease in both pre-pro-B and pro-B cells. Alternatively, depletion of the G protein α subunit, using the Osx-Cre system, impaired pro-B cell development. Although the promoters used in these mouse models were first thought to be specifically activated in OBs, recent results show that they are already activated in osteo-progenitors, including PSS cells [57,58,73]. Accordingly, *Bglap-Cre/Cxcl12*$^{lox/-}$ mice that lack CXCL12 expression in OBs did not show any B cell phenotype, while pre-pro-B cells were impaired in *Osx-Cre/Cxcl12*$^{lox/-}$ mice. This demonstrates the importance of CXCL12 expression in osteo-progenitors for pre-pro-B cell development [52].

IL7 is essentially expressed by a subset of PSS cells and at very low levels by BMECs [73,75]. Specific depletion of IL7 from PSS cells using *LepR-Cre/Il7*$^{lox/-}$ mice induced a strong decrease of B cell progenitors from the BLP stage, while depletion in BMEC using *Tie2-Cre/Il7*$^{lox/-}$ mice induced a low but significant decrease only from the pro-B cell stage [73]. These results indicate that pro-B cells are more affected by low IL7 fluctuations than BLP and are consistent with the fact that pro-B cell proliferation requires high levels of IL7 [94].

Altogether, these results suggest that the niche sustaining pre-pro-B and pro-B cell homeostasis is located in the peri-sinusoidal region, where both CXCL12 and IL7 levels are high (Figure 3). However, its precise location and the nature of the stromal cells involved in the control of their development remain to be defined.

4.3. The Pre-B Cell Niche

The expression of the pre-BCR at the large pre-BII stage is a crucial checkpoint allowing the selection of functional Igμ chains, the amplification of cells expressing such chains—ensuring higher Ig diversity—, and the induction of IgL recombination. The SLC plays a crucial role in pre-BCR signaling. Indeed, deletion of either λ5 or VpreB leads to a severe block at the pre-BI/large pre-BII transition, and deletion of both results in a complete block of differentiation [101–103]. Pre-BCR signaling relies on ligand-independent and ligand-dependent mechanisms, which both implicate the extra loop (EL) of λ5. Interactions between adjacent λ5-EL and VpreB-EL or between λ5-EL and a glycosylated chain linked to the Igμ constant region at position N46 were proposed to induce self-aggregation of the pre-BCR, resulting in tonic signaling [104,105]. This tonic signal is, however, increased by inducing pre-BCR cross-linking with an anti IgM antibody, suggesting the existence of a ligand for the pre-BCR [104]. Heparan sulfate proteoglycans present at the surface of stromal cells have the capacity to bind λ5-EL [106]. Interestingly, pretreatment of pre-B cells with heparan sulfate improves pre-BCR signaling induced by anti IgM cross-linking [107]. Galectin-1 (GAL1), an S-type lectin which binds β-galactoside glycoconjugates through its carbohydrate recognition domain (CRD), is a ligand for the pre-BCR and binds λ5-EL through direct protein–protein contacts [108,109]. GAL1 is secreted by stromal cells and acts as a docking protein by interacting with both the pre-BCR and glycosylated chains of integrins at the surface of pre-B cells to form a complex lattice [110,111]. Clustering of the pre-BCR is further increased by the interaction of the pre-B cell integrins with their ligands expressed by stromal cells, resulting in pre-BCR signaling. Accordingly, the inactivation of GAL1 expression by stromal cells in vitro and in vivo impairs large pre-BII cell proliferation and differentiation [111]. More recently, we identified GAL1-expressing stromal cells in the BM. Such cells are different from IL7-expressing cells and are not localized in peri-sinusoidal regions [75]. Most importantly, large pre-BII cells are in close contact with these GAL1+ cells. While pro-B cells need high levels of IL7 for their proliferation, large pre-BII cells are sensitive to low levels of IL7 [94], suggesting that the transition from the IL7+ to the GAL1+ stromal cell niche plays an important role in B cell development.

4.4. Immature B Cell Niches

The main features at the immature B cell stage are the negative selection of auto-reactive BCR and their egress to the periphery where they complete their maturation. As compared to pro-B and pre-B cells, immature B cells express low levels of CXCR4 and have a decreased capacity to adhere to VCAM1 in a CXCL12-induced manner [91,112]. Moreover, when mice transgenic for a hen egg lysosome (HEL)-specific BCR were treated with HEL to simulate self-antigen engagement, immature B cells upregulated CXCR4, resulting in reduced egress to the periphery. This suggests that immature B cells are retained in the BM if their BCR is auto-reactive. In line with this result, cannabinoid receptor 2 was shown to be involved in the retention of immature B cells in sinusoids and to favor receptor editing [113]. Furthermore, immature B cells are protected from BCR-mediated apoptosis and have the capacity to reinitiate the recombination-activating genes *RAG1* and *RAG2* when incubated with cells of the BM microenvironment expressing DX5 and low levels of Thy1 [114,115]. Altogether, these results show that endothelial and/or stromal cells play an important role in the retention of auto-reactive immature B cells in the BM and in receptor editing.

4.5. Recirculating B Cell and Plasma Cell Niches

BM also represents a privileged homing site for plasma cells (PC) and mature B cells. In particular, the BM represents a reservoir for long-lived PC recently requalified as "memory plasma cells" [116]. In addition, the capacity of mature B cells and plasma blasts to home to the BM is likely critical to ensure protection of the hematopoietic system and of maturing immune cells against pathogens. The CXCR4/CXCL12 axis plays a crucial role in the homing of mature B cells (called recirculating B cells in the BM) and plasma blasts, as demonstrated using CXCR4-deficient cells [48,90,117]. Both subsets localize in peri-sinusoidal regions and even in direct contact with CXCLC12-expressing PSS cells in the case of PC [42,118] (Figure 3). Cells of hematopoietic origin are part of the supportive niche and deliver survival cues to PC and recirculating B cells. Dendritic cells (DC) forming clusters in peri-vascular regions are located in close proximity to recirculating B cells [119]. Upon specific depletion of these DC using *CD11c-DTR* transgenic mice, wild-type but not *Bcl2* transgenic recirculating B cells are lost, demonstrating a role for the DC in their survival. Monocytes, eosinophils, and megakaryocytes are also involved in the survival of long-lived PC through the secretion of A proliferation-inducing ligand (APRIL) and IL6 [120–122]. Interestingly, monocytes and eosinophils express CXCR4 and the α4β1 integrin, which participate in the positioning of B cells, including PC close to CXCL12-expressing cells [42,91,122]. The complexity of long-lived PC niches, in which hematopoietic cells provide survival signals, while stromal cells represent an anchoring site, is intriguing. Indeed, long-lived PC are sessile and scattered throughout the BM in contact with stromal cells [123]. PC anchoring to stromal cells is likely to occur via the α4β1 and αLβ2 integrins, since treatment with a combination of monoclonal antibodies (mAbs) against these integrins (clone PS/2 and M17/4, respectively) depletes PC from the BM [124]. Whether a unique BM stromal cell subset expresses the different ligands for α4β1 and

Int. J. Mol. Sci. **2018**, *19*, 2353

αLβ2 (fibronectin and VCAM-1 for the former and ICAM-1 for the latter) remains to be demonstrated. However, these results argue for a model in which plasma blasts use CXCL12-driven migration to reach a limited number of niches made of stromal cells and IL6/APRIL-secreting eosinophils, before switching off migration and anchoring to stromal cells. This would be in agreement with studies showing that eosinophil depletion reduces by 70% the number of PC and that generation of new plasma blasts is accompanied by long-lived PC mobilization from the BM [121,125]. Therefore, BM stromal cells not only are able to give direct signals to differentiating hematopoietic cells, but also act as regulators of long-lived PC.

5. Concluding Remarks

At the time of stochastic versus instructive models of lymphocyte commitment and differentiation in the early nineties, pioneer studies by Rolink and others clearly established that BM stromal cells play a key role in the generation and regeneration of the B-lymphocyte lineage [126]. On one hand, this prompted many teams to work with co-culture conditions in order to generate large quantities of mature cells starting from few progenitors. On the other hand, this paved the way toward the search for stromal cells supporting long-term hematopoiesis [127]. More recently, and thanks to the use of reporter mice, the progress made in understanding BM organization and BM stromal cell heterogeneity has been tremendous. The influence of the BM microenvironment on pathologies affecting hematopoietic progenitors has benefited from the important advances in normal HSC niche characterization [128,129]. Resistance and relapse in the case of B cell acute lymphoblastic leukemia, the pathological equivalent of differentiating B cells, also concerns a great proportion of patients, most particularly adults. It is now clear that part of the resistance to treatment is related to protective cues transmitted by stromal cells [130]. Therefore, it is now crucial to translate our current understanding of mouse BM organization to human physiological and pathological situations.

Author Contributions: Writing—Original Draft Preparation, M.A.-L. and S.J.C.M.; Writing—Review & Editing, M.A.-L. and S.J.C.M.; Figure Preparation: S.J.C.M.

Funding: This work was supported by grants from the ARC Foundation (PJA#20161204555), the ANR (OSTEOVALYMPH, ANR-17-CE14-0019), and the Excellence Initiative of Aix-Marseille University-A*MIDEX, a French "Investissements d'Avenir" program.

Acknowledgments: We are grateful to our colleagues and to the members of the team for stimulating discussions. We apologize to those whose work was not cited because of space limitation.

Conflicts of Interest: The authors declare no conflict of interest.

References

1. Till, J.E.; Mc, C.E. A direct measurement of the radiation sensitivity of normal mouse bone marrow cells. *Radiat. Res.* **1961**, *14*, 213–222. [CrossRef] [PubMed]
2. Siminovitch, L.; McCulloch, E.A.; Till, J.E. The distribution of colony-forming cells among spleen colonies. *J. Cell. Comp. Physiol.* **1963**, *62*, 327–336. [CrossRef] [PubMed]
3. Spangrude, G.J.; Heimfeld, S.; Weissman, I.L. Purification and characterization of mouse hematopoietic stem cells. *Science* **1988**, *241*, 58–62. [CrossRef] [PubMed]
4. Ikuta, K.; Weissman, I.L. Evidence that hematopoietic stem cells express mouse c-kit but do not depend on steel factor for their generation. *Proc. Natl. Acad. Sci. USA* **1992**, *89*, 1502–1506. [CrossRef] [PubMed]
5. Morrison, S.J.; Weissman, I.L. The long-term repopulating subset of hematopoietic stem cells is deterministic and isolatable by phenotype. *Immunity* **1994**, *1*, 661–673. [CrossRef]
6. Wilson, A.; Laurenti, E.; Oser, G.; van der Wath, R.C.; Blanco-Bose, W.; Jaworski, M.; Offner, S.; Dunant, C.F.; Eshkind, L.; Bockamp, E.; et al. Hematopoietic stem cells reversibly switch from dormancy to self-renewal during homeostasis and repair. *Cell* **2008**, *135*, 1118–1129. [CrossRef] [PubMed]

7.   Kiel, M.J.; Yilmaz, O.H.; Iwashita, T.; Terhorst, C.; Morrison, S.J. Slam family receptors distinguish hematopoietic stem and progenitor cells and reveal endothelial niches for stem cells. *Cell* 2005, *121*, 1109–1121. [CrossRef] [PubMed]

8.   Adolfsson, J.; Borge, O.J.; Bryder, D.; Theilgaard-Monch, K.; Astrand-Grundstrom, I.; Sitnicka, E.; Sasaki, Y.; Jacobsen, S.E. Upregulation of Flt3 expression within the bone marrow Lin⁻Sca1⁺c-kit⁺ stem cell compartment is accompanied by loss of self-renewal capacity. *Immunity* 2001, *15*, 659–669. [CrossRef]

9.   Adolfsson, J.; Mansson, R.; Buza-Vidas, N.; Hultquist, A.; Liuba, K.; Jensen, C.T.; Bryder, D.; Yang, L.; Borge, O.J.; Thoren, L.A.; et al. Identification of Flt3⁺ lympho-myeloid stem cells lacking erythro-megakaryocytic potential a revised road map for adult blood lineage commitment. *Cell* 2005, *121*, 295–306. [CrossRef] [PubMed]

10. Pietras, E.M.; Reynaud, D.; Kang, Y.A.; Carlin, D.; Calero-Nieto, F.J.; Leavitt, A.D.; Stuart, J.M.; Gottgens, B.; Passegue, E. Functionally distinct subsets of lineage-biased multipotent progenitors control blood production in normal and regenerative conditions. *Cell Stem Cell* 2015, *17*, 35–46. [CrossRef] [PubMed]

11. Arinobu, Y.; Mizuno, S.; Chong, Y.; Shigematsu, H.; Iino, T.; Iwasaki, H.; Graf, T.; Mayfield, R.; Chan, S.; Kastner, P.; et al. Reciprocal activation of GATA-1 and PU.1 marks initial specification of hematopoietic stem cells into myeloerythroid and myelolymphoid lineages. *Cell Stem Cell* 2007, *1*, 416–427. [CrossRef] [PubMed]

12. Oguro, H.; Ding, L.; Morrison, S.J. Slam family markers resolve functionally distinct subpopulations of hematopoietic stem cells and multipotent progenitors. *Cell Stem Cell* 2013, *13*, 102–116. [CrossRef] [PubMed]

13. Guo, G.; Luc, S.; Marco, E.; Lin, T.W.; Peng, C.; Kerenyi, M.A.; Beyaz, S.; Kim, W.; Xu, J.; Das, P.P.; et al. Mapping cellular hierarchy by single-cell analysis of the cell surface repertoire. *Cell Stem Cell* 2013, *13*, 492–505. [CrossRef] [PubMed]

14. Paul, F.; Arkin, Y.; Giladi, A.; Jaitin, D.A.; Kenigsberg, E.; Keren-Shaul, H.; Winter, D.; Lara-Astiaso, D.; Gury, M.; Weiner, A.; et al. Transcriptional heterogeneity and lineage commitment in myeloid progenitors. *Cell* 2015, *163*, 1663–1677. [CrossRef] [PubMed]

15. Wilson, N.K.; Kent, D.G.; Buettner, F.; Shehata, M.; Macaulay, I.C.; Calero-Nieto, F.J.; Sanchez Castillo, M.; Oedekoven, C.A.; Diamanti, E.; Schulte, R.; et al. Combined single-cell functional and gene expression analysis resolves heterogeneity within stem cell populations. *Cell Stem Cell* 2015, *16*, 712–724. [CrossRef] [PubMed]

16. Nestorowa, S.; Hamey, F.K.; Pijuan Sala, B.; Diamanti, E.; Shepherd, M.; Laurenti, E.; Wilson, N.K.; Kent, D.G.; Gottgens, B. A single-cell resolution map of mouse hematopoietic stem and progenitor cell differentiation. *Blood* 2016, *128*, e20–e31. [CrossRef] [PubMed]

17. Rodriguez-Fraticelli, A.E.; Wolock, S.L.; Weinreb, C.S.; Panero, R.; Patel, S.H.; Jankovic, M.; Sun, J.; Calogero, R.A.; Klein, A.M.; Camargo, F.D. Clonal analysis of lineage fate in native haematopoiesis. *Nature* 2018, *553*, 212–216. [CrossRef] [PubMed]

18. Laurenti, E.; Gottgens, B. From haematopoietic stem cells to complex differentiation landscapes. *Nature* 2018, *553*, 418–426. [PubMed]

19. Kondo, M.; Weissman, I.L.; Akashi, K. Identification of clonogenic common lymphoid progenitors in mouse bone marrow. *Cell* 1997, *91*, 661–672. [CrossRef]

20. Inlay, M.A.; Bhattacharya, D.; Sahoo, D.; Serwold, T.; Seita, J.; Karsunky, H.; Plevritis, S.K.; Dill, D.L.; Weissman, I.L. Ly6d marks the earliest stage of B-cell specification and identifies the branchpoint between B-cell and T-cell development. *Genes Dev.* 2009, *23*, 2376–2381. [CrossRef] [PubMed]

21. Rolink, A.; Melchers, F. B-cell development in the mouse. *Immunol. Lett.* 1996, *54*, 157–161. [CrossRef]

22. Hardy, R.R.; Hayakawa, K. B cell development pathways. *Annu. Rev. Immunol.* 2001, *19*, 595–621. [CrossRef] [PubMed]

23. Hardy, R.R.; Carmack, C.E.; Shinton, S.A.; Kemp, J.D.; Hayakawa, K. Resolution and characterization of pro-B and pre-pro-B cell stages in normal mouse bone marrow. *J. Exp. Med.* 1991, *173*, 1213–1225. [CrossRef] [PubMed]

24. Rolink, A.; Grawunder, U.; Winkler, T.H.; Karasuyama, H.; Melchers, F. Il-2 receptor α chain (CD25, TAC) expression defines a crucial stage in pre-B cell development. *Int. Immunol.* 1994, *6*, 1257–1264. [CrossRef] [PubMed]

25. Nutt, S.L.; Kee, B.L. The transcriptional regulation of B cell lineage commitment. *Immunity* 2007, *26*, 715–725. [CrossRef] [PubMed]

26. Urbanek, P.; Wang, Z.Q.; Fetka, I.; Wagner, E.F.; Busslinger, M. Complete block of early B cell differentiation and altered patterning of the posterior midbrain in mice lacking Pax5BSAP. *Cell* **1994**, *79*, 901–912. [CrossRef]

27. Nutt, S.L.; Heavey, B.; Rolink, A.G.; Busslinger, M. Commitment to the B-lymphoid lineage depends on the transcription factor Pax5. *Nature* **1999**, *401*, 556–562. [CrossRef] [PubMed]

28. Schaniel, C.; Bruno, L.; Melchers, F.; Rolink, A.G. Multiple hematopoietic cell lineages develop in vivo from transplanted Pax5-deficient pre-B I-cell clones. *Blood* **2002**, *99*, 472–478. [CrossRef] [PubMed]

29. Sakaguchi, N.; Melchers, F. λ5, a new light-chain-related locus selectively expressed in pre-B lymphocytes. *Nature* **1986**, *324*, 579–582. [CrossRef] [PubMed]

30. Pillai, S.; Baltimore, D. Formation of disulphide-linked μ2ω2 tetramers in pre-B cells by the 18K ω-immunoglobulin light chain. *Nature* **1987**, *329*, 172–174. [CrossRef] [PubMed]

31. Kudo, A.; Melchers, F. A second gene, vpreB in the λ5 locus of the mouse, which appears to be selectively expressed in pre-B lymphocytes. *EMBO J.* **1987**, *6*, 2267–2272. [PubMed]

32. Karasuyama, H.; Kudo, A.; Melchers, F. The proteins encoded by the VpreB and λ5 pre-B cell-specific genes can associate with each other and with mu heavy chain. *J. Exp. Med.* **1990**, *172*, 969–972. [CrossRef] [PubMed]

33. Nemazee, D. Mechanisms of central tolerance for B cells. *Nat. Rev. Immunol.* **2017**, *17*, 281–294. [CrossRef] [PubMed]

34. Laiosa, C.V.; Stadtfeld, M.; Graf, T. Determinants of lymphoid-myeloid lineage diversification. *Annu. Rev. Immunol.* **2006**, *24*, 705–738. [CrossRef] [PubMed]

35. Laslo, P.; Pongubala, J.M.; Lancki, D.W.; Singh, H. Gene regulatory networks directing myeloid and lymphoid cell fates within the immune system. *Semin. Immunol.* **2008**, *20*, 228–235. [CrossRef] [PubMed]

36. Dexter, T.M.; Allen, T.D.; Lajtha, L.G. Conditions controlling the proliferation of haemopoietic stem cells in vitro. *J. Cell. Physiol.* **1977**, *91*, 335–344. [CrossRef] [PubMed]

37. Penn, P.E.; Jiang, D.Z.; Fei, R.G.; Sitnicka, E.; Wolf, N.S. Dissecting the hematopoietic microenvironment. IX. Further characterization of murine bone marrow stromal cells. *Blood* **1993**, *81*, 1205–1213. [PubMed]

38. Shackney, S.E.; Ford, S.S.; Wittig, A.B. Kinetic-microarchitectural correlations in the bone marrow of the mouse. *Cell Tissue Kinet.* **1975**, *8*, 505–516. [CrossRef] [PubMed]

39. Lord, B.I. The architecture of bone marrow cell populations. *Int. J. Cell Cloning* **1990**, *8*, 317–331. [CrossRef] [PubMed]

40. Calvi, L.M.; Adams, G.B.; Weibrecht, K.W.; Weber, J.M.; Olson, D.P.; Knight, M.C.; Martin, R.P.; Schipani, E.; Divieti, P.; Bringhurst, F.R.; et al. Osteoblastic cells regulate the haematopoietic stem cell niche. *Nature* **2003**, *425*, 841–846. [CrossRef] [PubMed]

41. Zhang, J.; Niu, C.; Ye, L.; Huang, H.; He, X.; Tong, W.G.; Ross, J.; Haug, J.; Johnson, T.; Feng, J.Q.; et al. Identification of the haematopoietic stem cell niche and control of the niche size. *Nature* **2003**, *425*, 836–841. [CrossRef] [PubMed]

42. Tokoyoda, K.; Egawa, T.; Sugiyama, T.; Choi, B.I.; Nagasawa, T. Cellular niches controlling B lymphocyte behavior within bone marrow during development. *Immunity* **2004**, *20*, 707–718. [CrossRef] [PubMed]

43. Sugiyama, T.; Kohara, H.; Noda, M.; Nagasawa, T. Maintenance of the hematopoietic stem cell pool by CXCL12-CXCR4 chemokine signaling in bone marrow stromal cell niches. *Immunity* **2006**, *25*, 977–988. [CrossRef] [PubMed]

44. Nagasawa, T.; Kikutani, H.; Kishimoto, T. Molecular cloning and structure of a pre-B-cell growth-stimulating factor. *Proc. Natl. Acad. Sci. USA* **1994**, *91*, 2305–2309. [CrossRef] [PubMed]

45. Ma, Q.; Jones, D.; Borghesani, P.R.; Segal, R.A.; Nagasawa, T.; Kishimoto, T.; Bronson, R.T.; Springer, T.A. Impaired B-lymphopoiesis, myelopoiesis, and derailed cerebellar neuron migration in CXCR4- and SDF-1-deficient mice. *Proc. Natl. Acad. Sci. USA* **1998**, *95*, 9448–9453. [CrossRef] [PubMed]

46. Nie, Y.; Han, Y.C.; Zou, Y.R. CXCR4 is required for the quiescence of primitive hematopoietic cells. *J. Exp. Med.* **2008**, *205*, 777–783. [CrossRef] [PubMed]

47. Aiuti, A.; Webb, I.J.; Bleul, C.; Springer, T.; Gutierrez-Ramos, J.C. The chemokine SDF-1 is a chemoattractant for human CD34+ hematopoietic progenitor cells and provides a new mechanism to explain the mobilization of CD34+ progenitors to peripheral blood. *J. Exp. Med.* **1997**, *185*, 111–120. [CrossRef] [PubMed]

48. Ma, Q.; Jones, D.; Springer, T.A. The chemokine receptor CXCR4 is required for the retention of B lineage and granulocytic precursors within the bone marrow microenvironment. *Immunity* **1999**, *10*, 463–471. [CrossRef]

49. Broxmeyer, H.E.; Orschell, C.M.; Clapp, D.W.; Hangoc, G.; Cooper, S.; Plett, P.A.; Liles, W.C.; Li, X.; Graham-Evans, B.; Campbell, T.B.; et al. Rapid mobilization of murine and human hematopoietic stem and progenitor cells with AMD3100, a CXCR4 antagonist. *J. Exp. Med.* **2005**, *201*, 1307–1318. [CrossRef] [PubMed]

50. Ding, L.; Morrison, S.J. Haematopoietic stem cells and early lymphoid progenitors occupy distinct bone marrow niches. *Nature* **2013**, *495*, 231–235. [CrossRef] [PubMed]

51. Asada, N.; Kunisaki, Y.; Pierce, H.; Wang, Z.; Fernandez, N.F.; Birbrair, A.; Ma'ayan, A.; Frenette, P.S. Differential cytokine contributions of perivascular haematopoietic stem cell niches. *Nat. Cell Biol.* **2017**, *19*, 214–223. [CrossRef] [PubMed]

52. Greenbaum, A.; Hsu, Y.M.; Day, R.B.; Schuettpelz, L.G.; Christopher, M.J.; Borgerding, J.N.; Nagasawa, T.; Link, D.C. CXCL12 in early mesenchymal progenitors is required for haematopoietic stem-cell maintenance. *Nature* **2013**, *495*, 227–230. [CrossRef] [PubMed]

53. Broudy, V.C. Stem cell factor and hematopoiesis. *Blood* **1997**, *90*, 1345–1364. [PubMed]

54. Ding, L.; Saunders, T.L.; Enikolopov, G.; Morrison, S.J. Endothelial and perivascular cells maintain haematopoietic stem cells. *Nature* **2012**, *481*, 457–462. [CrossRef] [PubMed]

55. Omatsu, Y.; Sugiyama, T.; Kohara, H.; Kondoh, G.; Fujii, N.; Kohno, K.; Nagasawa, T. The essential functions of adipo-osteogenic progenitors as the hematopoietic stem and progenitor cell niche. *Immunity* **2010**, *33*, 387–399. [CrossRef] [PubMed]

56. Mendez-Ferrer, S.; Michurina, T.V.; Ferraro, F.; Mazloom, A.R.; Macarthur, B.D.; Lira, S.A.; Scadden, D.T.; Ma'ayan, A.; Enikolopov, G.N.; Frenette, P.S. Mesenchymal and haematopoietic stem cells form a unique bone marrow niche. *Nature* **2010**, *466*, 829–834. [CrossRef] [PubMed]

57. Mizoguchi, T.; Pinho, S.; Ahmed, J.; Kunisaki, Y.; Hanoun, M.; Mendelson, A.; Ono, N.; Kronenberg, H.M.; Frenette, P.S. Osterix marks distinct waves of primitive and definitive stromal progenitors during bone marrow development. *Dev. Cell* **2014**, *29*, 340–349. [CrossRef] [PubMed]

58. Zhou, B.O.; Yue, R.; Murphy, M.M.; Peyer, J.G.; Morrison, S.J. Leptin-receptor-expressing mesenchymal stromal cells represent the main source of bone formed by adult bone marrow. *Cell Stem Cell* **2014**, *15*, 154–168. [CrossRef] [PubMed]

59. Adams, G.B.; Chabner, K.T.; Alley, I.R.; Olson, D.P.; Szczepiorkowski, Z.M.; Poznansky, M.C.; Kos, C.H.; Pollak, M.R.; Brown, E.M.; Scadden, D.T. Stem cell engraftment at the endosteal niche is specified by the calcium-sensing receptor. *Nature* **2006**, *439*, 599–603. [CrossRef] [PubMed]

60. Lam, B.S.; Cunningham, C.; Adams, G.B. Pharmacologic modulation of the calcium-sensing receptor enhances hematopoietic stem cell lodgment in the adult bone marrow. *Blood* **2011**, *117*, 1167–1175. [CrossRef] [PubMed]

61. Sugimura, R.; He, X.C.; Venkatraman, A.; Arai, F.; Box, A.; Semerad, C.; Haug, J.S.; Peng, L.; Zhong, X.B.; Suda, T.; et al. Noncanonical Wnt signaling maintains hematopoietic stem cells in the niche. *Cell* **2012**, *150*, 351–365. [CrossRef] [PubMed]

62. Kunisaki, Y.; Bruns, I.; Scheiermann, C.; Ahmed, J.; Pinho, S.; Zhang, D.; Mizoguchi, T.; Wei, Q.; Lucas, D.; Ito, K.; et al. Arteriolar niches maintain haematopoietic stem cell quiescence. *Nature* **2013**, *502*, 637–643. [CrossRef] [PubMed]

63. Kusumbe, A.P.; Ramasamy, S.K.; Adams, R.H. Coupling of angiogenesis and osteogenesis by a specific vessel subtype in bone. *Nature* **2014**, *507*, 323–328. [CrossRef] [PubMed]

64. Ito, K.; Hirao, A.; Arai, F.; Takubo, K.; Matsuoka, S.; Miyamoto, K.; Ohmura, M.; Naka, K.; Hosokawa, K.; Ikeda, Y.; et al. Reactive oxygen species act through p38 MAPK to limit the lifespan of hematopoietic stem cells. *Nat. Med.* **2006**, *12*, 446–451. [CrossRef] [PubMed]

65. Takubo, K.; Goda, N.; Yamada, W.; Iriuchishima, H.; Ikeda, E.; Kubota, Y.; Shima, H.; Johnson, R.S.; Hirao, A.; Suematsu, M.; et al. Regulation of the HIF-1α level is essential for hematopoietic stem cells. *Cell Stem Cell* **2010**, *7*, 391–402. [CrossRef] [PubMed]

66. Nombela-Arrieta, C.; Pivarnik, G.; Winkel, B.; Canty, K.J.; Harley, B.; Mahoney, J.E.; Park, S.Y.; Lu, J.; Protopopov, A.; Silberstein, L.E. Quantitative imaging of haematopoietic stem and progenitor cell localization and hypoxic status in the bone marrow microenvironment. *Nat. Cell Biol.* **2013**, *15*, 533–543. [CrossRef] [PubMed]

67. Itkin, T.; Gur-Cohen, S.; Spencer, J.A.; Schajnovitz, A.; Ramasamy, S.K.; Kusumbe, A.P.; Ledergor, G.; Jung, Y.; Milo, I.; Poulos, M.G.; et al. Distinct bone marrow blood vessels differentially regulate haematopoiesis. *Nature* **2016**, *532*, 323–328. [CrossRef] [PubMed]

68. Mendelson, A.; Frenette, P.S. Hematopoietic stem cell niche maintenance during homeostasis and regeneration. *Nat. Med.* **2014**, *20*, 833–846. [CrossRef] [PubMed]

69. Crane, G.M.; Jeffery, E.; Morrison, S.J. Adult haematopoietic stem cell niches. *Nat. Rev. Immunol.* **2017**, *17*, 573–590. [CrossRef] [PubMed]

70. Lo Celso, C.; Fleming, H.E.; Wu, J.W.; Zhao, C.X.; Miake-Lye, S.; Fujisaki, J.; Cote, D.; Rowe, D.W.; Lin, C.P.; Scadden, D.T. Live-animal tracking of individual haematopoietic stem/progenitor cells in their niche. *Nature* **2009**, *457*, 92–96. [CrossRef] [PubMed]

71. Dias, S.; Silva, H., Jr.; Cumano, A.; Vieira, P. Interleukin-7 is necessary to maintain the B cell potential in common lymphoid progenitors. *J. Exp. Med.* **2005**, *201*, 971–979. [CrossRef] [PubMed]

72. Terashima, A.; Okamoto, K.; Nakashima, T.; Akira, S.; Ikuta, K.; Takayanagi, H. Sepsis-induced osteoblast ablation causes immunodeficiency. *Immunity* **2016**, *44*, 1434–1443. [CrossRef] [PubMed]

73. Cordeiro Gomes, A.; Hara, T.; Lim, V.Y.; Herndler-Brandstetter, D.; Nevius, E.; Sugiyama, T.; Tani-Ichi, S.; Schlenner, S.; Richie, E.; Rodewald, H.R.; et al. Hematopoietic stem cell niches produce lineage-instructive signals to control multipotent progenitor differentiation. *Immunity* **2016**, *45*, 1219–1231. [CrossRef] [PubMed]

74. Zhang, J.; Link, D.C. Targeting of mesenchymal stromal cells by *Cre*-recombinase transgenes commonly used to target osteoblast lineage cells. *J. Bone Miner. Res.* **2016**, *31*, 2001–2007. [CrossRef] [PubMed]

75. Mourcin, F.; Breton, C.; Tellier, J.; Narang, P.; Chasson, L.; Jorquera, A.; Coles, M.; Schiff, C.; Mancini, S.J. Galectin-1-expressing stromal cells constitute a specific niche for pre-BII cell development in mouse bone marrow. *Blood* **2011**, *117*, 6552–6561. [CrossRef] [PubMed]

76. Wang, L.; Benedito, R.; Bixel, M.G.; Zeuschner, D.; Stehling, M.; Savendahl, L.; Haigh, J.J.; Snippert, H.; Clevers, H.; Breier, G.; et al. Identification of a clonally expanding haematopoietic compartment in bone marrow. *EMBO J.* **2013**, *32*, 219–230. [CrossRef] [PubMed]

77. Whitlock, C.A.; Witte, O.N. Long-term culture of B lymphocytes and their precursors from murine bone marrow. *Proc. Natl. Acad. Sci. USA* **1982**, *79*, 3608–3612. [CrossRef] [PubMed]

78. Namen, A.E.; Lupton, S.; Hjerrild, K.; Wignall, J.; Mochizuki, D.Y.; Schmierer, A.; Mosley, B.; March, C.J.; Urdal, D.; Gillis, S. Stimulation of B-cell progenitors by cloned murine interleukin-7. *Nature* **1988**, *333*, 571–573. [CrossRef] [PubMed]

79. Namen, A.E.; Schmierer, A.E.; March, C.J.; Overell, R.W.; Park, L.S.; Urdal, D.L.; Mochizuki, D.Y. B cell precursor growth-promoting activity. Purification and characterization of a growth factor active on lymphocyte precursors. *J. Exp. Med.* **1988**, *167*, 988–1002. [CrossRef] [PubMed]

80. Kincade, P.W.; Lee, G.; Pietrangeli, C.E.; Hayashi, S.; Gimble, J.M. Cells and molecules that regulate B lymphopoiesis in bone marrow. *Annu. Rev. Immunol.* **1989**, *7*, 111–143. [CrossRef] [PubMed]

81. Rolink, A.; Kudo, A.; Karasuyama, H.; Kikuchi, Y.; Melchers, F. Long-term proliferating early pre B cell lines and clones with the potential to develop to surface Ig-positive, mitogen reactive B cells in vitro and in vivo. *EMBO J.* **1991**, *10*, 327–336. [PubMed]

82. Rolink, A.G. B-cell development and pre-B-1 cell plasticity in vitro. *Methods Mol. Biol.* **2004**, *271*, 271–281. [PubMed]

83. Rolink, A.; Grawunder, U.; Haasner, D.; Strasser, A.; Melchers, F. Immature surface Ig+ B cells can continue to rearrange kappa and λ l chain gene loci. *J. Exp. Med.* **1993**, *178*, 1263–1270. [CrossRef] [PubMed]

84. Rolink, A.; Streb, M.; Nishikawa, S.; Melchers, F. The c-kit-encoded tyrosine kinase regulates the proliferation of early pre-B cells. *Eur. J. Immunol.* **1991**, *21*, 2609–2612. [CrossRef] [PubMed]

85. Jacobsen, K.; Osmond, D.G. Microenvironmental organization and stromal cell associations of B lymphocyte precursor cells in mouse bone marrow. *Eur. J. Immunol.* **1990**, *20*, 2395–2404. [CrossRef] [PubMed]

86. Jacobsen, K.; Miyake, K.; Kincade, P.W.; Osmond, D.G. Highly restricted expression of a stromal cell determinant in mouse bone marrow in vivo. *J. Exp. Med.* **1992**, *176*, 927–935. [CrossRef] [PubMed]

87. Nagasawa, T.; Hirota, S.; Tachibana, K.; Takakura, N.; Nishikawa, S.; Kitamura, Y.; Yoshida, N.; Kikutani, H.; Kishimoto, T. Defects of B-cell lymphopoiesis and bone-marrow myelopoiesis in mice lacking the CXC chemokine PBSF/SDF-1. *Nature* **1996**, *382*, 635–638. [CrossRef] [PubMed]

88. Zou, Y.R.; Kottmann, A.H.; Kuroda, M.; Taniuchi, I.; Littman, D.R. Function of the chemokine receptor CXCR4 in haematopoiesis and in cerebellar development. *Nature* **1998**, *393*, 595–599. [CrossRef] [PubMed]

89. Egawa, T.; Kawabata, K.; Kawamoto, H.; Amada, K.; Okamoto, R.; Fujii, N.; Kishimoto, T.; Katsura, Y.; Nagasawa, T. The earliest stages of B cell development require a chemokine stromal cell-derived factor/pre-B cell growth-stimulating factor. *Immunity* **2001**, *15*, 323–334. [CrossRef]

90. Nie, Y.; Waite, J.; Brewer, F.; Sunshine, M.J.; Littman, D.R.; Zou, Y.R. The role of CXCR4 in maintaining peripheral B cell compartments and humoral immunity. *J. Exp. Med.* **2004**, *200*, 1145–1156. [CrossRef] [PubMed]

91. Glodek, A.M.; Honczarenko, M.; Le, Y.; Campbell, J.J.; Silberstein, L.E. Sustained activation of cell adhesion is a differentially regulated process in B lymphopoiesis. *J. Exp. Med.* **2003**, *197*, 461–473. [CrossRef] [PubMed]

92. Arroyo, A.G.; Yang, J.T.; Rayburn, H.; Hynes, R.O. Differential requirements for α4 integrins during fetal and adult hematopoiesis. *Cell* **1996**, *85*, 997–1008. [CrossRef]

93. Park, S.Y.; Wolfram, P.; Canty, K.; Harley, B.; Nombela-Arrieta, C.; Pivarnik, G.; Manis, J.; Beggs, H.E.; Silberstein, L.E. Focal adhesion kinase regulates the localization and retention of pro-B cells in bone marrow microenvironments. *J. Immunol.* **2013**, *190*, 1094–1102. [CrossRef] [PubMed]

94. Marshall, A.J.; Fleming, H.E.; Wu, G.E.; Paige, C.J. Modulation of the IL-7 dose-response threshold during pro-B cell differentiation is dependent on pre-B cell receptor expression. *J. Immunol.* **1998**, *161*, 6038–6045. [PubMed]

95. Corcoran, A.E.; Smart, F.M.; Cowling, R.J.; Crompton, T.; Owen, M.J.; Venkitaraman, A.R. The interleukin-7 receptor α chain transmits distinct signals for proliferation and differentiation during B lymphopoiesis. *EMBO J.* **1996**, *15*, 1924–1932. [PubMed]

96. Lu, L.; Chaudhury, P.; Osmond, D.G. Regulation of cell survival during B lymphopoiesis: Apoptosis and Bcl-2/Bax content of precursor B cells in bone marrow of mice with altered expression of IL-7 and recombinase-activating gene-2. *J. Immunol.* **1999**, *162*, 1931–1940. [PubMed]

97. Von Freeden-Jeffry, U.; Vieira, P.; Lucian, L.A.; McNeil, T.; Burdach, S.E.; Murray, R. Lymphopenia in interleukin (IL)-7 gene-deleted mice identifies IL-7 as a nonredundant cytokine. *J. Exp. Med.* **1995**, *181*, 1519–1526. [CrossRef] [PubMed]

98. Peschon, J.J.; Morrissey, P.J.; Grabstein, K.H.; Ramsdell, F.J.; Maraskovsky, E.; Gliniak, B.C.; Park, L.S.; Ziegler, S.F.; Williams, D.E.; Ware, C.B.; et al. Early lymphocyte expansion is severely impaired in interleukin 7 receptor-deficient mice. *J. Exp. Med.* **1994**, *180*, 1955–1960. [CrossRef] [PubMed]

99. Alves, N.L.; Richard-Le Goff, O.; Huntington, N.D.; Sousa, A.P.; Ribeiro, V.S.; Bordack, A.; Vives, F.L.; Peduto, L.; Chidgey, A.; Cumano, A.; et al. Characterization of the thymic IL-7 niche in vivo. *Proc. Natl. Acad. Sci. USA* **2009**, *106*, 1512–1517. [CrossRef] [PubMed]

100. Mazzucchelli, R.I.; Warming, S.; Lawrence, S.M.; Ishii, M.; Abshari, M.; Washington, A.V.; Feigenbaum, L.; Warner, A.C.; Sims, D.J.; Li, W.Q.; et al. Visualization and identification of IL-7 producing cells in reporter mice. *PLoS ONE* **2009**, *4*, e7637. [CrossRef] [PubMed]

101. Kitamura, D.; Kudo, A.; Schaal, S.; Muller, W.; Melchers, F.; Rajewsky, K. A critical role of λ5 protein in B cell development. *Cell* **1992**, *69*, 823–831. [CrossRef]

102. Mundt, C.; Licence, S.; Shimizu, T.; Melchers, F.; Martensson, I.L. Loss of precursor B cell expansion but not allelic exclusion in VpreB1/VpreB2 double-deficient mice. *J. Exp. Med.* **2001**, *193*, 435–445. [CrossRef] [PubMed]

103. Shimizu, T.; Mundt, C.; Licence, S.; Melchers, F.; Martensson, I.L. VpreB1/vpreB2/λ5 triple-deficient mice show impaired B cell development but functional allelic exclusion of the IgH locus. *J. Immunol.* **2002**, *168*, 6286–6293. [CrossRef] [PubMed]

104. Ohnishi, K.; Melchers, F. The nonimmunoglobulin portion of λ5 mediates cell-autonomous pre-B cell receptor signaling. *Nat. Immunol.* **2003**, *4*, 849–856. [CrossRef] [PubMed]

105. Ubelhart, R.; Bach, M.P.; Eschbach, C.; Wossning, T.; Reth, M.; Jumaa, H. N-linked glycosylation selectively regulates autonomous precursor BCR function. *Nat. Immunol.* **2010**, *11*, 759–765. [CrossRef] [PubMed]

106. Bradl, H.; Wittmann, J.; Milius, D.; Vettermann, C.; Jack, H.M. Interaction of murine precursor B cell receptor with stroma cells is controlled by the unique tail of λ5 and stroma cell-associated heparan sulfate. *J. Immunol.* **2003**, *171*, 2338–2348. [CrossRef] [PubMed]

107. Milne, C.D.; Corfe, S.A.; Paige, C.J. Heparan sulfate and heparin enhance ERK phosphorylation and mediate preBCR-dependent events during B lymphopoiesis. *J. Immunol.* **2008**, *180*, 2839–2847. [CrossRef] [PubMed]

108. Gauthier, L.; Rossi, B.; Roux, F.; Termine, E.; Schiff, C. Galectin-1 is a stromal cell ligand of the pre-B cell receptor (BCR) implicated in synapse formation between pre-B and stromal cells and in pre-BCR triggering. *Proc. Natl. Acad. Sci. USA* **2002**, *99*, 13014–13019. [CrossRef] [PubMed]

109. Elantak, L.; Espeli, M.; Boned, A.; Bornet, O.; Bonzi, J.; Gauthier, L.; Feracci, M.; Roche, P.; Guerlesquin, F.; Schiff, C. Structural basis for galectin-1-dependent pre-B cell receptor (pre-BCR) activation. *J. Biol. Chem.* **2012**, *287*, 44703–44713. [CrossRef] [PubMed]

110. Rossi, B.; Espeli, M.; Schiff, C.; Gauthier, L. Clustering of pre-B cell integrins induces galectin-1-dependent pre-B cell receptor relocalization and activation. *J. Immunol.* **2006**, *177*, 796–803. [CrossRef] [PubMed]

111. Espeli, M.; Mancini, S.J.; Breton, C.; Poirier, F.; Schiff, C. Impaired B-cell development at the pre-BII-cell stage in galectin-1-deficient mice due to inefficient pre-BII/stromal cell interactions. *Blood* **2009**, *113*, 5878–5886. [CrossRef] [PubMed]

112. Beck, T.C.; Gomes, A.C.; Cyster, J.G.; Pereira, J.P. CXCR4 and a cell-extrinsic mechanism control immature B lymphocyte egress from bone marrow. *J. Exp. Med.* **2014**, *211*, 2567–2581. [CrossRef] [PubMed]

113. Pereira, J.P.; An, J.; Xu, Y.; Huang, Y.; Cyster, J.G. Cannabinoid receptor 2 mediates the retention of immature B cells in bone marrow sinusoids. *Nat. Immunol.* **2009**, *10*, 403–411. [CrossRef] [PubMed]

114. Sandel, P.C.; Monroe, J.G. Negative selection of immature B cells by receptor editing or deletion is determined by site of antigen encounter. *Immunity* **1999**, *10*, 289–299. [CrossRef]

115. Sandel, P.C.; Gendelman, M.; Kelsoe, G.; Monroe, J.G. Definition of a novel cellular constituent of the bone marrow that regulates the response of immature B cells to B cell antigen receptor engagement. *J. Immunol.* **2001**, *166*, 5935–5944. [CrossRef] [PubMed]

116. Chang, H.D.; Tokoyoda, K.; Radbruch, A. Immunological memories of the bone marrow. *Immunol. Rev.* **2018**, *283*, 86–98. [CrossRef] [PubMed]

117. Hargreaves, D.C.; Hyman, P.L.; Lu, T.T.; Ngo, V.N.; Bidgol, A.; Suzuki, G.; Zou, Y.R.; Littman, D.R.; Cyster, J.G. A coordinated change in chemokine responsiveness guides plasma cell movements. *J. Exp. Med.* **2001**, *194*, 45–56. [CrossRef] [PubMed]

118. Cariappa, A.; Mazo, I.B.; Chase, C.; Shi, H.N.; Liu, H.; Li, Q.; Rose, H.; Leung, H.; Cherayil, B.J.; Russell, P.; et al. Perisinusoidal B cells in the bone marrow participate in T-independent responses to blood-borne microbes. *Immunity* **2005**, *23*, 397–407. [CrossRef] [PubMed]

119. Sapoznikov, A.; Pewzner-Jung, Y.; Kalchenko, V.; Krauthgamer, R.; Shachar, I.; Jung, S. Perivascular clusters of dendritic cells provide critical survival signals to B cells in bone marrow niches. *Nat. Immunol.* **2008**, *9*, 388–395. [CrossRef] [PubMed]

120. Winter, O.; Moser, K.; Mohr, E.; Zotos, D.; Kaminski, H.; Szyska, M.; Roth, K.; Wong, D.M.; Dame, C.; Tarlinton, D.M.; et al. Megakaryocytes constitute a functional component of a plasma cell niche in the bone marrow. *Blood* **2010**, *116*, 1867–1875. [CrossRef] [PubMed]

121. Chu, V.T.; Frohlich, A.; Steinhauser, G.; Scheel, T.; Roch, T.; Fillatreau, S.; Lee, J.J.; Lohning, M.; Berek, C. Eosinophils are required for the maintenance of plasma cells in the bone marrow. *Nat. Immunol.* **2011**, *12*, 151–159. [CrossRef] [PubMed]

122. Belnoue, E.; Tougne, C.; Rochat, A.F.; Lambert, P.H.; Pinschewer, D.D.; Siegrist, C.A. Homing and adhesion patterns determine the cellular composition of the bone marrow plasma cell niche. *J. Immunol.* **2012**, *188*, 1283–1291. [CrossRef] [PubMed]

123. Zehentmeier, S.; Roth, K.; Cseresnyes, Z.; Sercan, O.; Horn, K.; Niesner, R.A.; Chang, H.D.; Radbruch, A.; Hauser, A.E. Static and dynamic components synergize to form a stable survival niche for bone marrow plasma cells. *Eur. J. Immunol.* **2014**, *44*, 2306–2317. [CrossRef] [PubMed]

124. DiLillo, D.J.; Hamaguchi, Y.; Ueda, Y.; Yang, K.; Uchida, J.; Haas, K.M.; Kelsoe, G.; Tedder, T.F. Maintenance of long-lived plasma cells and serological memory despite mature and memory B cell depletion during CD20 immunotherapy in mice. *J. Immunol.* **2008**, *180*, 361–371. [CrossRef] [PubMed]

125. Odendahl, M.; Mei, H.; Hoyer, B.F.; Jacobi, A.M.; Hansen, A.; Muehlinghaus, G.; Berek, C.; Hiepe, F.; Manz, R.; Radbruch, A.; et al. Generation of migratory antigen-specific plasma blasts and mobilization of resident plasma cells in a secondary immune response. *Blood* **2005**, *105*, 1614–1621. [CrossRef] [PubMed]

126. Rolink, A.; Melchers, F. Generation and regeneration of cells of the B-lymphocyte lineage. *Curr. Opin. Immunol.* **1993**, *5*, 207–217. [CrossRef]

127. Kodama, H.; Nose, M.; Yamaguchi, Y.; Tsunoda, J.; Suda, T.; Nishikawa, S. In vitro proliferation of primitive hemopoietic stem cells supported by stromal cells: Evidence for the presence of a mechanism(s) other than that involving c-kit receptor and its ligand. *J. Exp. Med.* **1992**, *176*, 351–361. [CrossRef] [PubMed]

128. Goulard, M.; Dosquet, C.; Bonnet, D. Role of the microenvironment in myeloid malignancies. *Cell. Mol. Life Sci.* **2018**, *75*, 1377–1391. [CrossRef] [PubMed]

129. Kramann, R.; Schneider, R.K. The identification of fibrosis-driving myofibroblast precursors reveals new therapeutic avenues in myelofibrosis. *Blood* **2018**, *131*, 2111–2119. [CrossRef] [PubMed]

130. Chiarini, F.; Lonetti, A.; Evangelisti, C.; Buontempo, F.; Orsini, E.; Cappellini, A.; Neri, L.M.; McCubrey, J.A.; Martelli, A.M. Advances in understanding the acute lymphoblastic leukemia bone marrow microenvironment: From biology to therapeutic targeting. *Biochim. Biophys. Acta* **2016**, *1863*, 449–463. [CrossRef] [PubMed]

International Journal of
Molecular Sciences

MDPI

Article

Identification of an Essential Cytoplasmic Region of Interleukin-7 Receptor α Subunit in B-Cell Development

Hirotake Kasai [1,†,‡], Taku Kuwabara [2,†], Yukihide Matsui [2,3,4], Koichi Nakajima [4] and Motonari Kondo [2,*]

[1] Department of Immunology, Duke University Medical Center, Durham, NC 27710, USA;
    hirotake.kasai@yamanashi.ac.jp
[2] Department of Molecular Immunology, Toho University School of Medicine, Tokyo 143-8540, Japan;
    kuwabara@med.toho-u.ac.jp (T.K.); ym04083@yahoo.co.jp (Y.M.)
[3] Toho University Graduate School of Medicine, Tokyo 143-8540, Japan
[4] Department of Urology, Toho University Omori Medical Center, Tokyo 143-8541, Japan;
    koichin@med.toho-u.ac.jp
*   Correspondence: motonari.kondo@med.toho-u.ac.jp; Tel.: +81-3-3762-4151
†   These authors contributed equally to this work.
‡   Current address: Department of Microbiology, Faculty of Medicine, Yamanashi University,
    Chuo 409-3898, Japan

Received: 29 June 2018; Accepted: 20 August 2018; Published: 25 August 2018

Abstract: Interleukin-7 (IL-7) is essential for lymphocyte development. To identify the functional subdomains in the cytoplasmic tail of the IL-7 receptor (IL-7R) α chain, here, we constructed a series of IL-7Rα deletion mutants. We found that IL-7Rα-deficient hematopoietic progenitor cells (HPCs) gave rise to B cells both in vitro and in vivo when a wild-type (WT) IL-7Rα chain was introduced; however, no B cells were observed under the same conditions from IL-7Rα-deficient HPCs with introduction of the exogenous IL-7Rα subunit, which lacked the amino acid region at positions 414–441 (d414–441 mutant). Signal transducer and activator of transcription 5 (STAT5) was phosphorylated in cells with the d414–441 mutant, similar to that in WT cells, in response to IL-7 stimulation. In contrast, more truncated STAT5 (tSTAT5) was generated in cells with the d414–441 mutant than in WT cells. Additionally, the introduction of exogenous tSTAT5 blocked B lymphopoiesis but not myeloid cell development from WT HPCs in vivo. These results suggested that amino acids 414–441 in the IL-7Rα chain formed a critical subdomain necessary for the supportive roles of IL-7 in B-cell development.

Keywords: interleukin-7; interleukin-7 receptor; B-cell development; signal transducer and activator of transcription 5

1. Introduction

Interleukin-7 (IL-7) provides signals via the IL-7 receptor (IL-7R), which are required during many stages of lymphocyte development [1,2]. IL-7R is composed of two different receptor subunits, IL-7Rα and common γ (γc) chains [3,4]. IL-7Rα is also a component of the thymic stromal lymphopoietin receptor complex, whereas γc is common to functional receptor complexes for IL-2, IL-4, IL-7, IL-9, IL-15, and IL-21 [5]. The developmental processes governed by IL-7R signaling include initiation of cell proliferation, protection from apoptotic cell death, and induction of lineage-specific events (i.e., gene rearrangement in antigen receptor loci) [6]. Mutations that interfere with IL-7R signaling cause profound immunodeficiency in both humans and mice [7–9], highlighting the central role IL-7 plays in lymphopoiesis.

Activation of signal transduction pathways via IL-7R begins with IL-7 binding to IL-7R complexes, followed by heterodimerization and conformational changes in IL-7Rα and γ$_c$ chains [10]. Next, Janus kinase 1 (JAK1) and JAK3 interphosphorylate, inducing kinase activity in both molecules. Activated JAK1 and JAK3 phosphorylate tyrosine residues on their substrates, including other kinases, adaptor molecules, and receptor subunits [11]. Phosphorylated tyrosine residues in the cytoplasmic domains of cytokine receptors can be docking sites for downstream signaling molecules with SH2 domains, such as phosphatidylinositol 3-kinase (PI3K) and signal transducer and activator of transcription 5 (STAT5) [11].

Four tyrosine residues are present in the cytoplasmic domain of mouse IL-7Rα. Among these four tyrosine residues, the third, Y449, is necessary for STAT5 binding to IL-7Rα [12,13]. Because the introduction of a constitutively active form of STAT5 in *IL-7Rα*$^{-/-}$ hematopoietic stem cells (HSCs) rescues impaired B-cell development [14], STAT5 is considered a critical signaling molecule for IL-7R signaling [12]. Indeed, B-cell development is completely shut down in 7RαYYFY knock-in mice, in which Y449 of IL-7Rα is substituted with phenylalanine (F449). In contrast, the number of early T-cell progenitors in 7RαYYFY knock-in mice is reduced, but the number of mature T cells in the periphery is close to normal. Therefore, other functional subdomains that play a role in the regulation of signal transduction via IL-7R should exist in the cytoplasmic tail of IL-7Rα.

In addition to the tyrosine residues, two motifs, namely Box1 and Box2, which are conserved in a number of cytokine receptors, are also recognized in the IL-7Rα chain [10,15,16]. These two subdomains are known to form binding sites of JAK family kinases. Therefore, if Box1 or Box2 is removed, the mutated cytokine receptors lack the ability to trigger signals upon binding of a cognate cytokine [10,15,16]. The acidic region, which was first identified in IL-2Rβ chain and forms a binding site for Src tyrosine kinases, such as Lck and Fyn, is also found in the IL-7Rα chain [17]. However, no other functional subdomains have been identified in the cytoplasmic tail of the IL-7Rα chain.

Therefore, in the present study, we constructed a series of IL-7Rα-deletion mutants to uncover hidden functional subdomains in the cytoplasmic tail of IL-7Rα chain. Our results provide important insights into B-cell development and IL-7R signal transduction.

2. Results

2.1. Identification of the IL-7Rα Cytoplasmic Regions that Are Necessary for IL-7-Mediated Cell Proliferation

We constructed a series of IL-7Rα deletion mutants to study the roles of different cytoplasmic regions in IL-7R signaling (Figure 1a). We divided the IL-7Rα cytoplasmic domain into nine segments and made deletion mutants lacking each segment. We also generated one mutant lacking all nine segments (dCyt). Two of the segments were chosen such that we could delete conserved motifs, such as Box1 (d280–307) and Box2 (d308–322) [18]. The d280–307 mutant lacked all amino acid residues between Box1 and Box2. Segments 379–396 and 397–413 were chosen so that only the first (Y390) or second (Y401) tyrosine residue was deleted (d379–396 and d397–413 mutants, respectively). The d414–441 mutant conserved all four tyrosine residues but lacked most of the amino acid residues between the second and third residues. Finally, we deleted the C terminal region of IL-7Rα, which contained the third (Y449) and fourth (Y456) tyrosine residues.

We screened each mutant using two different criteria. The first was the ability to transduce growth signals in an IL-2-dependent mouse T-cell line (CTLL-2 cells). The other was the potential to support B-cell development from hematopoietic progenitor cells (HPCs). In this case, we purified c-Kit$^+$ lineage$^-$ Sca-1$^+$ (KLS) HPCs from the bone marrow of *IL-7Rα*$^{-/-}$ mice and introduced wild-type (WT) and mutant IL-7Rα chains using a retroviral system. We used the nonfunctional IL-7Rα YYFY mutant as a negative control and the WT IL-7Rα chain as a positive control. After mutant IL-7Rα chains were introduced into *IL-7Rα*$^{-/-}$ HPCs, we cultured the *IL-7Rα*$^{-/-}$ HPCs with exogenous WT or mutant IL-7Rα subunits on OP9 cells in the presence of stem cell factor (SCF), Flt3 ligand, and IL-7 for 6 days. Then, we examined the presence of CD19$^+$ cells after culture.

As was the case with other cytokine receptor subunits, Box1 and Box2 were indispensable for IL-7Rα function based on the IL-7-mediated proliferation of CTLL-2 cells (see d272–279 and d308–322 mutants in Figure 1b). In addition, the d280–307 mutant, which lacked the region between Box1 and Box2, was not functional. Thus, dBox1, d280–307, and dBox2 mutants were not functional, perhaps due to the lack of association with JAK1 [19,20]. The d442–459 mutant was nonfunctional as well, presumably because of the lack of STAT5 binding in the absence of Y449. Although the d323–356 mutant showed lower functionality than the wild-type (WT) IL-7Rα chain, other mutants (i.e., d357–378, d379–396, d397–413, and d414–441) induced cell proliferation at a level comparable to that of the WT IL-7Rα chain (Figure 1b).

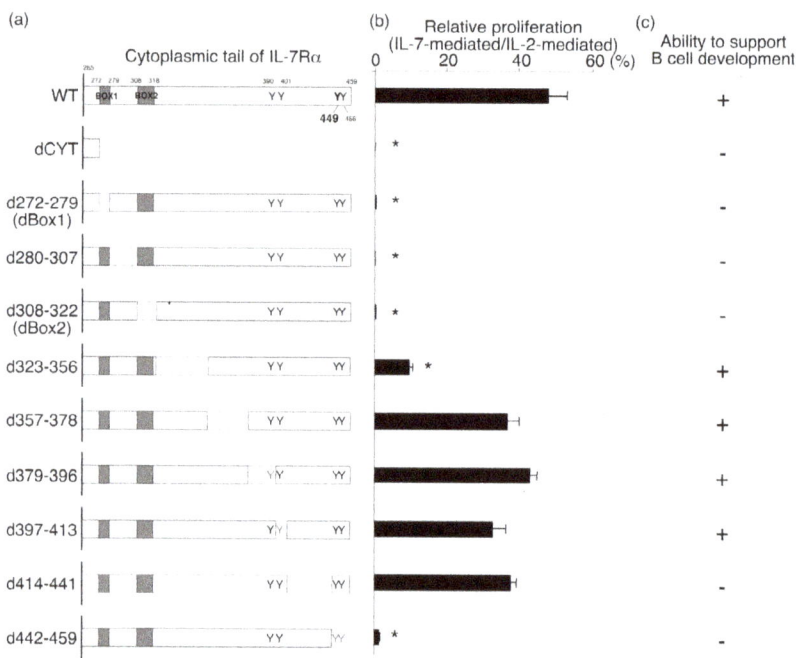

Figure 1. Identification of functional subdomains in the cytoplasmic tail of the IL-7Rα subunit. (**a**) Schematic structure of IL-7Rα deletion mutants. The positions of four tyrosine residues and Box1/2 in the intracytoplasmic domain of IL-7Rα are also indicated. The functionality of each mutant was tested by analysis of IL-7-mediated cell proliferation with CTLL-2 cells (**b**) and by appearance of CD19+ cells from IL-7Rα$^{-/-}$ c-Kit+ lineage$^-$ Sca-1+ (KLS) cells after the in vitro cultures (**c**). The data shown in (**b**) are representative of three independent experiments and normalized to the proliferation rate of each line in the presence of IL-2 to minimize variations in input cell numbers. Then, the proliferation rate of CTLL-2 cells with each mutant was compared to the rate of the cells with wild-type (WT) IL-7Rα chain (* $p < 0.05$).

2.2. The Amino Acids Region from Positions 414 to 441 of the IL-7Rα Subunit Was Necessary for B-Cell Development

There was no discrepancy between the ability to transduce growth signals and the potential to support B-cell development except for with the d414–441 mutant (Figures 1 and 2). Stagnation of B-cell development in IL-7Rα$^{-/-}$ HPCs with the d414–441 mutant occurred at the transition stage from pre-proB (B220+CD19$^-$) to proB (B220+CD19+) cells (Figure 2), where IL-7 stimulation was indispensable for the stage transition [14]. We also examined B-cell differentiation potential in

KLS cells with the d414–441 mutant in vivo by injecting the cells into 400 rad-irradiated *recombinant activating gene 2 (RAG2)*$^{-/-}$ mice. We found that no B cells were derived from *IL-7Rα*$^{-/-}$ HPCs with the d414–441 mutant, as was the case for *IL-7Rα*$^{-/-}$ HPCs with the nonfunctional IL-7Rα YYFY mutant at 5 weeks after injection (Figure 3). These results indicated that the amino acid region from positions 414 to 441 of IL-7Rα played a critical role in IL-7-dependent B-lymphocyte development.

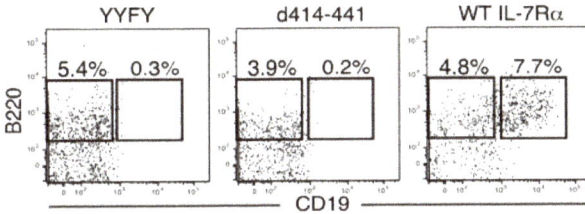

Figure 2. The d414–441 mutant did not support IL-7-mediated stage transition from the pre-proB (B220$^+$CD19$^-$) to proB stage (B220$^+$CD19$^+$) during B-cell development. KLS cells from *IL-7Rα*$^{-/-}$ bone marrow were infected with retroviruses expressing the YYFY mutant (negative control), d414–441 mutant, or WT IL-7Rα (positive control). Two days after infection, green fluorescent protein (GFP)-positive cells were purified by fluorescence-assisted cell sorting (FACS) and cultured on OP9 stromal cell layers in the presence of stem cell factor (SCF), Flt3 ligand, and IL-7 for 6 days. Cells were then stained with anti-B220 and anti-CD19 antibodies and analyzed by FACS.

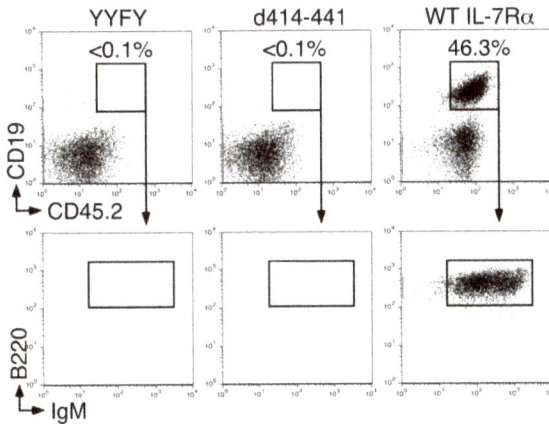

Figure 3. The d414–441 mutant did not support B-cell development in vivo. We infected *IL-7Rα*$^{-/-}$ KLS cells (CD45.2) with recombinant viruses, as indicated in the figure. GFP-positive cells were purified and injected into 400 rad-irradiated *RAG2*$^{-/-}$ (CD45.1) mice intravenously. Four weeks after injection, splenocytes from host mice were stained with anti-CD45.2, anti-CD19, anti-B220, and IgM antibodies and analyzed by FACS.

2.3. The Truncated Form of STAT5 (tSTAT5) Was Upregulated in CTLL-2 Cells with the d414–441 Mutant Compared with that in WT Cells after IL-7 Stimulation

The d414–441 mutant could transduce growth signals in CTLL-2 cells as efficiently as WT IL-7Rα (Figure 1) but did not support B-lymphocyte development (Figures 2 and 3). Because STAT5 plays a critical role in IL-7R signaling, we examined the phosphorylation status of STAT5 in CTLL-2 cells with the d414–441 mutant after IL-7 stimulation. As shown in Figure 4, the level of STAT5 phosphorylation in the d414–441 mutant was comparable to that in WT cells. Moreover, tSTAT5 was obviously upregulated in CTLL-2 cells with the d414–441 mutant than in cells expressing WT IL-7Rα (Figure 4).

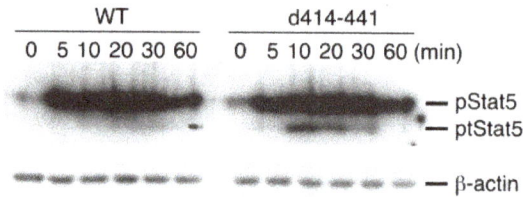

Figure 4. Signal transducer and activator of transcription 5 (STAT5) phosphorylation in response to IL-7 treatment in CTLL-2 cells expressing WT IL-7Rα and d414–441 mutant. Cells were starved of IL-2 for 8 h and stimulated with 100 ng/mL IL-7 for the indicated times. Whole cell lysates were subjected to sodium dodecyl sulfate polyacrylamide gel electrophoresis, and STAT5 activation was analyzed with immunoblotting using anti-phospho-STAT5 antibodies. Expression of β-actin was also examined with anti-β-actin antibodies as a loading control.

tSTAT5 can be generated by partial proteolysis after stimulation with cytokines, including IL-2 and IL-3. We previously demonstrated that tSTAT5 is a dominant-negative form of STAT5 [21]. Therefore, we examined whether B-cell development from HPCs was blocked in the presence of tSTAT5. For this purpose, we purified KLS cells as HPCs from WT mouse bone marrow. After the introduction of tSTAT5 (or control) in WT HPCs using a retroviral system, we injected these cells into RAG2$^{-/-}$ mice. We then examined spleen cells in RAG2$^{-/-}$ mice with HPCs with or without tSTAT5 at 5 weeks after injection. We found that B-cell development was severely impaired in the presence of tSTAT5, although the number of Mac-1$^+$ myeloid cells was not changed (Figure 5). These results suggested that the amino acid region from positions 414 to 441 of the IL-7Rα chain may form a docking site for the molecule, which inhibits the generation of tSTAT5 in cells after IL-7 stimulation.

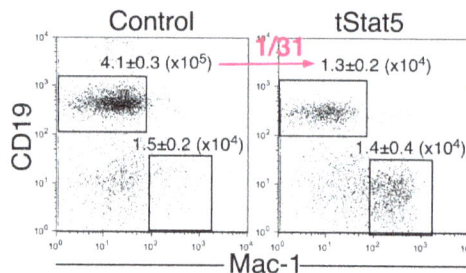

Figure 5. Truncated STAT5 (tSTAT5) had inhibitory functions in B-cell development. KLS cells from WT mice were infected with control or tSTAT5 retroviruses. These cells were injected into 400 rad-irradiated RAG2$^{-/-}$ mice. Recipient-derived cells in the spleens of host mice were analyzed at 5 weeks after injection. The numbers in the plots were the means of B and myeloid cell numbers from three mice.

3. Discussion

Activation of cytokine receptor signals is triggered by JAKs, resulting in phosphorylation of tyrosine residues of various signal molecules and the cytoplasmic tail of receptor subunits [22]. Therefore, among various protein modifications, phosphorylation is a main driver of signal cascades via cytokine receptors upon cognate ligand binding. However, other protein modifications are necessary for proper cytokine receptor signal transduction. For example, a number of molecules are acetylated in the cytoplasm, playing a role in positive regulation of interferon receptor signals [23]. Acetylated STAT1 and STAT2 exhibit enhanced transcriptional activity upon activation by tyrosine phosphorylation. Recently, we demonstrated that acetylation of JAK1, JAK3, and STAT5 occurs immediately in T cells in an IL-2-dependent manner [21]. This acetylation occurs via CREB binding protein (CBP), which relocates from the nucleus to the cytoplasm upon IL-2 stimulation. Acetylated STAT5 is a target of STAT5 protease, resulting in limited proteolysis and generation of tSTAT5. In this study, we demonstrated that tSTAT5 generation was increased after IL-7 stimulation if the 414–441 region in the IL-7Rα subunit was deleted. Therefore, we propose that the 414–441 region in the IL-7Rα subunit may form a binding site for inhibitors of CBP function. As shown in Figure 6, the 414–441 region contains multiple proline residues, which may serve as a docking site for the SH3 domain, although a conventional PxxP motif is absent. Accordingly, it is difficult to hypothesize which molecules may associate with the 414–441 region just based on the amino acid sequence.

```
        414        420                    430                  441
        * * * * P * P * * * Q * G I L - - * P * * Q * Q P I * T S *
mouse   P V P V P Q P L P F Q S G I L - - I P V S Q R Q P I S T S S
rat     N G T V P Q P F P L Q S G I L - - I P V S Q G Q P I S T S S
human   N S T L P P P F S L Q S G I L T L N P V A Q G Q P I L T S L
cat     N G P G P T P F P F Q S G I L T L N P A A Q G Q P I L T S L
cow     N S S L P P P F P F Q P G I L T L N P V A Q G Q P I L T S L
```

Figure 6. Comparison of the 414–441 region with the corresponding sites of IL-7Rα in various species. The 414-441 region contains proline residues (in red). These amino acids may provide docking site for the SH3 domain.

Various deletion mutants and point mutants of cytokine receptor subunits have been generated to identify the functional subdomains in the cytoplasmic tails of receptor subunits [24–26]. As a result, multiple signal pathways have been shown to be activated by different regions of the cytoplasmic tails of cytokine receptors. These studies have also shown that proliferation and differentiation signals are independent of one another. Accordingly, the d414–441 IL-7Rα mutant retained the potential to stimulate cell proliferation but lacked the ability to support B-cell development from HPCs. However, it is unclear why there was such a discrepancy despite the observation that STAT5 plays roles in both cell proliferation and support of B-cell development. In B-cell development, the first checkpoint at which IL-7 stimulation is required is the transition from the pre-proB cell population to the proB cell population. Expression of the transcription factor early B-cell factor (EBF) is indispensable for this stage transition [27]. We previously demonstrated that IL-7 stimulation is necessary for upregulation of EBF before entry to the proB cell stage [14]. Moreover, there was a threshold for EBF expression that was sufficient for the transition to the proB cell population from more immature cells. In addition, B-cell progenitors need to be stimulated with IL-7 before the pre-proB cell stage to enable sufficient EBF expression in response to IL-7 for the transition to the proB stage [28]. Therefore, one possible explanation for the discrepancy between proliferation and support of B-cell development by the d414–441 mutant may be related to the sensitivity of the mutant to the strength of STAT5 activity. Transcription levels of EBF could be more sensitive to STAT5 activity than proliferation. Furthermore, IL-7-induced EBF expression is mediated directly by STAT5, whereas cell proliferation is regulated by not only STAT5 but also other signaling components, such as PI3K, which associates with IL-7Rα at the region containing Y449 [29]. Since lack of a hypothetical molecule associated with the 414–441 region of the IL-7Rα chain may diminish IL-7R function in B-cell development, it is possible that the associated molecule plays a role in the regulation of B-cell number in vivo. Further investigations are necessary to determine why IL-7-stimulated cell proliferation and B-cell development have different requirements for the 414–441 region of IL-7Rα.

Notably, tSTAT5 inhibited IL-2-mediated proliferation of CTLL-2 cells, in contrast to IL-7-mediated cell growth. In all of our experiments, IL-2 was found to stimulate cell proliferation more strongly than IL-7; IL-7-driven cell proliferation levels which were only approximately 40% of that induced by IL-2. Therefore, cell proliferation in response to IL-2 may be more sensitive to negative effects, such as the presence of tSTAT5, than IL-7-mediated stimulation. Additionally, the effects of deletion of amino acids 414–441 from IL-7Rα may be different between T-cell development and B lymphopoiesis, as was the case in 7RαYYFY knock-in mice. Further studies using d414–441 mutant knock-in mice are needed to obtain insights into possible differential roles of the 414–441 region of IL-7Rα in T- and B-cell development. These future studies are expected to highlight the importance of acetylation in the regulation of signal transduction via cytokine receptors.

4. Materials and Methods

4.1. Mice

IL-7Rα$^{-/-}$ and RAG2$^{-/-}$ (CD45.1) mice on a C57Bl/6 background were bred, maintained under a specific pathogen-free environment at the Duke University Medical Center Animal Care Facility and the animal facility at Toho University School of Medicine, and used at 8–12 weeks of age. All studies and procedures were approved by the Duke University Animal Care and Use Committee (A246-07-09, 24 September 2009) and Toho University Administrative Panel for Animal Care (18-54-311, 1 April 2018) and Recombinant DNA (18-54-303, 1 April 2018).

4.2. Construction of Mutant IL-7Rα Subunits and Retrovirus Production

IL-7Rα deletion mutants shown in Figure 1 were generated by polymerase chain reaction (PCR) using full-length mouse IL-7Rα cDNA as a template and the following primers: 5′-tcaaggagg atgggatcc-3′ (common forward primer) and 5′-tacatagttgttccagagttttcttgacaggtttattcttttttttccat-3′ (272-279, reverse primer), 5′-cttcaacgcctttcacctcatgagtatgatcggggagactaggccatacg-3′ (280-307, reverse primer), 5′-gtgcaggaagatcattgggcagaaactggcagtccaggaaactttcggga-3′ (308-322, reverse primer), 5′-ctcttctaactgtttctggtgggctactttccacctcgtccctggcttca-3′ (323-356, reverse primer), 5′-tagaggaaaggagt ggaggggcattgctgactgaagtctcaggcgagcggtt-3′ (367-378, reverse primer), 5′-agtcttgatacacagg aggcttatta ttgcaggtactcagatttctagcc-3′ (379-396, reverse primer), 5′-atggttgagggacagggacagggaccctatttctgtcaccat ctctgtagtca-3′ (397-413, reverse primer), 5′-catacgcttcttcttgattcagtacgacatttgtgtttccagagtttggc-3′ (414-441, reverse primer), and 5′-tgaggaagtggagatgggctg-3′ (442-459, reverse primer). The IL-7Rα YYFY mutant, in which Y449 was substituted with F, was generated by site-directed mutagenesis. Mutants were confirmed by sequencing. All cDNAs were subcloned between the 5′-long terminal repeat and internal ribosome entry site (IRES) in the Murine Stem Cell Virus (MSCV)-IRES-green fluorescent protein (GFP) vector. IL-7Rα and tSTAT5 in the MSCV-IRES-GFP vector were described previously [14,21,30,31]. Retroviruses were prepared as described in [14].

4.3. Establishment of CTLL-2 Transfectants

CTLL-2 IL-2-dependent mouse T cells were cultured in complete medium (RPMI 1640 with 10% fetal calf serum [FCS] and 50 μM 2-mercaptoethanol) supplemented with 2 ng/mL hIL-2. CTLL-2 transfectants stably expressing WT or mutant IL-7Rα subunits were established with retroviral systems, followed by purification of GFP$^+$ cells by fluorescence-assisted cell sorting (FACS) as described previously [32]. All GFP$^+$ cells expressed exogenously introduced genes, as shown by staining for cell surface IL-7Rα.

4.4. Proliferation Assays

CTLL-2 transfectants were maintained in complete medium supplemented with hIL-2. After washing three times with phosphate-buffered saline, 5×10^4 cells in complete medium were cultured in each well of a 96-well plate in the presence of IL-7 (10 ng/mL) at 37 °C for 48 h. [^{3}H]-thymidine (1 μCi) was added to the culture 4 h before harvesting. Cells were harvested with an automatic cell harvester on a glass filter (Harvester 96; TOMTEC, Hamden, CT, USA). Radioactivity was determined USA).

4.5. Cell Sorting and FACS Analysis

Antibodies used in FACS sorting and analyses were as follows: phycoerythrin (PE)- or biotin-conjugated anti-IL-7Rα (A7R34); PE/Cy5- or allophycocyanin (APC)-conjugated anti-B220 (RA3-6B2); PE-anti-CD19 (6D5); fluorescein isothiocyanate (FITC)-, PE-, or PE-/Cy5-conjugated anti-Mac-1 (M1/70); PE/Cy5-conjugated anti-CD3 (145-2C11); anti-CD4 (RM4-5); anti-CD8 (53-6.7); anti-Gr-1 (RB6-8C5); anti-TER119 antibodies; and APC-conjugated anti-c-Kit (2B8); all antibodies were purchased from eBioscience (San Diego, CA, USA), Tombo (San Diego, CA, USA), or BD Bioscience (Mountan View, CA, USA). Alexa Fluor 594-anti-Sca-1 antibodies were prepared in our laboratory using standard procedures. Biotin-conjugated antibodies were visualized with PE-streptavidin (eBioscience).

The HPCs used in this paper were KLS cells, in which HSCs were highly enriched [33,34]. For cell surface phenotyping, cells were incubated with normal rat IgG (Sigma, St. Louis, MO, USA) and then fluorescence- or biotin-conjugated antibodies on ice for 20 min. If necessary, cells were further incubated with PE-streptavidin after washing with staining medium (Hanks' Balanced Salt solution (HBSS) with 2% FCS and 0.02% NaN$_3$). FACS analysis was performed using a FACSVantage with the DiVa option equipped with 488-nm argon, 599-nm dye, and 408-nm krypton lasers (BD Bioscience Flow Cytometry Systems) at the FACS facility of Duke University Comprehensive Cancer Center. A FACSAriaIII at the FACS facility of Toho University School of Medicine was also used for cell sorting. In addition, an LSRFortessa X-20 (BD Bioscience) was used for analyses. Data were analyzed with FlowJo software (BD Bioscience). Dead cells were excluded from analyses and sorting as cells showing positive staining with propidium iodide or 7-AAD.

4.6. Immunoblotting

CTLL-2 and its derivatives were cultured in complete medium without IL-2 for 6 h. These factor-starved cells were stimulated with a saturating amount of IL-7 (100 ng/mL) for the indicated times (Figure 4) at 37 °C. Cells were centrifuged and solubilized with lysis buffer (1% Triton X-100, 50 mM Tris-Cl, 300 mM NaCl, and 5 mM ethylenediaminetetraacetic acid) with protease inhibitor cocktail and phosphatase inhibitor cocktail (Sigma). After centrifugation, cell lysates were subjected to sodium dodecyl sulfate polyacrylamide gel electrophoresis and electrophoretically transferred to Immobilon-FL membranes (Millipore, Burlington, MA, USA). After blocking with 3% bovine serum albumin (BSA) in TBS-T (10 mM Tris-HCl, 150 mM NaCl, 0.1% Tween 20, pH = 7.5), membranes were incubated with anti-phospho-STAT5 (Tyr694) antibodies (Cell Signaling Technology, Danvers, MA, USA) in 1% BSA in TBS-T. After washing, membranes were further incubated with Alexa Fluor 680-conjugated anti-rabbit immunoglobulin (Molecular Probes, Eugene, OR, USA). Membranes were analyzed using an Odyssey infrared imaging system (LI-COR, Lincoln, NE, USA).

4.7. In Vitro and In Vivo Differentiation Assays

In vitro culture of HPCs was performed as described previously [28]. In brief, after retroviral infection, HPCs were cultured on OP9 cells in the presence of IL-7, Flt3 ligand, and SCF for 6 days. In vivo injections were performed as described in Reference [28] as well. In the experiment shown in Figure 3, HPCs were purified from *IL-7Ra*$^{-/-}$ mice (CD45.2) and were intravenously injected into 400 rad-irradiated *RAG2*$^{-/-}$ (CD45.1) mice. For the investigation shown in Figure 5, HPCs were purified from C57Bl/6 mice (CD45.2) and intravenously injected into 400 rad-irradiated *RAG2*$^{-/-}$ mice (CD45.1).

Author Contributions: M.K. conceived the study. H.K., T.K., and Y.M. designed and performed the experiments. K.N. supervised the study. H.K., T.K., Y.M. and M.K. analyzed the data and prepared the figures. T.K. and M.K. prepared the manuscript.

Funding: This work was supported by the Japan Society for the Promotion of Science Grant-in-Aid for Scientific Research (grant nos. 25460600 and 17K08892) to T.K.; a grant from the TAKEDA Science Foundation (to T.K.); a Strategic Research Foundation Grant-aided Project for Private School at Heisei 26th (S1411015) from the Ministry of Education, Culture, Sports, Science and Technology (to M.K.); and a Grant-in Aid for Private University Research Branding Project from the Ministry of Education, Culture, Sports, Science and Technology (to M.K.).

Acknowledgments: We would like to thank Editage (www.editage.jp) for English language editing. We would also like to dedicate this paper to Ton Rolink, who was a pioneer and established the knowledge base in the field of early hematopoiesis and B-cell development.

Conflicts of Interest: The authors declare no conflicts of interest.

Abbreviations

CBP	CREB-binding protein
CD	Cluster of differentiation
EBF	Early B cell factor
ETP	Early T cell progenitor
γ_c	Common γ
HPC	Hematopoietic progenitor cell
Lck	Lymphocyte-specific protein tyrosine kinase
IL	Interleukin
JAK	Janus kinase
KLS	c-Kit$^+$ Lineage$^-$ Sca-1$^+$
RAG	Recombination activating gene
Src	Rat sarcoma
SCF	Stem cell factor
STAT	Signal transducer and activator of transcription

References

1. Hofmeister, R.; Khaled, A.R.; Benbernou, N.; Rajnavolgyi, E.; Muegge, K.; Durum, S.K. Interleukin-7: Physiological roles and mechanisms of action. *Cytokine Growth Factor Rev.* **1999**, *10*, 41–60. [CrossRef]
2. Ceredig, R.; Rolink, A.G. The key role of IL-7 in lymphopoiesis. *Semin. Immunol.* **2012**, *24*, 159–164. [CrossRef] [PubMed]
3. Sugamura, K.; Asao, H.; Kondo, M.; Tanaka, N.; Ishii, N.; Ohbo, K.; Nakamura, M.; Takeshita, T. The interleukin-2 receptor gamma chain: Its role in the multiple cytokine receptor complexes and T cell development in XSCID. *Annu. Rev. Immunol.* **1996**, *14*, 179–205. [CrossRef] [PubMed]
4. Leonard, W.J. The molecular basis of X-linked severe combined immunodeficiency: Defective cytokine receptor signaling. *Annu. Rev. Med.* **1996**, *47*, 229–239. [CrossRef] [PubMed]
5. Rochman, Y.; Spolski, R.; Leonard, W.J. New insights into the regulation of T cells by γ_c family cytokines. *Nat. Rev. Immunol.* **2009**, *9*, 480–490. [CrossRef] [PubMed]
6. Corfe, S.A.; Paige, C.J. The many roles of IL-7 in B cell development; mediator of survival, proliferation and differentiation. *Semin. Immunol.* **2012**, *24*, 198–208. [CrossRef] [PubMed]
7. Puel, A.; Ziegler, S.F.; Buckley, R.H.; Leonard, W.J. Defective IL7R expression in T$^-$B$^+$NK$^+$ severe combined immunodeficiency. *Nat. Genet.* **1998**, *20*, 394–397. [CrossRef] [PubMed]
8. Peschon, J.J.; Morrissey, P.J.; Grabstein, K.H.; Ramsdell, F.J.; Maraskovsky, E.; Gliniak, B.C.; Park, L.S.; Ziegler, S.F.; Williams, D.E.; Ware, C.B.; et al. Early lymphocyte expansion is severely impaired in interleukin 7 receptor-deficient mice. *J. Exp. Med.* **1994**, *180*, 1955–1960. [CrossRef] [PubMed]
9. Von Freeden-Jeffry, U.; Vieira, P.; Lucian, L.A.; McNeil, T.; Burdach, S.E.; Murray, R. Lymphopenia in interleukin (IL)-7 gene-deleted mice identifies IL-7 as a nonredundant cytokine. *J. Exp. Med.* **1995**, *181*, 1519–1526. [CrossRef] [PubMed]
10. Kittipatarin, C.; Khaled, A.R. Interlinking interleukin-7. *Cytokine* **2007**, *39*, 75–83. [CrossRef] [PubMed]

11. Leonard, W.J. Role of Jak kinases and STATs in cytokine signal transduction. *Int. J. Hematol.* **2001**, *73*, 271–277. [CrossRef] [PubMed]

12. Jiang, Q.; Benbernou, N.; Chertov, O.; Khaled, A.R.; Wooters, J.; Durum, S.K. IL-7 induces tyrosine phosphorylation of clathrin heavy chain. *Cell Signal* **2004**, *16*, 281–286. [CrossRef]

13. Lin, J.X.; Migone, T.S.; Tsang, M.; Friedmann, M.; Weatherbee, J.A.; Zhou, L.; Yamauchi, A.; Bloom, E.T.; Mietz, J.; John, S.; et al. The role of shared receptor motifs and common Stat proteins in the generation of cytokine pleiotropy and redundancy by IL-2, IL-4, IL-7, IL-13, and IL-15. *Immunity* **1995**, *2*, 331–339. [CrossRef]

14. Kikuchi, K.; Lai, A.Y.; Hsu, C.L.; Kondo, M. IL-7 receptor signaling is necessary for stage transition in adult B cell development through up-regulation of EBF. *J. Exp. Med.* **2005**, *201*, 1197–1203. [CrossRef] [PubMed]

15. Tanner, J.W.; Chen, W.; Young, R.L.; Longmore, G.D.; Shaw, A.S. The conserved box 1 motif of cytokine receptors is required for association with JAK kinases. *J. Biol. Chem.* **1995**, *270*, 6523–6530. [CrossRef] [PubMed]

16. Jiang, Q.; Li, W.Q.; Hofmeister, R.R.; Young, H.A.; Hodge, D.R.; Keller, J.R.; Khaled, A.R.; Durum, S.K. Distinct regions of the interleukin-7 receptor regulate different Bcl2 family members. *Mol. Cell Biol.* **2004**, *24*, 6501–6513. [CrossRef] [PubMed]

17. Page, T.H.; Lali, F.V.; Foxwell, B.M. Interleukin-7 activates p56^{lck} and p59^{fyn}, two tyrosine kinases associated with the p90 interleukin-7 receptor in primary human T cells. *Eur. J. Immunol.* **1995**, *25*, 2956–2960. [CrossRef] [PubMed]

18. Murakami, M.; Narazaki, M.; Hibi, M.; Yawata, H.; Yasukawa, K.; Hamaguchi, M.; Taga, T.; Kishimoto, T. Critical cytoplasmic region of the interleukin 6 signal transducer gp130 is conserved in the cytokine receptor family. *Proc. Natl. Acad. Sci. USA* **1991**, *88*, 11349–11353. [CrossRef] [PubMed]

19. Zhu, M.H.; Berry, J.A.; Russell, S.M.; Leonard, W.J. Delineation of the regions of interleukin-2 (IL-2) receptor β chain important for association of Jak1 and Jak3. Jak1-independent functional recruitment of Jak3 to Il-2Rβ. *J. Biol. Chem.* **1998**, *273*, 10719–10725. [CrossRef] [PubMed]

20. Palmer, M.J.; Mahajan, V.S.; Trajman, L.C.; Irvine, D.J.; Lauffenburger, D.A.; Chen, J. Interleukin-7 receptor signaling network: An integrated systems perspective. *Cell Mol. Immunol.* **2008**, *5*, 79–89. [CrossRef] [PubMed]

21. Kuwabara, T.; Kasai, H.; Kondo, M. Acetylation Modulates IL-2 Receptor Signaling in T. Cells. *J. Immunol.* **2016**, *197*, 4334–4343. [CrossRef] [PubMed]

22. Leonard, W.J.; O'Shea, J.J. Jaks and STATs: Biological implications. *Annu. Rev. Immunol.* **1998**, *16*, 293–322. [CrossRef] [PubMed]

23. Tang, X.; Gao, J.S.; Guan, Y.J.; McLane, K.E.; Yuan, Z.L.; Ramratnam, B.; Chin, Y.E. Acetylation-dependent signal transduction for type I interferon receptor. *Cell* **2007**, *131*, 93–105. [CrossRef] [PubMed]

24. Ishihara, K.; Hirano, T. Molecular basis of the cell specificity of cytokine action. *Biochim. Biophys. Acta* **2002**, *1592*, 281–296. [CrossRef]

25. Minami, Y.; Taniguchi, T. IL-2 signaling: Recruitment and activation of multiple protein tyrosine kinases by the components of the IL-2 receptor. *Curr. Opin. Cell Biol.* **1995**, *7*, 156–162. [CrossRef]

26. Okuda, K.; Foster, R.; Griffin, J.D. Signaling domains of the βc chain of the GM-CSF/IL-3/IL-5 receptor. *Ann. N. Y. Acad. Sci.* **1999**, *872*, 305–312. [CrossRef] [PubMed]

27. Lin, H.; Grosschedl, R. Failure of B-cell differentiation in mice lacking the transcription factor EBF. *Nature* **1995**, *376*, 263–267. [CrossRef] [PubMed]

28. Kikuchi, K.; Kasai, H.; Watanabe, A.; Lai, A.Y.; Kondo, M. IL-7 specifies B cell fate at the common lymphoid progenitor to pre-proB transition stage by maintaining early B cell factor expression. *J. Immunol.* **2008**, *181*, 383–392. [CrossRef] [PubMed]

29. Rothenberg, E.V. Transcriptional control of early T and B cell developmental choices. *Annu. Rev. Immunol.* **2014**, *32*, 283–321. [CrossRef] [PubMed]

30. Kondo, M.; Scherer, D.C.; Miyamoto, T.; King, A.G.; Akashi, K.; Sugamura, K.; Weissman, I.L. Cell-fate conversion of lymphoid-committed progenitors by instructive actions of cytokines. *Nature* **2000**, *407*, 383–386. [CrossRef] [PubMed]

31. Hsu, C.L.; King-Fleischman, A.G.; Lai, A.Y.; Matsumoto, Y.; Weissman, I.L.; Kondo, M. Antagonistic effect of CCAAT enhancer-binding protein-α and Pax5 in myeloid or lymphoid lineage choice in common lymphoid progenitors. *Proc. Natl. Acad. Sci. USA* **2006**, *103*, 672–677. [CrossRef] [PubMed]

32. Hsu, C.L.; Kikuchi, K.; Kondo, M. Activation of mitogen-activated protein kinase kinase (MEK)/extracellular signal regulated kinase (ERK) signaling pathway is involved in myeloid lineage commitment. *Blood* **2007**, *110*, 1420–1428. [CrossRef] [PubMed]

33. Adolfsson, J.; Mansson, R.; Buza-Vidas, N.; Hultquist, A.; Liuba, K.; Jensen, C.T.; Bryder, D.; Yang, L.; Borge, O.J.; Thoren, L.A.; et al. Identification of Flt3$^+$ lympho-myeloid stem cells lacking erythro-megakaryocytic potential a revised road map for adult blood lineage commitment. *Cell* **2005**, *121*, 295–306. [CrossRef] [PubMed]

34. Christensen, J.L.; Weissman, I.L. Flk-2 is a marker in hematopoietic stem cell differentiation: A simple method to isolate long-term stem cells. *Proc. Natl. Acad. Sci. USA* **2001**, *98*, 14541–14546. [CrossRef] [PubMed]

International Journal of
Molecular Sciences

MDPI

Article

Mining the Plasma Cell Transcriptome for Novel Cell Surface Proteins

Stephanie Trezise [1,2,†], Alexander Karnowski [1,2,3,†], Pasquale L. Fedele [1,2,4],
Sridurga Mithraprabhu [5,6], Yang Liao [1,2], Kathy D'Costa [1], Andrew J. Kueh [1,2],
Matthew P. Hardy [3], Catherine M. Owczarek [3], Marco J. Herold [1,2], Andrew Spencer [5,6],
Wei Shi [1,7], Simon N. Willis [1,2], Stephen L. Nutt [1,2,*] and Lynn M. Corcoran [1,2]

[1]   The Walter and Eliza Hall Institute of Medical Research, Parkville, VIC 3052, Australia;
      trezise.s@wehi.edu.au (S.T.); fedele.p@wehi.edu.au (P.L.F.); liao@wehi.edu.au (Y.L.);
      dcostakj@gmail.com (K.D.); kueh@wehi.edu.au (A.J.K.); herold@wehi.edu.au (M.J.H.);
      shi@wehi.edu.au (W.S.); willis@wehi.edu.au (S.N.W.); corcoran@wehi.edu.au (L.M.C.)
[2]   Department of Medical Biology, University of Melbourne, Parkville, VIC 3010 Australia
[3]   CSL Limited, Parkville, VIC 3010, Australia; alexander.karnowski@csl.com.au (A.K.);
      Matt.Hardy@csl.com.au (M.P.H.); Catherine.Owczarek@csl.com.au (C.M.O.)
[4]   Haematology Department, Monash Health, Clayton, VIC 3168, Australia
[5]   Department of Clinical Haematology, Alfred Health, Melbourne 3004, Australia;
      durga.mithraprabhu@monash.edu (S.M.); aspencer@netspace.net.au (A.S.)
[6]   Australian Centre for Blood Diseases, Monash University, Melbourne 3004, Australia
[7]   Department of Computing and Information Systems, University of Melbourne,
      Parkville, VIC 3010, Australia
*   Correspondence: nutt@wehi.edu.au
†   These authors contributed equally to this study.

Received: 28 June 2018; Accepted: 20 July 2018; Published: 24 July 2018

Abstract: Antibody Secreting Cells (ASCs) are a fundamental component of humoral immunity, however, deregulated or excessive antibody production contributes to the pathology of autoimmune diseases, while transformation of ASCs results in the malignancy Multiple Myeloma (MM). Despite substantial recent improvements in treating these conditions, there is as yet no widely used ASC-specific therapeutic approach, highlighting a critical need to identify novel methods of targeting normal and malignant ASCs. Surface molecules specifically expressed by the target cell population represent ideal candidates for a monoclonal antibody-based therapy. By interrogating the ASC gene signature that we previously defined we identified three surface proteins, Plpp5, Clptm1l and Itm2c, which represent potential targets for novel MM treatments. *Plpp5*, *Clptm1l* and *Itm2c* are highly and selectively expressed by mouse and human ASCs as well as MM cells. To investigate the function of these proteins within the humoral immune system we have generated three novel mouse strains, each carrying a loss-of-function mutation in either *Plpp5*, *Clptm1l* or *Itm2c*. Through analysis of these novel strains, we have shown that Plpp5, Clptm1l and Itm2c are dispensable for the development, maturation and differentiation of B-lymphocytes, and for the production of antibodies by ASCs. As adult mice lacking either protein showed no apparent disease phenotypes, it is likely that targeting these molecules on ASCs will have minimal on-target adverse effects.

Keywords: plasma cell; antibody; CLPTM1L; ITM2C; PLPP5; membrane protein

1. Introduction

The differentiation of mature B cells into Antibody Secreting Cells (ASCs) is an essential part of the adaptive immune response and underlies virtually all current vaccination strategies. The antibodies that these cells produce are important for the elimination of infecting pathogens and the persistent secretion of these antibodies after pathogen clearance provides long-term protection against re-infection. However, the generation of self-reactive antibodies is the driver of many autoimmune diseases, including Systemic Lupus Erythematosus and Sjörgrens Syndrome and we currently lack effective methods of targeting the long-lived ASC population [1]. Furthermore, Multiple Myeloma (MM) is an ASC malignancy that constitutes approximately 10% of all hematological malignancies [2]. MM derives from the clonal expansion of a transformed post germinal center plasma cell, leading to bone marrow plasmacytosis and production of a monoclonal immunoglobulin. In addition to the consequences of effacement of normal bone marrow hematopoiesis resulting in anemia and other cytopenias, patients with MM develop specific end-organ sequale including renal failure, hypercalcemia and bone lesions [3]. Despite recent advances in treatment for this disease, through the use of proteasome inhibitors and immunomodulatory drugs, MM remains incurable and the median survival from diagnosis is only 5–6 years [2]. Recently, monoclonal antibody (mAb)-based therapies have been at the forefront of novel treatments for MM.

mAb-based immunotherapies are a highly specific and effective method of targeting a given cell type for depletion. The power of this therapy in the treatment of hematological cancers was first demonstrated by Rituximab, directed against CD20, a cell surface protein that is expressed on mature B cells, however, Rituximab is ineffective against MM and ASCs, as they no longer express CD20 [4,5]. There are several mAb therapies targeting MM cells that are currently in clinical use including Elotuzumab, and Daratumumab, which target Signaling lymphocytic activation molecule F7 (SLAMF7) [6,7] and CD38 [8], respectively. Unfortunately, the development of clones resistant to these therapies ultimately means that cancer regrowth is inevitable. Developing new targeted therapies for MM to use in combination with existing ones increases the likelihood that all clones could be targeted during treatment and would allow for more complete clearance of the cancer.

To increase our understanding of ASC biology and to enable us to identify novel therapeutic strategies to target pathogenic ASC, including MM, we have previously examined the transcriptional changes that occur during B cell differentiation in the mouse [9]. This study allowed us to define a robust ASC gene signature, a collection of genes consistently upregulated during the process of terminal differentiation. It is highly likely that this signature contains surface molecules, which are selectively expressed on ASCs and will consequently be suitable targets for an ASC directed immunotherapy. In agreement with this conclusion, all the current targets of mAb therapy for MM are found within the ASC gene signature, however, there is also a large collection of genes whose function remains either undetermined or that had not previously been associated with ASCs. We believe that some of these genes will present potential novel ways of targeting ASCs and MM.

We have identified three genes within the ASC gene signature, phospholipid phosphatase 5 (*Plpp5*), cleft-lip and palate transmembrane protein 1-like (*Clptm1l*) and integral membrane protein type 2 C (*Itm2c*), that are candidates for a mAb-based therapy to target pathogenic plasma cells. All three proteins display surface expression, and their highly conserved human homologues are expressed in both healthy ASCs and in MM cells. To identify possible side-effects of targeting these three proteins, we generated three novel stains of mice, each carrying a loss-of-function mutation in one of the three candidate genes. This work details the first analysis of *Plpp5*, *Clptm1l* or *Itm2c* function within immune cells.

2. Results

2.1. Identification of Candidate Cell Surface Proteins in Anibody Secreting Cells

We have previously generated gene expression profiles for mature B cells and ASC populations and identified a subset of genes, termed the ASC gene signature, which are upregulated during the process of B cell terminal differentiation [9]. From this signature, we searched the current literature for proteins with evidence of surface localization, resulting in a shortened list of 39 genes encoding membrane spanning proteins for which there is some evidence for cell surface localization (Figure 1A).

In addition to the established markers of plasma cells, including *Sdc1* (*Cd138*) and *Slamf7*, there were many genes with no known association with ASC biology, thus representing potential novel targets for an ASC specific therapy. From this shortened list we selected three genes for further investigation, *Plpp5*, *Clptm1l* and *Itm2c*. All three of these genes are poorly characterized and currently have no published association with any immune cell type. We next interrogated the Immgen Consortium database (www.immgen.org) to examine the expression of these three genes across a range of immune cell populations (Figure 1B) [10]. *Plpp5* and *Clptm1l* displayed high expression almost exclusively in ASC populations, while *Itm2c* was also highly expressed in dendritic cells. The selective expression of these genes suggests that they are candidates for a possible ASC-specific therapy.

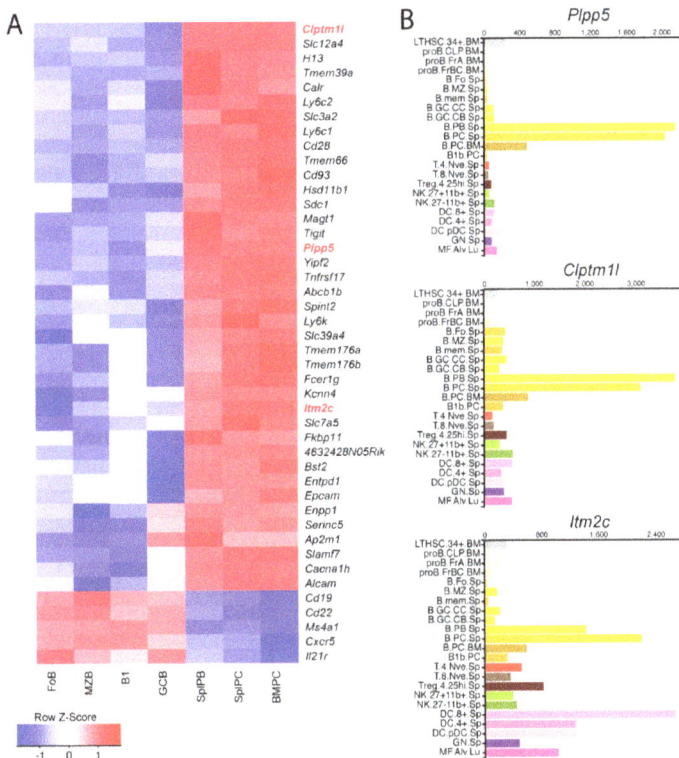

Figure 1. Identification of genes encoding novel surface proteins in mouse ASCs. (**A**) Expression profiles of genes within the ASC gene signature that encode transmembrane proteins that are either known or predicted to be expressed on the plasma membrane. The expression of five additional genes encoding cell surface proteins expressed in B cells, but not plasma cells is shown for comparison. The positions of *Plpp5*, *Clptm1l* and *Itm2c* are highlighted in red. Expression is represented as a Z-score as defined by the legend; (**B**) expression of *Plpp5*, *Clptm1l* and *Itm2c* in selected mouse immune cell populations. Data obtained from the Immgen Consortium. Expression value normalized by DEseq2. Immgen nomenclature: BM, bone marrow; Sp, splenic; PC, peritoneal cavity; Lu, lung; LTHSC.34+, CD34+ long-term hematopoietic stem cell; proB.CLP, common lymphoid progenitor; proB.FrA, pre-pro-B cell; proB.FrBC, pro-B cell; B.Fo, Follicular B cell; B.MZ, MZ B cell; B.mem, memory B cell; B.GC.CC, GC centrocyte; B.GC.CB, GC centroblast; B.PB., Plasmablast; B.PC, Plasma cell; T.4.Nve, naïve CD4+ T cell; T.8.Nve, naïve CD8+ T cell; Treg.4.25^{hi}, CD25^{hi} Treg; NK.27+11b−, CD27+ Cd11b− NK cell; DC.8+, CD8+ Dendritic Cell (DC); DC.4+, CD4+ DC; DC.pDC, plasmacytoid DC; GN, neutrophil; MF.Alv, alveolar macrophage.

2.2. Plpp5, Clptm1l and Itm2c Are Highly Conserved between Mice and Humans

Having identified Plpp5, Clptm1l and Itm2c as candidate ASC markers in the mouse, we next examined whether their sequences and expression patterns were conserved in humans. We performed pairwise sequence analysis of the mouse and human amino acid sequences for each of PLPP5, CLPTM1L and ITM2C, and found that they have sequence identity of 87.9%, 92.8%, and 92.9% respectively (Figure 2A–C). To determine whether *PLPP5*, *CLPTM1L* and *ITM2C* have similar expression patterns in mice and humans, we examined the expression of each gene in human B cell and ASC populations (Figure 2D). The pattern of expression of *CLPTM1L* and *ITM2C* during the terminal differentiation of both mouse and human B cells was very similar; low expression in B cell subsets, which increased markedly in ASC populations. *CLPTM1L* and *ITM2C* displayed the same pattern of expression as *SLC3A2*, *SDC1* and *SLAMF7*, which are current targets of ASC-directed immunotherapies. The expression of *PLPP5* differed between mice and humans, with expression in both naïve B cells and ASCs in humans while expression in mice was exclusive to ASCs. To determine whether the expression of these genes within human immune cell populations mirrored expression in the mouse we interrogated the BLUEPRINT consortium RNAseq database (http://www.blueprint-epigenome.eu) and observed that *PLPP5*, *CLPTM1L* and *ITM2C* expression was similarly restricted to B cells and ASCs (Figure 2E) [11]. The high degree of sequence identity and similar expression patterns suggests that it is likely that these genes serve a similar function in both mice and humans ASCs.

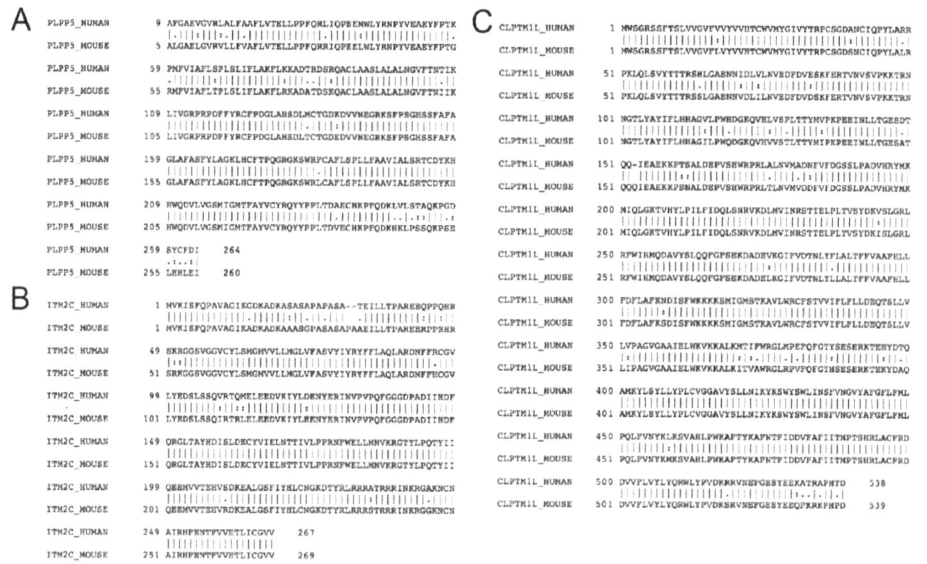

Figure 2. *Cont.*

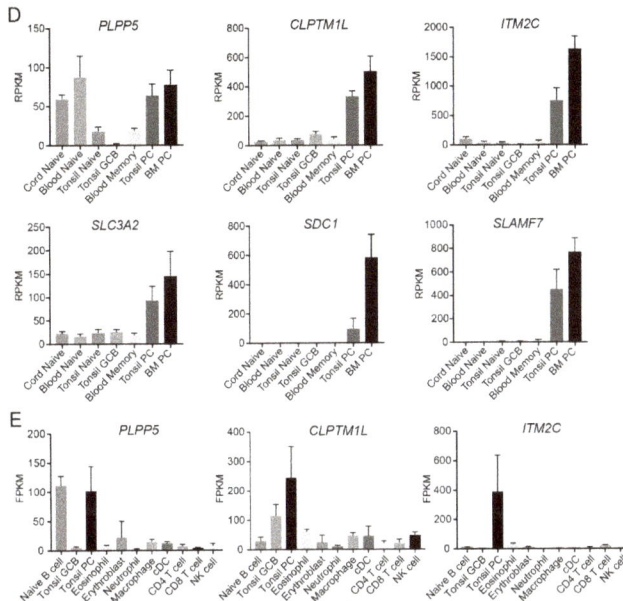

Figure 2. Expression of the human homologues in ASCs. Alignment of mouse and human amino acid sequences for (**A**) PLPP5, (**B**) ITM2C, and (**C**) CLPTM1L. Symbols indicate conserved (I), highly similar (:), and similar (.) residues. (**D**) RNAseq showing the expression of *PLPP5, CLPTM1L, ITM2C* and three established clinical candidates *SLC3A2, SDC1* and *SLAMF7* in purified human B cell and ASC populations. Data are the mean reads per kilobase per million reads (RPKM) ± SD from 3–5 donors each. Cord naïve B cell (CD19$^+$ CD20$^+$ HLA-DR$^+$ IgG$^-$), blood naïve B cell (CD19$^+$ IgD$^+$ CD27$^-$), tonsil naïve B cell (CD19$^+$ CD38$^-$ CD27$^-$), tonsil germinal center (GCB) B cell (CD19$^+$ CD38$^+$ CD27$^+$), blood memory B cell (CD19$^+$ IgD$^-$ CD27$^+$ CD38$^{-/low}$), tonsil plasma cell/plasmablast (PC, CD19$^+$ CD27^{++} CD38^{++}), bone marrow plasma cell (BM PC, CD138^{++} CD38$^+$). (**E**) RNAseq showing the expression of *PLPP5, CLPTM1L* and *ITM2C* in human immune cell populations. Data obtained from the BLUEPRINT consortium (www.blueprint-epigenome.eu). Data are the mean fragments per kilobase per million reads (FPKM) ± SD. Blood naïve B cell ($n = 5$, CD19$^+$ CD20$^+$ CD23$^+$ CD38^{low}), tonsil GCB cell ($n = 3$, CD19$^+$ CD20$^+$ CD38^{medium}), tonsil PC ($n = 21$, CD20^{medium} CD38^{++}), blood eosinophil ($n = 2$, CD66$^+$ CD16$^-$), blood erythroblast ($n = 8$, CD36$^+$ CD71$^+$ CD235a$^+$), blood neutrophil ($n = 16$, CD66b$^+$ CD16$^+$), cultured macrophages ($n = 18$) and conventional dendritic cell cDC ($n = 3$), blood CD4$^+$ T cell ($n = 10$, CD3$^+$ CD4$^+$ CD45RA$^+$), blood CD8$^+$ T cell ($n = 2$, CD3$^+$ CD8$^+$ CD45RA$^+$), blood natural killer (NK) cell ($n = 4$, CD3$^-$ CD56$^+$ CD16^{dim}).

2.3. Confirmation of Plpp5, Clptm1l and Itm2c Surface Expression and Prediction of Membrane Topology

Since there has been only limited analysis of the protein products of *PLPP5, CLPTM1L* and *ITM2C* we conducted an in silico analysis with several tools (Phobius, Spoctopus, TMHMM2 and TMMOD [12–15] to predict putative signal peptides, trans-membrane (TM) domains and membrane topology. The majority of the tools predicted murine and human CLPTM1L and PLPP5 being type IIIb membrane proteins, lacking signal peptides with six putative TM domains. Murine and human ITM2C was predicted to be a type II transmembrane protein without a signal peptide and one putative TM domain. To confirm surface expression and orientation we generated amino- and/or carboxy-terminal tagged expression constructs for each protein. Flow cytometric analysis of transiently expressed CLPTM1L and PLPP5, using tag-specific antibodies showed only surface staining of the N-terminal FLAG tag of PLPP5 (Figure 3A). However, both tags were detected in an intra-cellular stain (Figure 3B). Using anti-CLPTM1L or PLPP5 specific polyclonal antibodies that were raised against peptides localized within the central to N-terminal region, we detected surface expression of both transmembrane proteins (Figure 3C).

Carboxy-terminal-tagged ITM2C was detected on the cell surface of transiently transfected FS-293F cells by surface staining with a tag-specific antibody and an ITM2C specific antibody (Figure 3D). However, we observed about 30% of tag-positive cells in the tag-specific surface stain, but almost 60% of ITM2C positive cells using the specific antibody. This discrepancy is likely due to protein cleavage of the carboxy-terminal end of ITM2C [16]. The surface staining of human CLPTM1L and ITM2C fits well with the predicted topology (Figure 3E,F) and a previous study [17], demonstrating that these proteins are expressed on the cell surface. The cell surface detection of the N-terminal tag of PLPP5 suggests that the predicted trans-membrane domain at the N-terminal end is not utilized and that a 57 amino acid N-terminal region that contains the polyclonal antibody-binding site is exposed at the cell surface (Figure 3G).

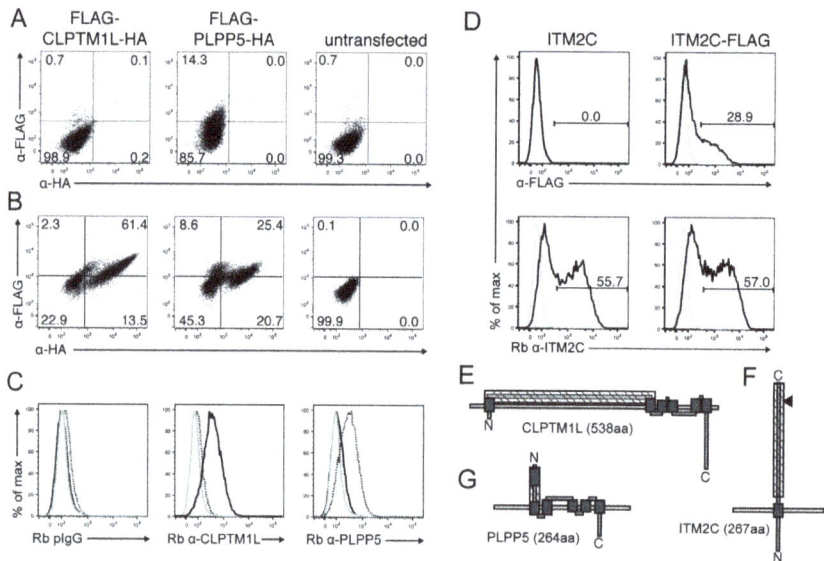

Figure 3. Cell surface expression of human PLPP5, CLPTM1L and ITM2C. FS-293F cells were transiently transfected with human CLPTM1L or PLPP5 expression constructs with N-terminal FLAG tags and C-terminal HA tags. (**A**) Cell surface or (**B**) Intra-cellular flow cytometric staining with anti-FLAG and anti-HA antibodies; (**C**) cell surface staining of transduced FS-293F cells with Rabbit (Rb) IgG control, Rb anti-CLPTM1L polyclonal (p)Ab and Rb anti-PLPP5 pAb (grey histograms: un-transfected cells, black lines: FLAG-Human-CLPTM1L-HA cells, dashed line: FLAG-Human-PLPP5-HA cells); (**D**) FS-293F cells were transiently transfected with a human ITM2C expression constructs with or without C-terminal FLAG tags. Flow cytometric analysis of surface anti-FLAG and Rb anti-ITM2C pAb staining (grey histograms: un-transfected cells, black lines: Human-ITM2C transfected cells); (**E–G**) predicted domain structure and cell surface localization of (**E**) CLPTM1L, (**F**) ITM2C and (**G**) PLPP5. (dark-grey boxes: putative TM-domains, hatched boxes: peptides used for pAb generation, black triangle Furin cleavage site).

2.4. PLPP5, CLPTM1L and ITM2C Are Highly Expressed in Multiple Myeloma

To determine whether PLPP5, CLPTM1L and ITM2C were potential targets in MM cells, we first analyzed the expression of each gene in a range of MM cells lines using qPCR (Figure 4A). The expression of *PLPP5*, *CLPTM1L* and *ITM2C* varied between different established human myeloma cell lines, however, all cell lines examined had detectable expression of at least one of these genes of interest.

We next determined the expression of *PLPP5*, *CLPTM1L*, *ITM2C* and *SLAMF7* (as a positive control) in primary MM and plasma cell (PC) samples (Figure 4B). All four genes showed significantly increased expression in MM samples compared to PCs. To confirm that there was a corresponding increase in protein expression we examined the expression of PLPP5 in a selection of MM cell lines (Figure 4C). In agreement with our qPCR data, all examined cell lines showed cell surface expression of PLPP5.

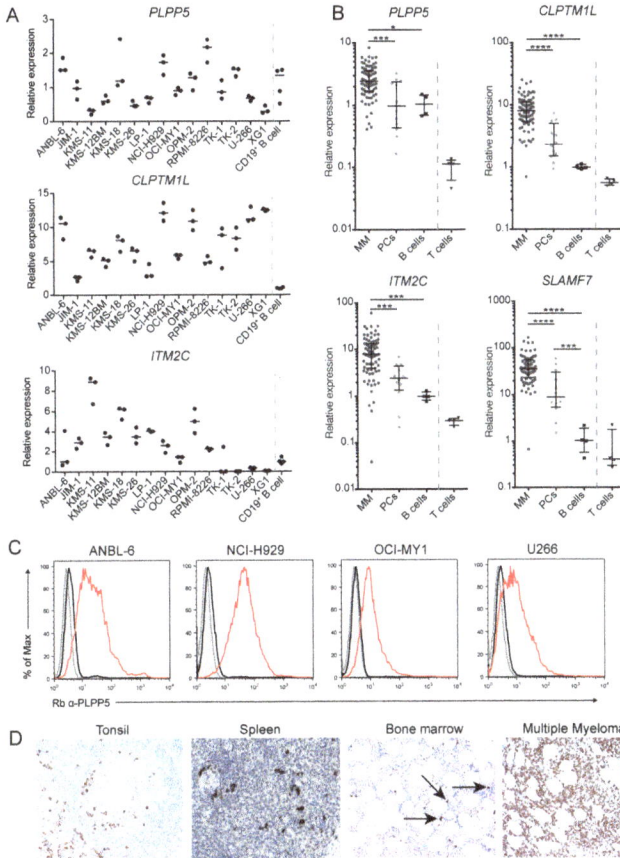

Figure 4. *PLPP5*, *CLPTM1L* and *ITM2C* are expressed in human Multiple Myeloma. (**A**) Expression of *PLPP5*, *CLPTM1L*, and *ITM2C* in human myeloma cell lines and primary human CD19$^+$ B cells as determined by qPCR. Expression data was normalized by the house-keeping genes *GAPDH*, *GPBP1* and *TPT1*. Bars show median normalized expression of experimental replicates ($n = 3$ myeloma cell lines, $n = 4$ for CD19$^+$ primary peripheral B cells); (**B**) expression of *PLPP5*, *CLPTM1L*, *ITM2C* and *SLAMF7* in primary bone marrow MM samples ($n = 83$), sorted plasma cells ($n = 12$), peripheral blood B cells ($n = 4$) and T cells ($n = 4$). Expression data was normalized as in (**A**). Bars show median expression and interquartile range. One-way ANOVA with Tukey's multiple comparison test on log transformed data from MM, plasma cells and B cells; * ($p < 0.05$), *** ($p < 0.001$), **** ($p < 0.0001$). (**C**) cell surface expression of PLPP5 in human myeloma cell lines was determined by flow cytometry. Grey histograms: unstained cells, black lines: anti-rabbit (Rb) IgG control, red lines: Rb anti-PLPP5; (**D**) expression of CLPTM1L in human tonsil, spleen and bone marrow trephine from patients with non-Hodgkin's Lymphoma (labeled, Bone marrow) or multiple myeloma. Spleen image was obtained from the Human Protein Atlas v18 (www.proteinatlas.org/ENSG00000049656-CLPTM1L/tissue/spleen#img). Images were captured at 100× magnification.

Finally, we used a commercially available anti-CLPTM1L antibody to perform immunohistochemistry on human tonsil and on bone marrow from patients with either non-Hodgkin's lymphoma (non-involved bone marrow) or MM (Figure 4D). Staining revealed that there were a small number of CLPTM1L-positive cells in the bone marrow and in the tonsil surrounding a follicle, which was the expected location and prevalence for ASCs. Our staining was in agreement with images obtained from the Human Protein Atlas [18], which showed a small number of anti-CLPTM1L-labeled cells in human spleen. Staining of bone marrow from a patient with MM revealed widespread CLPTM1L-positive cells, further supporting our qRNA analysis showing high *CLPTM1L* expression in MM. Together, these results confirm that *PLPP5*, *CLPTM1L* and *ITM2C* are expressed in human ASCs and MM, and are therefore candidates for ASC-directed immunotherapy.

2.5. Generation of Plpp5, Clptm1l and Itm2c Knock-Out Mouse Models

As noted above PLPP5, CLPTM1L and ITM2C are poorly characterized proteins, whose functions are unknown, although mutation in *CLPTM1L* has been associated with several human cancers (see Discussion). To interrogate the roles of these proteins within ASCs we generated knock-out mouse strains, each carrying loss-of-function mutations in either *Plpp5*, *Clptm1l* or *Itm2c*.

The *Plpp5* deficient mice were generated using embryonic stem (ES) cells obtained from the KOMP Repository (Figure 5A). The ES cells carried an IRES-LacZ cassette replacing exons 2–6 of the *Plpp5* gene. The ES cells were injected into day 3.5 blastocysts to generate chimeric mice, which were then bred to generate heterozygous and ultimately homozygous progeny. PCR amplification using primer pairs that spanned the insertion site confirmed the targeting vector was inserted correctly (Figure 5B). To demonstrate that the insertion of the targeting vector disrupted the *Plpp5* locus, we measured the abundance of *Plpp5* mRNA transcripts in lipopolysaccharide (LPS) stimulated wild type (WT), $Plpp5^{+/-}$ and $Plpp5^{-/-}$ B cell cultures (Figure 5C). This analysis confirmed that there was no detectable *Plpp5* mRNA in the culture from $Plpp5^{-/-}$ mice.

We generated *Clptm1l* mutant mice using ES cells obtained from the EUCOMM Consortium (Figure 5D). The construction of the targeting vector meant that the inactivation of *Clptm1l* could be performed conditionally ("flox" allele) or constitutively ("del" allele). As before, the ES cells were used to generate chimeric mice. These mice were crossed to Flp recombinase-expressing mice to excise the Neo-LacZ selection cassette and produce the functional $Clptm1l^{flox}$ allele. $Clptm1l^{flox/+}$ mice were crossed to a strain expressing Cre recombinase in the germline to generate the $Clptm1l^{del}$ allele. The excision of exon three creates a frame shift mutation with a stop codon immediately downstream. The correct structure of each allele was confirmed by PCR (Figure 5E). WT, $Clptm1l^{del/+}$ and $Clptm1l^{del/del}$ B cells were stimulated with LPS and the abundance of $Clptm1l$ mRNA in each culture was measured to confirm the absence of $Clptm1l$ expression in del/del mice (Figure 5F). Curiously, we observed the abundance of $Clptm1l$ mRNA was greater in the $Clptm1l^{del/+}$ cells than in the WT cell (Figure 5F). To investigate a potential consequence of this finding, our subsequent studies also included $Clptm1l^{del/+}$ mice.

We generated an *Itm2c* null mouse de novo using CRISPR/Cas9-mediated gene deletion. Guide RNAs were designed to target intronic sequences flanking exons 2 and 5, resulting in the deletion of the intervening exons (Figure 5G). This region was amplified by PCR to confirm that the desired resection event had occurred (Figure 5H). As before, we stimulated naïve B cells from WT, $Itm2c^{+/-}$ and $Itm2c^{-/-}$ mice with LPS and measured the abundance of *Itm2c* mRNA in each culture (Figure 5I). *Itm2c* transcripts were not detectable in the cultures produced from $Itm2c^{-/-}$ cells.

$Plpp5^{-/-}$, and $Itm2c^{-/-}$ mice were viable and showed no overt physical defects or disease phenotype. $Plpp5^{-/-}$ and $Itm2c^{-/-}$ pups were born at the expected frequencies from $+/- \times +/-$ matings, suggesting that the loss of either gene did not have a detrimental impact on viability. In contrast, $Clptm1l^{del/del}$ mice were born at a significantly lower than expected frequency from heterozygous mating pairs (Table 1). Additionally, fewer than 15% of the $Clptm1l^{del/del}$ pups that were born, survived past day 2 (Table 2). Curiously, those $Clptm1l^{del/del}$ pups that survived the neo-natal period appeared healthy and indistinguishable from their littermates. Together, this suggests that Clptm1l has a function that is important for the survival of embryonic and neonatal mice, which warrants further investigation in the future. Both male and female $Plpp5^{-/-}$, $Clptm1l^{del/del}$ and $Itm2c^{-/-}$ mice were able to produce viable pups, suggesting that none of these genes play an important role in fertility.

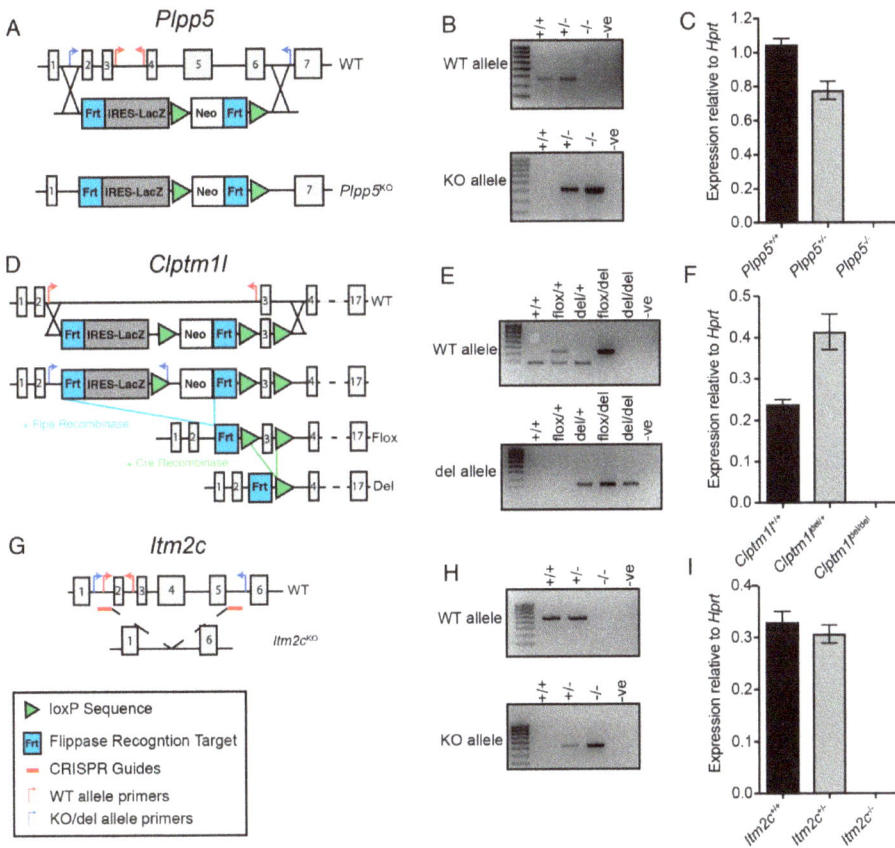

Figure 5. Generation of *Plpp5*, *Clptm1l* and *Itm2c* deficient mouse models. (**A**) Strategy for the generation of the non-functional (−) *Plpp5* allele; (**B**) gel electrophoresis of PCR products amplified from WT, $Plpp5^{+/-}$ and $Plpp5^{-/-}$ pups. −ve, negative control water only; (**C**) mRNA expression of *Plpp5* in WT, $Plpp5^{+/-}$ and $Plpp5^{-/-}$ B cell cultures following 4 days of LPS stimulation; (**D**) strategy for the generation of the conditional (flox) and non-functional (del) *Clptm1l* alleles; (**E**) gel electrophoresis of PCR products amplified from WT, $Clptm1l^{flox/+}$, $Clptm1l^{del/+}$, $Clptm1l^{flox/del}$ and $Clptm1l^{del/del}$ pups; (**F**) mRNA expression of *Clptm1l* in WT, $Clptm1l^{del/+}$ and $Clptm1l^{del/del}$ B cell cultures following 4 days of LPS stimulation; (**G**) strategy for the generation of the non-functional (−) *Itm2c* allele using CRISPR. The location of the small guide RNAs are indicated; (**H**) gel electrophoresis of PCR products amplified from WT, $Itm2c^{+/-}$ and $Itm2c^{-/-}$ pups; (**I**) mRNA expression of *Itm2c* in WT, $Itm2c^{+/-}$ and $Itm2c^{-/-}$ B cell cultures following 4 days of LPS stimulation. All mRNA expression was determined using qRT-PCR and normalized to *Hprt* expression. Data are the mean ± SD of technical triplicates. Genomic structures in (**A**,**D**,**G**) indicate the numbered exons of each gene (gray boxes) and the mutant alleles generated. The targeting constructs for *Plpp5* and *Clptm1l* are shown below the WT genes. The positions of genotyping primers, LoxP and Frt sites are indicated.

Table 1. Genotypes of offspring from $Clptm1l^{del/+}$ with $Clptm1l^{del/+}$ mating.

Genotype:	$Clptm1l^{+/+}$	$Clptm1l^{del/+}$	$Clptm1l^{del/del}$
Expected	13.25	26.5	13.25
Observed	21	29	3 *

$* p = 0.00175.$

Table 2. Survival of $Clptm1l^{del/del}$ pups.

Total Number Born	Number Survived Past Day 2
37	5

2.6. Characterization of Plpp5, Clptm1l and Itm2c Deficient Mice

RNA-seq data from the Immgen consortium showed that *Plpp5*, *Clptm1l* and *Itm2c* were not highly expressed in hematopoietic stem cells, lymphoid progenitors or developing B cell populations (Figure 1B), making it unlikely that the loss of either protein would impact B cell development and maturation. To ensure that B cell development was normal we analyzed B cell maturation stages in the spleen and bone marrow of each mutant strain. As expected, we did not observe differences in the frequency or number of precursor ($B220^+IgM^-$), immature ($B220^{lo}IgM^+$), or recirculating mature ($B220^{hi}IgM^+$) B cells in the bone marrow of WT mice compared with $Plpp5^{-/-}$, $Clptm1l^{del/del}$ or $Itm2c^{-/-}$ mice (Supplementary Figure S1A,B). Furthermore, we did not detect any changes in the frequency or total number or naïve mature ($B220^+IgD^{hi}IgM^+$) or immature ($B220^+IgD^{lo}IgM^{hi}$) B cell in the spleens of *Plpp5*, *Clptm1l* or *Itm2c* deficient mice compared to WT mice (Supplementary Figure S1C,D).

To determine whether the loss of *Plpp5*, *Clptm1l* or *Itm2c* influenced the generation of ASCs in vivo, we analyzed the steady state ASC populations in both the BM and spleen in each of the mutant mouse strains. We crossed each strain to Blimp-1-GFP mice [19], which allowed us to differentiate between the PC and the plasmablast (PB) populations in the spleen. We did not observe any marked difference in the frequency of BM PCs ($CD138^+Blimp-1-GFP^+$), Spl PCs ($CD138^+Blimp-1-GFP^{hi}$) or Spl PBs ($CD138^+Blimp-1-GFP^{int}$) between WT and the $Plpp5^{-/-}$, $Clptm1l^{del/+}$, $Clptm1l^{del/del}$ or $Itm2c^{-/-}$ mice. (Figure 6A,B). We also did not detect a significant difference in the serum concentrations of IgM, IgG1, IgG2b, IgG2c, IgG3 or IgA in the mice deficient in *Plpp5*, *Clptm1l* or *Itm2c* when compared with WT controls (Figure 6C).

To determine whether the loss of Plpp5, Clptm1l or Itm2c would alter the response to stimulation, we isolated naïve B cells from mice of each genotype and cultured them under multiple T cell-dependent (CD40 Ligand (CD40L) + Interleukin (IL)-4 ± IL-5), T cell-independent (LPS) or mixed (LPS + IL-4) stimulation conditions. We did not observe any difference in the differentiation of B cells to ASCs from any of the mutant mice under any of the conditions tested (Supplementary Figure S2A). Additionally, there was no observable defect in the ability of *Plpp5*, *Clptm1l* or *Itm2c* deficient B cells to undergo immunoglobulin class-switch recombination under any of the tested stimulation conditions (Supplementary Figure S2B). Interestingly, we observed a significant increase in the proportion of $Clptm1l^{del/+}$ B cells that had undergone class-switch recombination following T cell-dependent stimulation when compared to WT cells. We next used retroviral transduction to ectopically express each gene in activated B cells and found that this premature expression did not impact on the efficiency of ASC differentiation or class-switch recombination in LPS stimulated B cells (Supplementary Figure S2C,D).

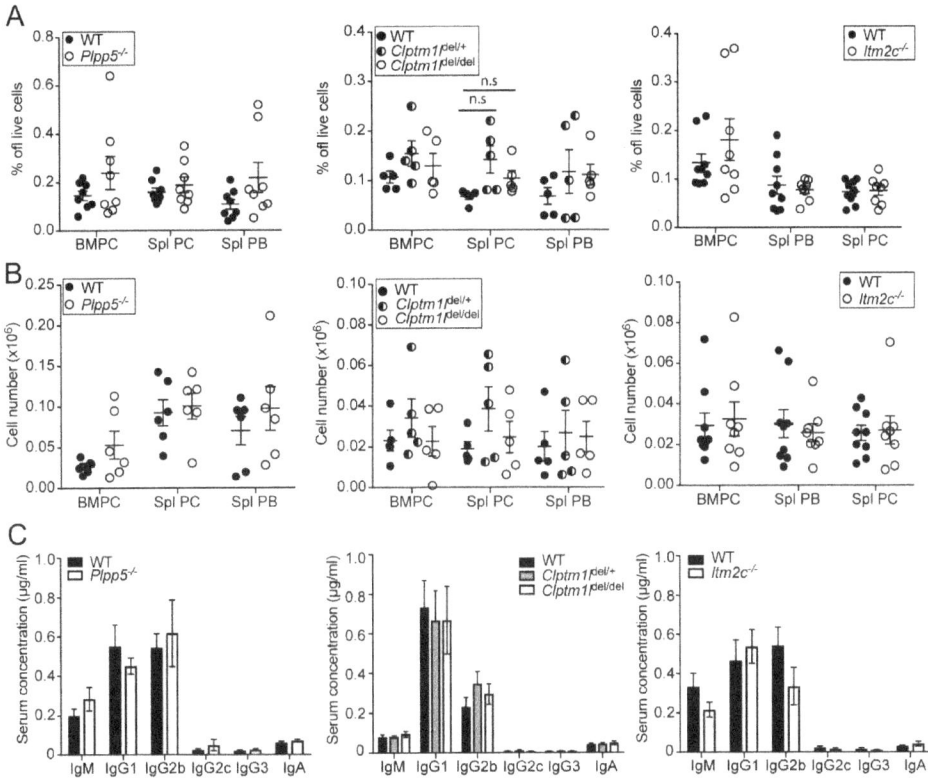

Figure 6. Characterization of ASC populations in $Plpp5^{-/-}$, $Clptm1l^{del/+}$, $Clptm1l^{del/del}$ and $Itm2c^{-/-}$ mice. (**A**) Frequency and; (**B**) total cell number of bone marrow plasma cells (BMPC), splenic plasma cells (Spl PC) and splenic plasmablasts (Spl PB) in $Plpp5^{-/-}$, $Clptm1l^{del/+}$, $Clptm1l^{del/del}$, $Itm2c^{-/-}$ and age-matched WT mice. Results are combined from two ($Clptm1l$) or three ($Plpp5$, $Itm2c$) independent experiments. Each dot represents a single mouse, horizontal lines show the means $\pm$ SEM; (**C**) serum concentration of immunoglobulin isotypes in $Plpp5^{-/-}$, $Clptm1l^{del/+}$, $Clptm1l^{del/del}$, $Itm2c^{-/-}$ and age- and sex-matched WT mice as determined by ELISA. Data are the mean of six samples from two independent experiments $\pm$ SEM. Statistical significance was analyzed using unpaired *t*-test, correcting for multiple comparisons. n.s., not significant ($p > 0.05$).

Finally, we examined the ability of *Plpp5*, *Clptm1l* and *Itm2c* deficient mice to form an antigen specific response following immunization. We immunized $Plpp5^{-/-}$, $Clptm1l^{del/del}$, $Itm2c^{-/-}$ and age-matched WT controls with 4(hydroxy-3-nitrophenyl) acetyl coupled to Keyhole Limpet Hemocyanin (NP-KLH) and determined the frequency of antigen-specific ASCs at multiple timepoints post-immunization (Figure 7). No significant difference was observed in the number of NP-specific ASCs in $Plpp5^{-/-}$ and $Itm2c^{-/-}$ mice when compared to WT controls at any examined timepoint. Unfortunately, due to the survival disadvantage in the $Clptm1l^{del/del}$ mice we were unable to obtain enough age-matched $Clptm1l^{del/del}$ mice to do multiple immunization timepoints. However, NP-specific ASCs were detectable in $Clptm1l^{del/del}$ mice 3 months post-immunization, indicating that *Clptm1l* deficient mice are capable of mounting a long-lived response to immunization. Together, these data demonstrate that *Plpp5*, *Clptm1l* and *Itm2c* are dispensable for the differentiation of B cells into ASCs, both in vivo and in vitro.

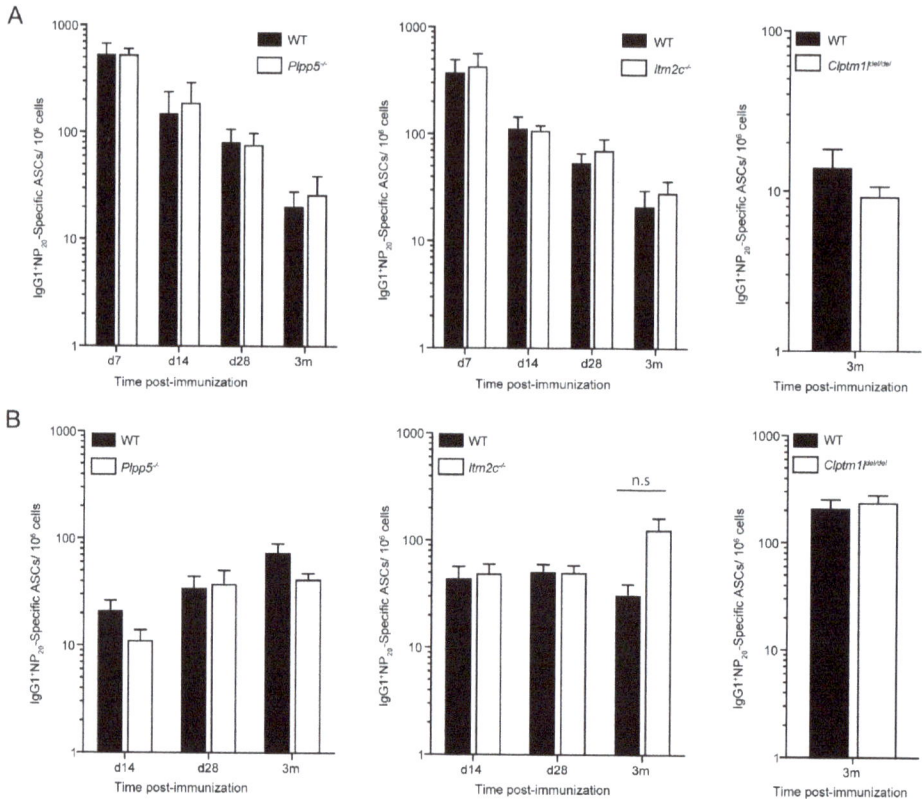

Figure 7. Characterization of the antigen-specific antibody response of *Plpp5*$^{-/-}$ *Clptm1l*$^{del/del}$ and *Itm2c*$^{-/-}$ mice. Mice were immunized intraperitoneally with NP-KLH in alum and, at the indicated time post-immunization, the number of NP20-specific IgG1+ ASCs in the (**A**) spleen and (**B**) BM was determined by ELISpot. *Plpp5* data are combined from three independent experiments at each time point where n = 3–6. *Itm2c* data are combined from 1 (day (d) 7) or 2 (d14, d28, 3 months (m)) independent experiments where n = 3–4. *Clptm1l* data are from 1 experiment where n = 4. Data are the mean ±SEM. Statistical significance was analyzed using unpaired t-test, correcting for multiple comparisons. n.s., not significant ($p > 0.05$).

3. Discussion

Despite recent improvements in the treatment of MM, it remains an incurable disease. The use of mAb therapies in the treatment of MM has been promising, with recent approvals for anti-SLAMF7 and anti-CD38 mAbs [20,21]. However, the presence of resistant MM cells means that patients often relapse. This highlights the continuing need for novel methods to target MM cells as combination therapies that kill MM cells through multiple approaches increases the likelihood that complete clearance of the cancer can be achieved. Surface molecules present exciting candidates for immunotherapies as they allow for the selective targeting of a cell population either through mAbs or, more recently, through chimeric antigen receptor T cells, which combine the specificity of mAbs with the cytotoxic capabilities of conventional CD8$^+$ T cells [20]. Therefore, we chose to focus on surface proteins that are expressed in ASCs and consequently MM. In this study we have examined three proteins, PLPP5, CLPTM1L and ITM2C, which are ASC surface proteins that we believe present promising candidates for an ASC-directed immunotherapy, but whose biological functions were largely unknown.

PLPP5 (also known as PPAPDC1B and HTPAP) encodes a lipid phosphatase that has been shown to be present on the plasma membrane and within the cytoplasm of PLPP5 overexpressing human hepatocellular carcinoma cell lines [22,23]. Although the function of PLPP5 has not been determined, it has been proposed to have an oncogenic role in breast cancer. *PLPP5* is estimated to be amplified in 10–15% of ductal breast carcinomas, and its knock-down in breast cancer cell lines causes an increase in apoptosis [24,25]. Similar observations have also been reported in pancreatic adenocarcinoma and small-cell lung cancer cell lines [25]. Within the mouse immune system we found that *Plpp5* expression was highly restricted to ASCs. Although its expression in ASCs and MM cell was conserved in human samples, we also observed considerable *PLPP5* expression in naïve, but not germinal center or memory B cells, suggesting that any mAb therapy based on PLPP5-binding and depletion would also target naïve B cells. As B cell depletion using Rituximab has been successfully used for many years in lymphoma [26] and some autoimmune contexts [27], the expression of PLPP5 in B cells is not likely to be an insurmountable obstacle for an anti-PLPP5-based therapy for MM or Ab mediated autoimmunity.

CLPTM1L is a close homologue of Cleft lip and palate transmembrane protein 1 (CLPTM1) and is predicted to contain six transmembrane domains [17]. There is conflicting data regarding the subcellular localization of CLPTM1L, with previous studies observing it exclusively in the mitochondria [28] or endoplasmic reticulum [29], while the data presented here and another recent study [17], have found evidence for localization to the plasma membrane. It is possible that the location of CLPTM1L is dependent on cell context or that aberrant localization is a result of the high levels of expression seen in the examined cancer cell lines. Mutations in *CLPTM1L* have been identified as risk factors in a range of cancers, including lung [30,31], pancreatic [32], colorectal [33], glioma [34] and testicular germ cell cancer [35], and the expression of *CLPTM1L* is significantly increased in cancerous cells when compared to healthy adjacent tissues [29,30,36]. This increase in expression in malignant cells was also evident in our analysis of primary MM samples compared with plasma cells. The knock-down of *CLPTM1L* with siRNA or targeting with an anti-CLPTM1L antibody increased the sensitivity of *CLPTM1L* overexpressing cancer cells to killing through genotoxic stress inducing agents [17,30]. Together, these observations suggest that targeting CLPTM1L in MM could potentially increase sensitivity to other treatments, however, this remains to be tested.

Curiously, we observed that *Clptm1l*[del/del] pups appeared to not only be born at lower than anticipated frequencies, but to have an initial survival disadvantage compared to their *Clptm1l*[+/+] and *Clptm1l*[del/+] littermates. *Clptm1l*[del/del] mice that survived past two days old showed no sign of physical defect or fitness disadvantage, suggesting that Clptm1l plays an important role in early life but is dispensable later on. Future investigation into the role of Clptm1l in adult mice should utilize the conditional strain that we have generated to avoid the complication of the poor neonatal survival that we have observed.

ITM2C is a member of the type 2 integral transmembrane family that has previously been reported to be localized to the plasma membrane [37], Golgi apparatus [38] and within lysosomes [39]. Previous work has focused on its role within the brain, as it is highly expressed in both the embryonic and adult mouse brain, as well as in adult human brain tissue [40,41]. Within the immune system, we observed strong expression of ITM2C in mouse and human ASCs, but also expression in dendritic cells and some macrophage populations. Whether these non-ASC expression domains will impact on the utility of any anti-ITM2C mAb therapy remains to be determined, but it is noteworthy that the two clinically approved mAb for MM, Daratumumab (CD38) and Elotuzumab (SLAMF7) target antigens expressed on ASC and multiple other cell types.

ITM2C is thought to have a role in the inhibition of the β-amyloid protein processing pathway based on the observations that it is capable of directly binding the β-amyloid precursor protein and *Itm2c* expression is inversely correlated with β-amyloid peptide production [38,42]. Due to the high expression of *Itm2c* within the embryonic mouse brain we were surprised to find that *Itm2c*[−/−] pups were viable and showed no evidence of a fitness disadvantage when compared to littermates.

Additionally, *Itm2c* is expressed in the testis during sexual maturation [43], however, *Itm2c*$^{-/-}$ male mice were able to produce pups indicating that the loss of *Itm2c* does not influence fertility.

Despite the increase in expression of *Plpp5*, *Clptm1l* and *Itm2c* during B cell differentiation, we did not identify any of these genes as being essential for either the generation of ASCs in steady state, either in vitro or in vivo, or for the ability of ASCs to secrete antibody. The presence of long-lived plasma cells in the spleen and bone marrow of all three KO mice suggests that these genes are also not required for the long-term survival of ASCs, or for homing to and retention within the bone marrow. This conclusion is supported by the presence of normal numbers of antigen-specific ASC after a T-dependent immunization. Whether this is due to functional redundancy with related proteins is at present unclear. However, it is noteworthy that the ITM2 family consists of three members (a, b, c) that show around 40% amino acid identity [40] and similar expression domains in late B cells (www.immgen.org) raising the possibility of redundancy between ITM2 family members.

This work describes the first investigation into the roles of PLPP5, CLPTM1L and ITM2C within the immune system. Furthermore, this work details the first time mutant mouse strains of each gene have been described and presents them as valuable tools for the investigation into the roles of these genes within other cell populations of interest, particularly in the context of cancer. With their surface localization and their high expression within both ASCs and MM samples, PLPP5. CLPTM1L and ITM2C represent enticing candidates for novel immunotherapies to target these cell populations. Adult mutant mice showed no apparent fitness defect or disease phenotype, which suggests that targeting these molecules will have minimal on target side effects. The next stage in the investigation of these proteins will be the generation of mAbs, first to allow for further investigation into the biology of PLPP5. CLPTM1L and ITM2C and then to determine their potential for therapeutic use.

4. Materials and Methods

4.1. Mice

Mice were bred and maintained on a C57BL/6 background and housed in a specific pathogen free facility. *Plpp5* mutant mice were generated using *Plpp5*$^{tm1(KOMP)Mbp}$ embryonic stem (ES) cells obtained from the KOMP Repository. *Clptm1l* mice were made using Clptm1l$^{tm1a\ (EUCOMM)Hmgu}$ ES cells obtained from the EUCOMM Consortium. In the case of *Itm2c*, we originally obtained targeted ES cells from EUCOMM (ID:69805, clones EDP0351_4_B07 and _F05). No germline transmission occurred with the B07 ES cells, while the F05 ES cells contained a gene duplication at the *Itm2c* locus and were thus not useful for to generate a loss-of-function allele. As an alternative, *Itm2c* mice were generated in house using CRISPR/Cas9. Guide sequences were AACTGCTAAAGAGGGTGGTC and GGTCGACATTCACTATAGTC. In the first generation, mice were generated that bore *Itm2c* deletions with micro heterogeneities around the deletion endpoints. As founders, we chose to mate a male and female that carried identical mutated sequences. Primers for genotyping were as follows: Plpp5WT-F (GTCTTAGTGTTGGCAAGTAGCTATGGG), Plpp5WT-R (CCATCTGCTTGG AGAAGAGTAAGCC), Plpp5KO-F (GCCTGTCAATCTTCCCCGTTTCCTCCCC), Plpp5KO-R (GGTGAGAGGAGAATTCTGGAATCCATCC), ClptWT-F (TCCTATTCATCACCCTGTGCCAGG), ClptWT-R (CCCACCTCTGTTAGAGCCTCAGACTAC), ClptDel-F (TCCTATTCATCACCCTGTGC CAGG), ClptDel-R (CTGATGGCGAGCTCAGACCATAACTTCG), Itm2cWT-F (AAATTCGGGCTG ATTGTTTG), Itm2cWT-R (CCTAAGAGCTCCTGGTGACG), Itm2cKO-F (TTCCCATGAACTCCTT GGTC), Itm2cKO-R (GCGAGGCAAGTGAGGTAGAC). Blimp-1-GFP [19], Flp [44] and Cre [45] recombinase expressing mice have been described previously. All animal experiments were conducted in accordance with protocols approved by the Walter and Eliza Hall Institute Animal Ethics Committee (2016.002, approved 31 March 2016).

4.2. Bioinformatic Analysis

RNAseq data for mouse immune cells derives from (GSE60927) [9] and the Immunological Genome Consortium database (www.immgen.org). RNA-seq from human hematopoietic cell populations derives from the BLUEPRINT consortium (www.blueprint-epigenome.eu) and Fedele et al. (unpublished). Pairwise sequence alignments between the mouse and human protein sequences were performed using EMBOSS Needle [46]. In silico analysis of the protein structures used the Phobius [13], Spoctopus [15], TMHMM2 [14] and TMMOD [12] tools.

4.3. Generation, Expression and Detection of Tagged Proteins

The coding sequence of the predominant isoform of human *CLPTM1L* (NP_110409) and *PLPP5* (NP_001096029) were synthesized at GeneArt (Thermo Fisher, Waltham, MA, USA) inframe with a 5′ FLAG-tag coding sequence (DYKDDDDK) and 3′ a HA-tag coding-sequence (YPYDVPDYA). *ITM2C* (NP_071862) coding sequence was synthesized with a 3′ FLAG-tag coding sequence. The coding sequences were cloned into a pCDNA 3.1-based expression vector and transiently transfected into FreeStyle 293-F (FS-293F) using 293fectin (Gibco, Thermo Fisher, Waltham, MA, USA) according to manufacturer's protocol. Expression was enhanced by adding Lucratone Lupin (Millipore-Sigma, Burlington, MA, USA) at 0.5% v/v final concentration 18 h post-transfection. FS293-S cells were cultured in Freestyle 293 Expression Media (Gibco). Cells were analyzed 48 h post-transfection.

For intra-cellular staining, cells were fixed and permeabilized using BDCytofix/Cytoperm system according to the manufacturer's guidelines (BD Biosciences, Franklin Lakes, NJ, USA). Cells were blocked with FcR Blocking reagent (130-059-901, Miltenyi Biotec, Bergisch Gladbach, Germany) and stained using the following antibodies: Mouse anti-HA-Alexa Fluro-488 mAb (2350, Cell Signaling Technology, Danvers, MA, USA), Rabbit anti-FLAG polyclonal (p)Ab (2368, Cell Signaling Technology), Rabbit anti-Human-CLPTM1L pAb (GTX116893, Genetex, Irvine, CA, USA), Rabbit anti Human-ITM2C pAb (GTX116904, Genetex) and Rabbit anti-Human-PLPP5 (LS-B4624, LifeSpan BioSciences, Seattle, WA, USA). Rabbit IgG isotype control (GTX35035, Genetex) was used in some experiments. Unconjugated rabbit pAb were detected with secondary antibody anti-Rabbit IgG Alexa Flour 647 (4414, Cell Signaling Technology). Cells were analyzed on FACSCanto (BD Biosciences).

4.4. Patient Samples

BM from MM patients (newly diagnosed, relapsed/refractory) or healthy donors was obtained following written informed consent as per Alfred Hospital Human Ethics Committee-approved protocol. Isolation of bone marrow mononuclear cells (BMMNC), determination of MM cell proportion and isolation of CD138+ cells from MM cells and plasma cells from healthy donors was performed as previously described [47]. Briefly, Ficoll Plaque Plus (GE Healthcare, Chicago, IL, USA) was utilized to isolate BMMNC as per manufacturer's guidelines. Red blood cells were removed using red blood cell lysis buffer (10 mmol/L $KHCO_3$, 150 mmol/L NH_4Cl and 0.1 mmol/L EDTA, pH 8.0) for 5 min at 37 °C followed by washing with sterile phosphate buffered saline (PBS). The proportion of MM or normal plasma cells (CD38+CD45−CD138+) in BMMNC isolated from each patient was determined through flow cytometric enumeration on a FACSCalibur Flow Cytometer (BD Biosciences). To isolate MM cells, anti-CD138 MACS beads were employed using manufacturer's guidelines (Miltenyi Biotec). CD138+ cells were selected through magnetic isolation using an MS-column (Miltenyi Biotec). For normal BM, flow cytometry was performed to ensure that proportion of plasma cells fell within the normal range (<3%). Normal plasma cells were isolated through the utilization of plasma cell isolation kit II (Miltenyi Biotec). Purified cells were stored in TRIzol (Life Technologies, Thermo Fisher, Waltham, MA, USA). Human B cells and T cell controls were purified from peripheral blood mononuclear cells (PBMCs) derived from buffy coats using anti-CD19-MACS beads or anti-CD3-MACS beads respectively. All samples were de-identified for this study.

Protocols were approved by the Alfred Hospital Human Ethics Committee, Monash Health and the Walter and Eliza Hall Institute Human Research Ethics Committees.

4.5. Flow Cytometry

4.5.1. Mouse

Single cell suspensions were labeled with the following mAbs: anti-CD138 (281-2), anti-B220 (RA3-6B2), anti-IgG1 (X56), all from BD Biosciences; anti-IgD (11–26c.2a) and anti-IgM (II/41), both from eBioScience (Affymetrix, Santa Clara, CA, USA). Cells were analyzed on FACSCanto Flow Cytometer (BD Biosciences).

4.5.2. Human

Human multiple myeloma cell lines were labeled with Rabbit anti-Human-PLPP5 pAb (LS-B4624, LifeSpan BioSciences) or Rabbit IgG isotype control (GTX35035, Genetex). Unconjugated pAb was detected with secondary antibody anti-Rabbit IgG Alexa Flour 647 (4414, Cell Signaling Technology). Cells were analyzed on FACSCanto (BD Biosciences).

4.6. Multiple Myeloma Cell Lines

Human multiple myeloma cell lines were grown in RPMI supplemented with 10% FCS, 1% L-Glutamine, 1% HEPES, 1% non-essential amino acids, 1% Sodium pyruvate, 50 μM β-mercaptoethanol. Recombinant Human IL-6 (R&D Systems, Minneapolis, MN, USA) was supplied when necessary. 1×10^6 cells were harvested in experimental triplicates or duplicates over subsequent passages for RNA purification.

4.7. Immunohistochemistry

Paraffin-embedded (de-identified) archival samples were de-waxed, and antigen retrieval was performed (120 C under 20 psi pressure, 3 min, 1 mM EDTA, pH 9.0). Blocking was with 5% FCS in PBS, 15 min. Primary rabbit anti-CLPTM1L antibody (HPA014791, Sigma-Aldrich, St. Louis, MO, USA) was added at 1:100, incubated at RT for 15 min, washed with 5% FCS in PBS, then the secondary antibody (goat anti-rabbit Ig-HRP, sc-2004, 1:500; Santa Cruz Biotechnology, Dallas, TX, USA) was added and incubated for 40 min, washed and developed using standard protocols.

4.8. B Cell Isolation and Cell Culture

Naïve splenic B cells were isolated using a B Cell Isolation Kit (Miltenyi Biotech,) and cultured in RPMI (supplemented with 10% FCS, 1% L-Glutamine, 1% HEPES, 1% non-essential amino acids, 1% Sodium pyruvate, 50 μM β-mercaptoethanol) with combinations of 20 μg/mL lipopolysaccharide (LPS, Sigma-Aldrich) and 100 ng/mL CD40 ligand (CD40L), 10 ng/mL Interleukin-4 (IL-4) and 5 ng/mL Interleukin-5 (IL-5) (all from R&D Systems).

4.9. RNA Isolation and qRT-PCR

4.9.1. Mouse

Naïve B cells were culture for 4 days in the presence of LPS before being resuspended in RLT lysis buffer (QIAGEN, Venlo, Netherlands) and transferred to a QIA shredder column (QIAGEN) for lysis and homogenization. RNA extraction was performed using an RNeasy Plus Mini Kit (QIAGEN). cDNA was generated using iScript reverse transcription supermix (BioRad, Hercules, CA, USA). qPCR was performed using Taqman probes (*Hprt*: Mm00446968_m1, *Clptm1l*: Mm00524746_m1, *Plpp5*: Mm01210970_m1, *Itm2c*: 00499081_m1) and TaqMan Universal Mastermix all from Applied Biosystems.

4.9.2. Human

Primary cells and myeloma cell lines were homogenized in RLT lysis buffer using the QIA shredder columns (QIAGEN). RNA extraction for myeloma cell lines and PBMC derived control B and T cells was performed using the RNeasy mini kit (QIAGEN) including a genomic DNA digestion step. RNA extraction from primary human plasma cells and MM cells was performed using a RNeasy micro kit (QIAGEN). cDNA was generated using SSIII First-Strand cDNA synthesis kit (Invitrogen, Thermo Fisher Scientific, Waltham, MA, USA). qPCR used Taqman probes (Applied Biosystems, Foster City, CA, USA): *CLPTM1L*: Hs00363947_m1, *ITM2C*: Hs00985194_g1, *PLPP5*: Hs00998335_g1, *SLAMF7*: HS00221793_m1, *GPBP1*: Hs00607556_m1, *GAPDH*: Hs02758991_g1, *TPT1*: Hs01044518_g1. Raw qPCR data was imported into LinReg [48] software to determine Cq and reaction specificities. Final analysis was conducted in qBASEplus (Biogazelle, Zwijnaarde, Belgium).

4.10. ELISA

Blood for serum antibody analysis was collected from all mice at 49 days of age. Plates were coated with anti-mouse IgM, IgG1, IgG2b, IgG2c, IgG3 or IgA (Southern Biotech, Birmingham, AL, USA) for 24 h before the addition of diluted serum. IgM, IgG1, IgG2b, IgG3 standards were obtained from Sigma-Aldrich, IgG2c standard was from Southern Biotech, and IgA standard was from Organon Tekcika–Cappel (Durham, NC, USA). Serum Ig was detected with anti-IgM-HRP, anti-IgG1-HRP, anti-IgG2b-HRP, anti-IgG2c-HRP, anti-IgG3-HRP, anti-IgA-biotin and streptavidin-HRP (Southern Biotech) and visualised using ABTS substrate (2,2′-Azinobis (3-ethylbenzthiazoline Sulfonic Acid); Sigma-Aldrich). All samples and standards were measured in duplicate.

4.11. Retroviral Transduction of B Cells

Plasmids containing pMD1-gag-pol, pCAG-Eco, and pMIG (expressing either GFP alone (empty vector) or murine *Plpp5*, *Itm2c* or *Clptm1l* and GFP) were transfected into 293T cells using the Calcium phosphate method. Retroviral supernatant was collected after 48 h and transferred to B cells that had been pre-stimulated with LPS for 24 h. Transduction was performed using a spin infection in the presence of polybrene (4 μg/mL). Transduced cells were then cultured in LPS for a further 3 days before analysis.

4.12. Immunization

4(hydroxy-3-nitrophenyl) acetyl coupled to Keyhole Limpet Hemocyanin (NP-KLH) (Biosearch Technologies, Petaluma, CA, USA) at concentration of 1 mg/mL was added to Imject Alum (Thermo Scientific) at a ratio of 1:1. 200 μL was injected into mice intraperitoneally. All mice were at least 7 weeks old at the time of immunization.

4.13. ELISpot

Multiscreen HA plates (Millipore-Sigma) were coated with 10 μg/mL NP20-BSA for 4 h before the addition of cells. Plates were then incubated for 14–18 h at 37 °C 10% CO_2. NP-specific antibody was detected with anti-IgG1-HRP and visualized using 3-amino-9-ethylCarbazole (Sigma-Aldrich).

4.14. Statistical Analysis

Gene expression in human MM, plasma cells and B cells were analyzed using one-way ANOVA with Tukey's multiple comparison test on log transformed data. Data from *Plpp5*$^{-/-}$ *Itm2c*$^{-/-}$ and *Clptm1l*$^{del/del}$ experiments were analyzed using unpaired t-tests using the Holm–Sidak method to correct for multiple comparisons.

Supplementary Materials: Supplementary materials can be found at http://www.mdpi.com/1422-0067/19/8/2161/s1.

Author Contributions: Conceptualization, S.T., A.K., S.N.W., S.L.N. and L.M.C.; Investigation, S.T. and A.K.; Methodology, P.L.F., S.M., Y.L., K.D., A.J.K., M.P.H., C.M.O., M.J.H., W.S. and L.M.C.; Supervision, M.J.H., A.S., W.S., S.N.W., S.L.N. and L.M.C.; Writing—original draft, S.T., A.K., S.L.N. and L.M.C.; Writing—review & editing, S.T., A.K., P.L.F., S.M., Y.L., A.J.K., M.J.H., Andrew Spencer, W.S., S.N.W., S.L.N. and L.M.C.

Funding: This project was supported by a Cancer Council Victoria Grant-in-Aid (1192662 to S.N. and L.C.), research grants from the National Health and Medical Research Council of Australia (1054925 to S.N. and L.C., 1058238 to S.N.). P.L.F. was supported by a Leukaemia Foundation of Australia Clinical PhD Scholarship and an RCPA Foundation Postgraduate Research Fellowship, S.N.W. by The Walter and Eliza Hall Trust Centenary Fellowship, W.S. was supported by a Walter and Eliza Hall Centenary Fellowship sponsored by CSL. This work was made possible through Victorian State Government Operational Infrastructure Support and NHMRC Independent Research Institute Infrastructure Support Scheme.

Acknowledgments: We thank Carola Vinuesa and Ilenia Papa (John Curtin School of Medical Research, Canberra Australia), Simon Harrison (Peter MacCallum Cancer Centre, Melbourne Australia), Andrew Roberts (Walter and Eliza Hall Institute, Melbourne Australia) and the Volunteer Blood Donor Registry (Walter and Eliza Hall) for samples and advice. This study makes use of data generated by the BLUEPRINT Consortium. A full list of the investigators who contributed to the generation of the data is available from www.blueprint-epigenome.eu. Funding for the project was provided by the European Union's Seventh Framework Programme (FP7/2007–2013) under grant agreement no 282510—BLUEPRINT.

Conflicts of Interest: A.K., M.H. and C.O. are employed by CSL Ltd., who funded part of the study. All other authors declare no conflict of interest.

References

1. Hiepe, F.; Radbruch, A. Plasma cells as an innovative target in autoimmune disease with renal manifestations. *Nat. Rev. Nephrol.* **2016**, *12*, 232–240. [CrossRef] [PubMed]
2. Siegel, R.L.; Miller, K.D.; Jemal, A. Cancer statistics, 2015. *CA Cancer J. Clin.* **2015**, *65*, 5–29. [CrossRef] [PubMed]
3. Kyle, R.A.; Rajkumar, S.V. Criteria for diagnosis, staging, risk stratification and response assessment of multiple myeloma. *Leukemia* **2009**, *23*, 3–9. [CrossRef] [PubMed]
4. Reff, M.; Carner, K.; Chambers, K.; Chinn, P.; Leonard, J.; Raab, R.; Newman, R.; Hanna, N.; Anderson, D. Depletion of B cells in vivo by a chimeric mouse human monoclonal antibody to CD20. *Blood* **1994**, *83*, 435–445. [PubMed]
5. DiLillo, D.J.; Hamaguchi, Y.; Ueda, Y.; Yang, K.; Uchida, J.; Haas, K.M.; Kelsoe, G.; Tedder, T.F. Maintenance of long-lived plasma cells and serological memory despite mature and memory B cell depletion during CD20 immunotherapy in mice. *J. Immunol.* **2008**, *180*, 361–371. [CrossRef] [PubMed]
6. Hsi, E.D.; Steinle, R.; Balasa, B.; Szmania, S.; Draksharapu, A.; Shum, B.P.; Huseni, M.; Powers, D.; Nanisetti, A.; Zhang, Y.; et al. CS1, a potential new therapeutic antibody target for the treatment of multiple myeloma. *Clin. Can. Res.* **2008**, *14*, 2775–2784. [CrossRef] [PubMed]
7. Tai, Y.-T.; Dillon, M.; Song, W.; Leiba, M.; Li, X.-F.; Burger, P.; Lee, A.I.; Podar, K.; Hideshima, T.; Rice, A.G.; et al. Anti-CS1 humanized monoclonal antibody HuLuc63 inhibits myeloma cell adhesion and induces antibody-dependent cellular cytotoxicity in the bone marrow milieu. *Blood* **2008**, *112*, 1329–1337. [CrossRef] [PubMed]
8. De Weers, M.; Tai, Y.-T.; van der Veer, M.S.; Bakker, J.M.; Vink, T.; Jacobs, D.C.H.; Oomen, L.A.; Peipp, M.; Valerius, T.; Slootstra, J.W.; et al. Daratumumab, a novel therapeutic human CD38 monoclonal antibody, induces killing of multiple myeloma and other hematological tumors. *J. Immunol.* **2011**, *186*, 1840–1848. [CrossRef] [PubMed]
9. Shi, W.; Liao, Y.; Willis, S.N.; Taubenheim, N.; Inouye, M.; Tarlinton, D.M.; Smyth, G.K.; Hodgkin, P.D.; Nutt, S.L.; Corcoran, L.M. Transcriptional profiling of mouse B cell terminal differentiation defines a signature for antibody-secreting plasma cells. *Nat. Immunol.* **2015**, *16*, 663–673. [CrossRef] [PubMed]
10. Immgen Consortium. Available online: www.Immgen.Org (accessed on 9 May 2018).
11. BLUEPRINT Consortium. Available online: www.blueprint-epigenome.eu/ (accessed on 18 July 2018).

12. Kahsay, R.Y.; Gao, G.; Liao, L. An improved hidden markov model for transmembrane protein detection and topology prediction and its applications to complete genomes. *Bioinformatics* **2005**, *21*, 1853–1858. [CrossRef] [PubMed]

13. Kall, L.; Krogh, A.; Sonnhammer, E.L. Advantages of combined transmembrane topology and signal peptide prediction—The phobius web server. *Nucleic Acids Res.* **2007**, *35*, W429–W432. [CrossRef] [PubMed]

14. Krogh, A.; Larsson, B.; von Heijne, G.; Sonnhammer, E.L. Predicting transmembrane protein topology with a hidden markov model: Application to complete genomes. *J. Mol. Biol.* **2001**, *305*, 567–580. [CrossRef] [PubMed]

15. Viklund, H.; Bernsel, A.; Skwark, M.; Elofsson, A. Spoctopus: A combined predictor of signal peptides and membrane protein topology. *Bioinformatics* **2008**, *24*, 2928–2929. [CrossRef] [PubMed]

16. Wickham, L.; Benjannet, S.; Marcinkiewicz, E.; Chretien, M.; Seidah, N.G. Beta-amyloid protein converting enzyme 1 and brain-specific type ii membrane protein BRI3: Binding partners processed by furin. *J. Neurochem.* **2005**, *92*, 93–102. [CrossRef] [PubMed]

17. Puskas, L.G.; Man, I.; Szebeni, G.; Tiszlavicz, L.; Tsai, S.; James, M.A. Novel anti-CRR9/CLPTM1l antibodies with antitumorigenic activity inhibit cell surface accumulation, PI3K interaction, and survival signaling. *Mol. Cancer Ther.* **2016**. [CrossRef] [PubMed]

18. Human Protein Atlas v18. Available online: https://www.Proteinatlas.Org/ensg00000049656-clptm1l/tissue/spleen#img (accessed on 8 May 2018).

19. Kallies, A.; Hasbold, J.; Tarlinton, D.M.; Dietrich, W.; Corcoran, L.M.; Hodgkin, P.D.; Nutt, S.L. Plasma cell ontogeny defined by quantitative changes in Blimp-1 expression. *J. Exp. Med.* **2004**, *200*, 967–977. [CrossRef] [PubMed]

20. Kumar, S.K.; Anderson, K.C. Immune therapies in multiple myeloma. *Clin. Cancer Res.* **2016**, *22*, 5453–5460. [CrossRef] [PubMed]

21. Sherbenou, D.W.; Behrens, C.R.; Su, Y.; Wolf, J.L.; Martin, T.G., 3rd; Liu, B. The development of potential antibody-based therapies for myeloma. *Blood Rev.* **2015**, *29*, 81–91. [CrossRef] [PubMed]

22. Takeuchi, M.; Harigai, M.; Momohara, S.; Ball, E.; Abe, J.; Furuichi, K.; Kamatani, N. Cloning and characterization of Dpp11 and Dppl2, representatives of a novel type of mammalian phosphatidate phosphatase. *Gene* **2007**, *399*, 174–180. [CrossRef] [PubMed]

23. Dai, C.; Dong, Q.Z.; Ren, N.; Zhu, J.J.; Zhou, H.J.; Sun, H.J.; Wang, G.; Zhang, X.F.; Xue, Y.H.; Jia, H.L.; et al. Downregulation of HTPAP transcript variant 1 correlates with tumor metastasis and poor survival in patients with hepatocellular carcinoma. *Cancer Sci.* **2011**, *102*, 583–590. [CrossRef] [PubMed]

24. Bernard-Pierrot, I.; Gruel, N.; Stransky, N.; Vincent-Salomon, A.; Reyal, F.; Raynal, V.; Vallot, C.; Pierron, G.; Radvanyi, F.; Delattre, O. Characterization of the recurrent 8p11-12 amplicon identifies Ppapdc1b, a phosphatase protein, as a new therapeutic target in breast cancer. *Cancer Res.* **2008**, *68*, 7165–7175. [CrossRef] [PubMed]

25. Mahmood, S.F.; Gruel, N.; Nicolle, R.; Chapeaublanc, E.; Delattre, O.; Radvanyi, F.; Bernard-Pierrot, I. Ppapdc1b and Whsc1l1 are common drivers of the 8p11-12 amplicon, not only in breast tumors but also in pancreatic adenocarcinomas and lung tumors. *Am. J. Pathol.* **2013**, *183*, 1634–1644. [CrossRef] [PubMed]

26. Engelhard, M. Anti-CD20 antibody treatment of non-hodgkin lymphomas. *Clin. Immunol.* **2016**, *172*, 101–104. [CrossRef] [PubMed]

27. Franks, S.E.; Getahun, A.; Hogarth, P.M.; Cambier, J.C. Targeting B cells in treatment of autoimmunity. *Curr. Opin. Immunol.* **2016**, *43*, 39–45. [CrossRef] [PubMed]

28. Ni, Z.; Chen, Q.; Lai, Y.; Wang, Z.; Sun, L.; Luo, X.; Wang, X. Prognostic significance of CLPTM1L expression and its effects on migration and invasion of human lung cancer cells. *Cancer Biomark.* **2016**, *16*, 445–452. [CrossRef] [PubMed]

29. Jia, J.; Bosley, A.D.; Thompson, A.; Hoskins, J.W.; Cheuk, A.; Collins, I.; Parikh, H.; Xiao, Z.; Ylaya, K.; Dzyadyk, M.; et al. CLPTM1L promotes growth and enhances aneuploidy in pancreatic cancer cells. *Cancer Res.* **2014**, *74*, 2785–2795. [CrossRef] [PubMed]

30. James, M.A.; Wen, W.; Wang, Y.; Byers, L.A.; Heymach, J.V.; Coombes, K.R.; Girard, L.; Minna, J.; You, M. Functional characterization of CLPTM1L as a lung cancer risk candidate gene in the 5p15.33 locus. *PLoS ONE* **2012**, *7*, e36116. [CrossRef] [PubMed]

31. Wauters, E.; Smeets, D.; Coolen, J.; Verschakelen, J.; De Leyn, P.; Decramer, M.; Vansteenkiste, J.; Janssens, W.; Lambrechts, D. The TERT-CLPTM1L locus for lung cancer predisposes to bronchial obstruction and emphysema. *Eur. Respir. J.* **2011**, *38*, 924–931. [CrossRef] [PubMed]

32. Petersen, G.M.; Amundadottir, L.; Fuchs, C.S.; Kraft, P.; Stolzenberg-Solomon, R.Z.; Jacobs, K.B.; Arslan, A.A.; Bueno-de-Mesquita, H.B.; Gallinger, S.; Gross, M.; et al. A genome-wide association study identifies pancreatic cancer susceptibility loci on chromosomes 13q22.1, 1q32.1 and 5p15.33. *Nat. Genet.* **2010**, *42*, 224–228. [CrossRef] [PubMed]

33. Peters, U.; Hutter, C.M.; Hsu, L.; Schumacher, F.R.; Conti, D.V.; Carlson, C.S.; Edlund, C.K.; Haile, R.W.; Gallinger, S.; Zanke, B.W.; et al. Meta-analysis of new genome-wide association studies of colorectal cancer risk. *Hum. Genet.* **2012**, *131*, 217–234. [CrossRef] [PubMed]

34. Zhao, Y.; Chen, G.; Zhao, Y.; Song, X.; Chen, H.; Mao, Y.; Lu, D. Fine-mapping of a region of chromosome 5p15.33 (TERT-CLPTM1L) suggests a novel locus in TERT and a CLPTM1L haplotype are associated with glioma susceptibility in a chinese population. *Int. J. Cancer* **2012**, *131*, 1569–1576. [CrossRef] [PubMed]

35. Turnbull, C.; Rapley, E.A.; Seal, S.; Pernet, D.; Renwick, A.; Hughes, D.; Ricketts, M.; Linger, R.; Nsengimana, J.; Deloukas, P.; et al. Variants near DMRT1, TERT and ATF7IP are associated with testicular germ cell cancer. *Nat. Genet.* **2010**, *42*, 604–607. [CrossRef] [PubMed]

36. Ni, Z.; Tao, K.; Chen, G.; Chen, Q.; Tang, J.; Luo, X.; Yin, P.; Tang, J.; Wang, X. CLPTM1L is overexpressed in lung cancer and associated with apoptosis. *PLoS ONE* **2012**, *7*, e52598. [CrossRef] [PubMed]

37. Martin, L.; Fluhrer, R.; Haass, C. Substrate requirements for SPPL2B-dependent regulated intramembrane proteolysis. *J. Biol. Chem.* **2009**, *284*, 5662–5670. [CrossRef] [PubMed]

38. Martins, F.; Rebelo, S.; Santos, M.; Cotrim, C.Z.; da Cruz e Silva, E.F.; da Cruz e Silva, O.A. BRI2 and BRI3 are functionally distinct phosphoproteins. *Cell. Signal.* **2016**, *28*, 130–144. [CrossRef] [PubMed]

39. Wu, H.; Liu, G.; Li, C.; Zhao, S. BRI3, a novel gene, participates in tumor necrosis factor-alpha-induced cell death. *Biochem. Biophys. Res. Commun.* **2003**, *311*, 518–524. [CrossRef] [PubMed]

40. Choi, S.C.; Kim, J.; Kim, T.H.; Cho, S.Y.; Park, S.S.; Kim, K.D.; Lee, S.H. Cloning and characterization of a type II integral transmembrane protein gene, Itm2c, that is highly expressed in the mouse brain. *Mol. Cells* **2001**, *12*, 391–397. [PubMed]

41. Vidal, R.; Calero, M.; Revesz, T.; Plant, G.; Ghiso, J.; Frangione, B. Sequence, genomic structure and tissue expression of human BRI3, a member of the Bri gene family. *Gene* **2001**, *266*, 95–102. [CrossRef]

42. Matsuda, S.; Matsuda, Y.; D'Adamio, L. BRI3 inhibits amyloid precursor protein processing in a mechanistically distinct manner from its homologue dementia gene BRI2. *J. Biol. Chem.* **2009**, *284*, 15815–15825. [CrossRef] [PubMed]

43. Rengaraj, D.; Gao, F.; Liang, X.H.; Yang, Z.M. Expression and regulation of type II integral membrane protein family members in mouse male reproductive tissues. *Endocrine* **2007**, *31*, 193–201. [CrossRef] [PubMed]

44. Farley, F.W.; Soriano, P.; Steffen, L.S.; Dymecki, S.M. Widespread recombinase expression using Flper (Flipper) mice. *Genesis* **2000**, *28*, 106–110. [CrossRef]

45. Schwenk, F.; Baron, U.; Rajewsky, K. A cre-transgenic mouse strain for the ubiquitous deletion of loxp-flanked gene segments including deletion in germ cells. *Nucleic Acids Res.* **1995**, *23*, 5080–5081. [CrossRef] [PubMed]

46. Rice, P.; Longden, I.; Bleasby, A. Emboss: The European molecular biology open software suite. *Trends Genet.* **2000**, *16*, 276–277. [CrossRef]

47. Mithraprabhu, S.; Kalff, A.; Chow, A.; Khong, T.; Spencer, A. Dysregulated class I histone deacetylases are indicators of poor prognosis in multiple myeloma. *Epigenetics* **2014**, *9*, 1511–1520. [CrossRef] [PubMed]

48. Ruijter, J.M.; Ramakers, C.; Hoogaars, W.M.; Karlen, Y.; Bakker, O.; van den Hoff, M.J.; Moorman, A.F. Amplification efficiency: Linking baseline and bias in the analysis of quantitative PCR data. *Nucleic Acids Res.* **2009**, *37*, e45. [CrossRef] [PubMed]

International Journal of
Molecular Sciences

MDPI

Review

Regulation of Energy Metabolism during Early B Lymphocyte Development

Sophia Urbanczyk [1], Merle Stein [2], Wolfgang Schuh [1], Hans-Martin Jäck [1],
Dimitrios Mougiakakos [3] and Dirk Mielenz [1,*]

[1] Division of Molecular Immunology, Nikolaus-Fiebiger-Center, Friedrich-Alexander-Universität Erlangen-Nürnberg (FAU), 91054 Erlangen, Germany; sophia.urbanczyk@uk-erlangen.de (S.U.); wolfgang.schuh@uk-erlangen.de (W.S.); hjaeck@gmail.com (H.-M.J.)

[2] Institute of Comparative Molecular Endocrinology (CME), University of Ulm, 89081 Ulm, Germany; merle.stein@uni-ulm.de

[3] Department of Internal Medicine V, University Hospital, Friedrich-Alexander-Universität Erlangen-Nürnberg (FAU), 91054 Erlangen, Germany; dimitrios.mougiakakos@uk-erlangen.de

* Correspondence: dirk.mielenz@fau.de or dirk.mielenz@uk-erlangen.de; Tel.: +49-9131-8539105; Fax: +49-9131-8539343

Received: 16 July 2018; Accepted: 25 July 2018; Published: 27 July 2018

Abstract: The most important feature of humoral immunity is the adaptation of the diversity of newly generated B cell receptors, that is, the antigen receptor repertoire, to the body's own and foreign structures. This includes the transient propagation of B progenitor cells and B cells, which possess receptors that are positively selected via anabolic signalling pathways under highly competitive conditions. The metabolic regulation of early B-cell development thus has important consequences for the expansion of normal or malignant pre-B cell clones. In addition, cellular senescence programs based on the expression of B cell identity factors, such as Pax5, act to prevent excessive proliferation and cellular deviation. Here, we review the basic mechanisms underlying the regulation of glycolysis and oxidative phosphorylation during early B cell development in bone marrow. We focus on the regulation of glycolysis and mitochondrial oxidative phosphorylation at the transition from non-transformed pro- to pre-B cells and discuss some ongoing issues. We introduce Swiprosin-2/EFhd1 as a potential regulator of glycolysis in pro-B cells that has also been linked to Ca^{2+}-mediated mitoflashes. Mitoflashes are bioenergetic mitochondrial events that control mitochondrial metabolism and signalling in both healthy and disease states. We discuss how Ca^{2+} fluctuations in pro- and pre-B cells may translate into mitoflashes in early B cells and speculate about the consequences of these changes.

Keywords: B lymphocyte development; metabolism; EFhd1; pre-BCR; mitochondria; mitoflash; oxidative phosphorylation; glycolysis

1. B Lymphocyte Development

B lymphocytes develop in adult vertebrates in the bone marrow (BM). They are derived from pluripotent stem cells and develop through the following stages: hematopoietic stem cells (HSCs); common lymphoid progenitors (CLPs); B cell-biased lymphoid progenitors (BLPs); and pre-pro-, pro- and pre-B cells. B cell precursors require cell contact and specific niches in the BM for their survival and growth [1]. Proliferative HSC and pre-pro-B cells, the earliest committed B lymphocyte progenitors, develop in the vicinity of sinusoids [2–4]. Pre-pro-B cells localize next to CXCL12 (also: SDF-1, stromal cell derived factor)-abundant reticular (CAR) cells, whereas pro-B cells are found adjacent to IL-7-expressing stromal cells, the majority of which are in close contact with the vasculature [5]. Pre-B cells localize near Galectin-1-expressing cells [1]. Each of these different niches possesses different oxygen tensions, indicating that there is a need to adapt mitochondrial respiration during different B cell developmental stages [6]. The specific characteristics of the niches required for early B lymphocyte development need further exploration and are at least partially and indirectly dependent on osteoblasts [2,4]. The active migration of cells towards their respective niches is induced by chemokines such as CXCL12 [3,5]. CXCL12 and CXCR4, the only receptor for CXCL12, are required for HSC and B cell development in a non-redundant manner [1,7–9], with CXCL12 eliciting an intracellular Ca^{2+} signal [7]. The first step of B lymphocyte development is controlled by the transcription factors (TFs) PU.1 and Ikaros (IKZF1), both of which are expressed in CLPs. Progenitors then commit to the B cell lineage by expressing E2A, EBF-1 and Pax-5 (reviewed in [10]). Pre-pro-B cells develop into pro-B cells (Figure 1), in which proliferation is supported by numerous factors, especially the cytokine interleukin 7 (IL-7) as well as CXCL12 and stem cell factor (SCF) [1]. Another important factor supporting early B cell development is Fms-like tyrosine kinase (Flt) 3 ligand [1]. B lymphocyte development follows defined stages that can be distinguished by the expression of cell surface markers, genetic rearrangements of Immunoglobulin (Ig) heavy and light chain loci and cell size and mitotic activity [11] (Figure 1; for detailed reviews see [1,12–14]).

In pro-B cells, binding of IL-7 to the IL-7 receptor drives the expression of the anti-apoptotic molecules Bcl-2 and myeloid-cell leukaemia sequence 1 (MCL1), enhancing survival and proliferation [3,15]. In vitro IL-7 induces proliferation in pro-B cells (Hardy fraction B and C) but not in further differentiated B cells [11]. This IL-7 dependency appears to be stronger in mice than in humans [16]. During development from pro-B cells to immature B cells, IL-7R is downregulated and responsiveness to IL-7 decreases [17,18]. A much smaller proportion of pre-B cells and immature B cells is found with higher concentrations of IL-7 but this is not due to the active suppression of differentiation. In fact, pro-B cells can also differentiate into pre-B cells and sIgM+ cells in the presence of higher IL-7 concentrations but these cells are outnumbered by proliferating pro-B cells [18]. The expression of Rag1 and 2 by pro-B cells allows diverse to joining (D-J) and variable D-J (VDJ) recombination of the gene segments that encode the μ heavy chain (μHC) of the B cell receptor (BCR) in pre-pro-B cells (Fraction A; Hardy et al. [11]) and pro-B cells (Hardy fraction B/C), respectively (Figure 1; reviewed in detail in [12–14]). After productive VDJ recombination, the newly formed μHC can pair with the surrogate light chain complex consisting of VpreB and λ5, resulting in pre-BCR expression and the appearance of large pre-B cells (Hardy fraction C/C') (Figure 1). Mice deficient in either of these Rag genes show a developmental B lymphocyte block and accumulate pro-B cells in the BM because rearrangement of μHC D-J and then VDJ elements cannot take place [19,20].

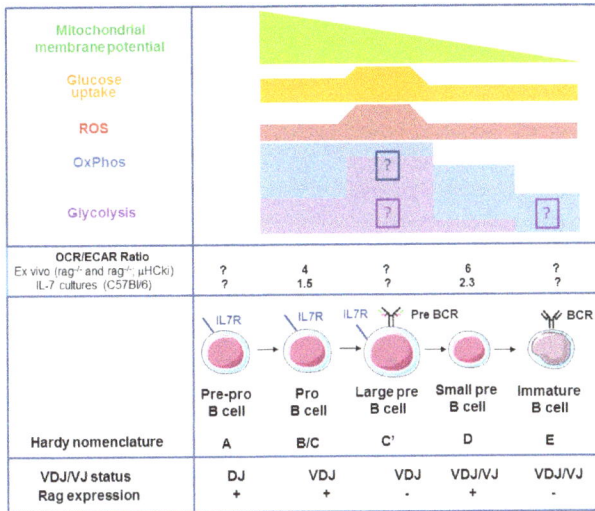

Figure 1. Relationship between oxidative phosphorylation and glycolysis during early murine B cell development. Summary of experimental determinations of the mitochondrial membrane potential $\Delta\psi_m$, glucose uptake, ROS production, Oxphos and glycolysis during B cell development from pro- to large and from small pre-B to immature B cells [21,22]. Extracellular flux analysis of pro- and small pre-B cells (mainly small) obtained from rag2$^{-/-}$ or rag2$^{-/-}$; µHC knock-in mice or from sorted wildtype pro- and pre-B cells (mainly small) obtained from IL-7 cultures revealed a decline in both glycolysis and OxPhos, with the decline in glycolysis being more pronounced [21]. Consequently, small pre-B cells show a higher OxPhos to glycolysis ratio in both systems, with a lower ratio observed in the IL-7 cultures. Immature B cells reveal an even lower rate of OxPhos [22]. Missing data are indicated by question marks. The increase in glycolysis and OxPhos in large pre-B cells is speculative (see boxed question marks in matching colours) and based on literature reviews.

Ectopic expression of the µHC on a Rag2$^{-/-}$ background in mice led to the development of phenotypic pre-B cells, while the introduction of µHC and lambda (λ)-LC led to the production of peripheral, monoclonal and immunoglobulin-secreting B cells [20]. The Pre-BCR elicits an increase in the cytosolic Ca^{2+} concentration [23–26] and acts as an inducible proliferative signal in pre-B cells with an expansion factor of 20–100 (approximately 4–6 cell divisions) [27]. Hence, B cell clones with an optimal pre-BCR signalling strength, based on µHC idiotype, will expand (pre-BCR signal 1) [28]. This defines the basis for the (mostly autoreactive) pre-immune BCR repertoire, which represents a direct link between metabolism, growth control and autoreactivity [29,30]. The mechanisms by which the pre-BCR induces this expansion signal at the structural level have been reviewed elsewhere [5,31,32]. Expression of the pre-BCR also inhibits further rearrangements of the V to DJ loci in the not yet re-arranged µHC allele (allelic exclusion) [28]. After the first round of clonal expansion, pre-B cells become quiescent again and decrease in size (pre-BCR signal 2). In these resting, small pre-B cells (Hardy fraction D), gene rearrangements occur in the V and J segments encoding the BCR light chain [14]. Successful VJ rearrangement gives rise to light chain protein, BCR expression and naïve, immature B cells (Hardy fraction E) (Figure 1). Immature B cells then complete development into resting mature follicular and marginal zone B cells in the spleen [33]. As outlined above, pro-B cells proliferate in response to IL-7, expand transiently into large pre-B cells upon early pre-BCR expression and then become quiescent as small pre-B cells again to allow VJ recombination to occur (reviewed in Clark et al. [14]) (Figure 1). The proximal signalling pathways that control these transitions have been reviewed in detail elsewhere [14,32,34] but several questions remain incompletely answered. For example, how is pro-B cell proliferation maintained homeostatically? How does IL-7 affect early B cell metabolism? How do pre-BCRs, IL-7 and nutrients control the transient expansion of large pre-B cells and the subsequent quiescence of small pre-B cells? The purpose of this review is to summarize what is currently known about metabolism during early B cell development.

2. Oxidative Phosphorylation and Glycolysis in Pro- and Pre-B Cells

The ultimate downstream biochemical events that supply cells with adenosine triphosphate (ATP) are glycolysis and mitochondrial oxidative phosphorylation (OxPhos) (reviewed in detail in the context of lymphocytes [35]). The oxidation of fatty acids (FAs), carbohydrates and amino acids is coupled to ATP synthesis in mitochondria by the proton gradient across the inner mitochondrial membrane (IMM). The proton gradient (ΔpH_m) across the IMM is established by the electron transport chain (ETC) by mitochondrial respiratory chain complexes I, III and IV, which pump protons from the matrix into the mitochondrial intermembrane space [35]. ΔpH and the mitochondrial membrane potential ($\Delta\psi m$) contribute independently to the proton motive force (Δp) that drives the synthesis of ATP via the ATP synthase complex (complex V) ($\Delta p = \Delta pH_m + \Delta\psi m$) [36]. The concentration of ATP relative to that of ADP and AMP is an indicator of the cellular energy status and is sensed by a kinase complex called adenosine monophosphate – activated protein kinase (AMPK). When the AMP/ATP ratio reaches a certain threshold, AMPK becomes activated to support catabolic pathways and ensure an ongoing energy supply. AMPK activity promotes mitochondrial biogenesis and autophagy and represses the mammalian target of Rapamycin (mTOR) pathway [37–39].

Inhibition experiments performed with 2-deoxyglucose (2-DG), a non-hydrolysable glucose analogue that blocks glycolysis, have shown that pro-/early/pre-B cells depend on the glycolytic pathway, whereas late (small) pre-B cells do not [40]. In contrast, a lack of glucose did not prevent the development of IgM-positive cells in vitro in total BM cultures [41]. It should be noted that 2-DG has off-target effects, including endoplasmic reticulum (ER) stress, autophagy induction, interference with mannose and reduced protein N-glycosylation (reviewed in detail in [42]). Hence, these findings need to be reconciled. However, the experiments performed by Kojima et al. revealed the existence of a metabolic checkpoint in early B cell development. This finding was corroborated by a genetic screen that revealed the existence of a metabolic checkpoint controlled by folliculin interacting protein 1 (Fnip1). Fnip1 forms a complex with AMPK [39] and in Fnip1$^{-/-}$ mice, B cell development is blocked at the large pre-B cell stage due to an imbalance in metabolism [41]. In WT BM B cell cultures derived from total BM cells grown in the presence of IL-7, SCF and Flt 3 ligand for 48 h, depleting the cells of glucose, glutamine or essential amino acids did not affect the number of developing IgM-positive B cells. However, Fnip1$^{-/-}$ B cell progenitors were sensitive to these depletions, indicating a state of energy exhaustion. Under the same experimental conditions, oligomycin (an inhibitor of ATP synthase activity in mitochondrial respiratory chain complex V) at 10 or 50 nM did not affect the appearance of-IgM positive WT B cells, while 10 nM oligomycin did alter the appearance of these cells in Fnip1$^{-/-}$ cultures. Extracellular flux analyses performed with a Seahorse analyser showed that pro-/pre-B cells responded to IL-7 by increasing their oxygen consumption rate (OCR; an indicator of oxidative phosphorylation/OxPhos) (We would like to add, as a technical note not related specifically to the cited publication [41], that measuring OCR in a Seahorse analyser does not provide information about the substrates fuelling the TCA cycle. Information about these substrates can be obtained by using labelled substrates or inhibitors. For instance, FA importation into mitochondria can be inhibited by Etomoxir but at high concentrations, Etomoxir exerts off-target effects, including inhibiting complex I of the electron transport chain [43]) and extracellular acidification rate (ECAR; an indicator of glycolysis) (ECAR measured in a Seahorse analyser represents a pH measurement. To ensure that an observed decrease in extracellular pH is due to an increase in lactate secretion that occurs as a consequence of glycolysis, it is recommended that lactate should be measured or ^{13}C-labeled glucose be tracked) [41]. However, Fnip1$^{-/-}$ pro/pre-B cells responded better. To determine which substrates fuel the observed increase in OCR, sorted pro-/pre-B cells were treated with 2-DG and Etomoxir. Both treatments reduced the OCR in WT cells but it was reduced even more in Fnip1$^{-/-}$ pro-/pre-B cells. These data suggested that pro-/pre-B cells utilize glucose and FA for OxPhos and that Fnip1 renders pro-/pre-B cells resilient to inhibition of glycolysis and of FA oxidation. Further experiments connected this metabolic checkpoint to the Fnip1:AMPK complex and the pro-/pre-B cell transition (Figure 2A). The anabolic ATP exhaustion observed in Fnip1$^{-/-}$ pro-/pre-B cells is likely mediated by an increase in *rps6ka1* expression [41]. A similar mechanism has been observed in transformed haploinsufficient Phosphatase and Tensin homologue (PTEN)$^{-/+}$ and PTEN$^{-/-}$ pre-B acute lymphoblastic leukaemia (ALL) cells [41,44]. While the experiments performed by Kojima et al. and Park et al. were seminal, measurements of OxPhos and glycolysis in discrete pro- and pre-B cell populations have not yet been performed under more defined conditions (e.g., medium with IL-7 only). Thus, we analysed metabolism in discrete pro- and pre-B cells (Figure 1) [21]. Mitochondrial mass relative to cell size is decreased in large pre-B cells but remains constant during later B cell development [21]. Pro-B cells exhibited the highest $\Delta\psi\mu$; $\Delta\psi\mu$ is then significantly lower in small pre-B cells and declines further during development. Reactive oxygen species (ROS) production, as measured by 2'-7'-dichlorodihydrofluorescein diacetate (DCDFA, a dye that does not specifically quantify mitochondrial ROS) and glucose uptake are highest in large pre-B cells but reduced in small pre-B cells, supporting the data described by Kojima and colleagues [40]. To measure glycolysis and OxPhos directly in pro- and pre-B cells, we established a μHC knock-in (ki) mouse model (33.C9μHCki) and crossed these mice to Rag1$^{-/-}$ mice [19] (Rag1$^{-/-}$;33.C9μHCki) [21]. Pre-B cells obtained from Rag1$^{-/-}$;33.C9μHCki mice are mainly small. Extracellular flux analyses performed with sorted primary pro- and pre-B cells obtained from this system revealed that in general, under normoxic conditions, OCR and ECAR were lower in Rag1$^{-/-}$;33.C9μHCki pre-B cells than pro-B cells. These data were confirmed by Zeng et al., who also analysed immature B cells, which have an OCR similar to that of small pre-B cells [22]. In contrast to Zeng et al. we also assessed glycolysis. In our experiments, glycolysis (evaluated by ECAR) was significantly reduced relative to OCR in small pre- versus pro-B cells, resulting in a higher OCR/ECAR ratio (Figure 1). However, the contributing mechanisms and consequences of the alterations in OCR/ECAR ratios and mitochondrial spare capacity observed in this system require more study. Nevertheless, we noted that the OCR/ECAR ratio was in general lower in IL-7 cultures, suggesting that IL-7 promotes glycolysis (Figure 1). In fact, IL-7 promotes glycolysis by activating Akt [13,45,46] and this might be important in IL-7-rich niches in BM [1,6]. IL-7 also appears to elevate mitochondrial spare capacity, perhaps via the pyruvate that is generated by glycolysis and directed towards the tricarbon (TCA) cycle (Figure 2A). The data described in Park et al. [41] do indeed imply that mixed pro-/pre-B cell cultures use pyruvate derived from glycolysis to fuel and maintain OxPhos but more experiments are needed to define the TCA substrates used in pro- and pre-B cells. In summary, pre-BCR expression ultimately promotes metabolic quiescence (pre-BCR signal 2) by reducing glycolysis (as defined by ECAR using a Seahorse analyser), resulting in an increased OCR/ECAR ratio. The decrease in glycolysis observed in small pre-B cells compared to pro-B cells is in agreement with the proposal that Akt is inactivated [32,47] (pre-BCR signal 2) and that glucose up-take [21] and responsiveness to IL-7 are reduced.

It appears that the reduction of metabolism observed in small pre-B cells is maintained in immature B cells [22]. Additionally, resting splenic B cells exhibit low metabolic activity [48] and consume FA to produce ATP via OxPhos, as shown by metabolic tracking of FA [48]. Upon BCR activation or by lipopolysaccharide (LPS) mediated TLR4 activation, normal but not anergic splenic murine B cells again upregulated both OxPhos and glycolysis in a Myc-dependent manner in a balanced ratio [48–50], concomitant with an increase in the glucose transporter *glut1*. The newly activated B cells then oxidize glutamine and pyruvate [48]. Interestingly, murine peritoneal B1 B cells are metabolically more active than follicular B cells and depend on glycolysis [51]. Glycolysis (analysed in inhibition experiments performed with dichloroacetate, an inhibitor of pyruvate dehydrogenase) also supported the secretion of antibodies from murine and human B cells both in vitro and in vivo [48]. Glucose taken up by plasma cells is mainly used for antibody glycosylation [52]. No information is available regarding differences between follicular and marginal zone B cells or whether small pre-B cells or immature B cells rely on FA. In addition, extracellular flux analyses of primary pro-and small pre-B cells failed to identify large pre-B cells for technical reasons. Large pre-B cells could, in the future, potentially be enriched from Irf4/8 double [53] or BLNK/SLP-65 knock-out mice [54], which fail to downregulate pre-BCR, show hyperproliferation of large pre-B cells and are prone to malignant transformation. It is likely that glycolysis is increased in large pre-B cells and this would provide more energy and more anaplerotic reactions for macromolecules and intermediate products, thereby protecting cells from ROS [55], in addition to more pyruvate to support mitochondrial ATP production (Figure 1). A genetic in vivo system that allows the mitochondrial respiratory chain to be manipulated could address this question. For instance, experiments performed in mice with an inducible deletion of the mitochondrial pyruvate importer Mcp2 (in *Mcp2*$^{fl/fl}$*;ROSA26 CreER* mice) revealed that plasma cells in the BM rely on this mechanism [52].

A caveat of all the studies mentioned so far is that ex vivo experiments are generally performed under normoxic conditions, while many parts of the BM and several of its niches are hypoxic [6]. Immune cells adapt to hypoxia by stabilizing HIF at the protein level [56]. Indeed, HIF1α deficiency impaired early B cell development in Rag2$^{-/-}$ blastocyst complementation chimeras by reducing the number of proliferating CD43$^-$HSA$^+$B220$^+$ cells [57]. On the other hand, B cell development in the BM of HIF1α$^{fl/fl}$ or HIF2α$^{fl/fl}$ mice crossed to Mb1-Cre mice is normal [58]. Although the CD43$^-$HSA$^+$B220$^+$ cells affected by HIF1α deficiency are likely proliferating pre B cells, there is the possibility that HIF1α mediated metabolic adaptations influence B cell development already before the mb-1 promotor is active. HIF-1α controls glycolysis in BM precursor B cells in a developmental stage-specific manner by regulating the genes that encode glucose transporters and the key glycolytic enzyme 6-phosphofructo-2-kinase/fructose-2,6-bishosphatase 3 [40]. Interestingly, HIF1α-deficient B cell progenitors compensated for defects in glycolytic enzymes by increasing the expression of respiratory chain-related genes and TCA-related genes, enabling more efficient pyruvate usage [40]. These data reveal that B cell progenitors are metabolically flexible and show a propensity to adapt to different oxygen tensions in the BM, thus ensuring survival and correct development. A very interesting point is that these adaptations in the BM due to loss of HIF1α appear to impact on B1 B cells and autoimmunity [57]. In accordance, recent elegant experiments have revealed that HIF1α is important for expansion of CD1d^{high}CD5$^+$ B cells via glycolysis and production of anti-inflammatory IL-10 by those B cells [58]. Further studies performed under hypoxic conditions are required to fully explore the physiological role of glycolysis and OxPhos in pro-B and pre-B cells. Another important experimental step was to establish a system that physiologically represents human early B-cell development [59]. Studying metabolic changes during early human B cell development in vitro could provide crucial evidence about the development of autoreactivity, malignant transformation and B cell repopulation following eradication of the bone marrow (e.g., by radiation or chemotherapy) [59].

3. Signalling Pathways Linking Membrane Receptor Signals to Glycolysis and Oxidative Phosphorylation in Pro- and Early Pre-B Cells

The genes that control mitochondria and glycolysis are targets of both pre-BCR and BCR [48,60]. This indicates that the metabolic machinery of B cells integrates signals down-stream of (pre-) BCRs and growth factors [47,48], thereby, connecting the μHC idiotype with metabolism. The IL-7 and pre-BCR signalling network in pro- and pre-B cells has been reviewed in detail elsewhere [13,34]. Briefly, pro-B cells receive signals related to survival and proliferation via IL-7, Janus kinase (JAK) and signal transducer of activation and transcription (STAT) factors that enforce B cell identity via the expression of Pax5 and Ebf1. Interestingly, the B cell identity-related TFs Pax5 and IKZF1 limit glucose uptake in normal B cells and thereby support a metabolic program that leads to metabolic exhaustion when an oncogene, such as Bcr-Abl, is activated [61]. IL-7 activates the phosphatidyl-inositol-3-kinase (PI3K) pathway, leading to the activation of extracellular regulated kinase (Erk) [62] and Akt and the inactivation of Foxo1 [15,63,64]. The early pre-BCR signal also engages the PI3K cascade via Syk. In particular, the PI3K and Erk pathways control proliferation during pre-B cell development [46,47,62]. It appears, however, that PI3K activity needs to be limited by PTEN, which controls IL-7R expression, to allow pro-B cell development [22]. In mammals, mTOR is downstream of PI3K signalling in B cells (reviewed in [39]). mTOR is a kinase complex that supports IL-7-induced anabolism via glycolysis and Myc [22] (Figure 2A). The mTORC1 complex is positively regulated by Raptor, whereas the mTORC2 complex is controlled by Rictor. Anabolic mTORC1 activity is counterbalanced by AMPK (reviewed in [39]). Because Fnip1 forms a complex with AMPK and a lack of Fnip1 leads to the hyperactivation of mTOR, Fnip1 mediates the inhibitory effects of AMPK on mTOR [41,65] (Figure 2A). The conditional deletion of Raptor in B cells in mb1-Cre mice [66] led to the B cell-specific inactivation of the mTORC1 complex [67].

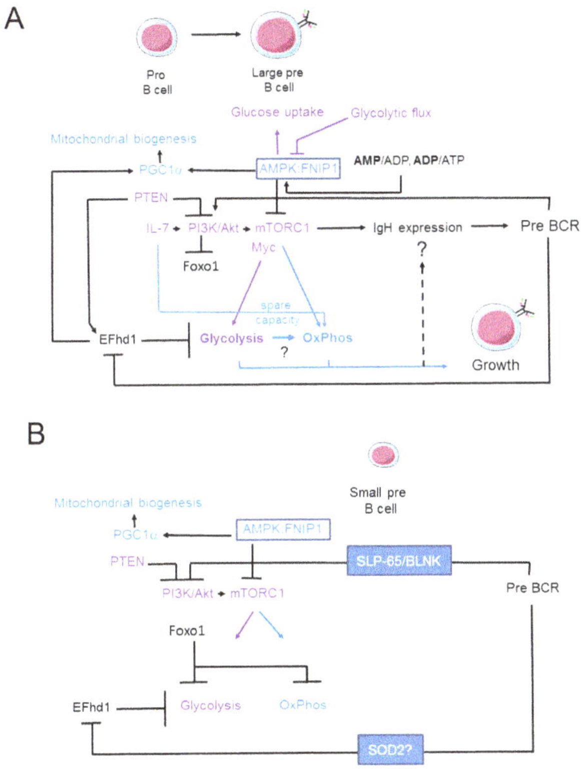

Figure 2. Relationship between oxidative phosphorylation and glycolysis during early murine B cell development. In pro- and large pre-B cells, growth is controlled by IL-7 and the pre-BCR, which lead to the activation of the PI3K/Akt/mTOR pathway. mTORC1 activates glycolysis and OxPhos and is required for the protein expression of μHC, which controls pre-BCR expression and signalling. Whether control of μHC protein expression occurs via glycolysis and/or OxPhos is not clear. Surface expression of the pre-BCR leads to the downregulation of EFhd1, a Ca^{2+}-binding protein localized on the inner mitochondrial membrane. EFhd1 suppresses glycolysis in transformed pro-B cells and is induced by PTEN. In primary pre-B cells, overexpression of EFhd1 induces PGC1α, which is also controlled by AMPK1, a negative regulator of anabolic pathways and a positive regulator of catabolic pathways. AMPK activity is controlled by ADP/AMP and ATP/ADP ratios and fructose 1,6 bisphosphate and is an indicator of glycolytic flux. AMPK activity is also modified by FNIP1. Loss of FNIP1 by genetic ablation leads to anabolic exhaustion in pro-/pre-B cells. IL-7 increases the mitochondrial spare capacity in pre-B cells. (**A**) The network in pro- and large pre-B cells with an activated PI3K pathway and inactivated Foxo1; and (**B**) the network in small pre-B cells, in which the PI3K and mTORC1 pathways are inactivated but Foxo1 is activated. It is unclear whether large pre-B cells depend solely on glycolysis or whether glycolytic pathways are mutually linked to OxPhos.

mTORC1-deficient pro- and perhaps early pre-B cells (Hardy fractions C and C′) developed in these mice, small pre-B cells (fraction D) were strongly reduced and later stages were absent. Although VDJ and VJ recombination were normal in the remaining C/C′ and D fraction cells, the expression of the IgM (μ) heavy chain was severely impaired, even when expressed as anti-hen egg lysozyme (HEL) transgenic BCR. Hence, B cell development was not rescued by the expression of anti-HEL BCR. It is unclear why mTORC1 activity is required for IgM expression. It is possible that mTORC1 controls IgM stability downstream of glycolysis and OxPhos, or it may support glycosylation via the hexosamine biosynthetic pathway and thereby stabilize the membrane expression of μHC within the pre-BCR complex (Figure 2A).

4. Swiprosin-2/EFhd1 as a Regulator of Glycolysis in Pro-B Cells

Swiprosin-2/EFhd1 (EFhd1) is a Ca^{2+}-binding protein that localizes to the IMM and consists of an N-terminal disordered region, two central Ca^{2+}-binding EF-hands and a C-terminal coiled-coil domain [68–71] (Figure 3). EFhd1 is a target gene of the TFs involved in B cell identity and controlling early B cell development. These TFs include Foxo1, Brg1 and Ebf1 in pro-B cells [72,73]. PTEN also promotes *efhd1* expression [44] (Figure 2A). EFhd1 becomes upregulated together with PGC-1α, uncoupling protein (UCP) 2 and other proteins involved in mitochondrial functions in the distal convoluted tubule cells of the kidney by inducing the deletion of the cytosolic Ca^{2+} buffer parvalbumin [74]. We showed that EFhd1 is expressed in primary mouse pro-B cells at the RNA and protein levels [21]. Surface expression of the pre-BCR resulted in the downregulation of EFhd1 in pro-B cells. Hence, IgM-positive B cells no longer express EFhd1 [21]. These data indicate that very early pre-B cells still express EFhd1 protein, although the half-life of EFhd1 is unknown (Figure 2A). The mechanism underlying the surface pre-BCR-mediated repression of EFhd1 is also unknown but it may involve tonic pre-BCR/BCR signals. Pro-B cells only tolerate a moderate amount of EFhd1 and pre-B cells need to downregulate EFhd1, probably to maintain mitochondrial ATP production during the pro- to pre-B cell transition [21].

As optimal pre-BCR signalling strength depends on efficient pairing of the newly generated μHC with VpreB and λ5, the downregulation of EFhd1 by the pre-BCR might link μHC signalling competence (the μHC repertoire) to metabolic fitness in pre-B cells. Along these lines, the Crispr-Cas9-mediated knock-out of EFhd1 as well as the its shRNA-mediated knock-down in the transformed pro-B cell line 38B9 resulted in increased glycolysis and a higher glycolytic rate and glycolytic spare capacity [21] (Figure 2A,B). We propose that the downregulation of EFhd1 in pro-B cells by pre-BCR signals is one of the mechanisms that drives pre-B cell expansion via glycolysis. We further speculate that the pre-BCR-mediated repression of EFhd1 represents a cellular readout for optimal pre-BCR signalling strength. The upregulation of PGC-1α by EFhd1 and its negative effect on glycolysis indicate that EFhd1 is a potential catabolic factor. It is thus of interest that the *efhd1* promotor is frequently methylated in tumour biopsies in colorectal cancer patients [75].

5. Signalling Pathways Linking Membrane Receptor Signals to Glycolysis and Oxidative Phosphorylation in Late Pre-B Cells

In parallel with decreasing IL-7 responsiveness, the pre-BCR initiates the expression of SLP-65/BLNK, allowing large pre-B cells to differentiate into small pre-B cells, by inhibiting the PI3K/Akt pathway and inducing a concomitant increase in FOXO1 activity [47,76,77] (Figure 2B). In small pre-B cells, PI3K and mTORC1 activity are strongly inhibited, enabling Foxo1 to suppress cell growth and facilitate LC rearrangement. EFhd1 is downregulated and its putative catabolic, that is, limiting effect on glycolysis may be overcome by a reduction in mTORC1 activity and an increase in Foxo1 activity (speculative model shown in Figure 2B). Together with Pax5, FOXO1 transactivates Rag1/2, IRF4 and p27, thereby inducing cell cycle arrest, while LC rearrangement takes place in small pre-B cells [13,73,78]. FOXO1 is also activated by ROS and upregulates superoxide dismutase 2 (*sod2*) [15]. The increase in the FOXO1 target gene, *sod2*, observed in pre-B cells [21] is compatible with the dephosphorylation and activation of FOXO1, which occurs as a consequence of reduced Akt activity [15] and might limit ROS. Interestingly, EFhd1 has been shown to be downregulated by SOD2 [79] (Figure 2B).

6. Putative Existence of Mitoflashes in Early B Cells—Regulated by EFhd1?

Careful measurement of mitochondrial pH (pH$_{mito}$) has revealed that pH$_{mito}$ is in dynamic equilibrium with cytosolic pH (reviewed in [36]). Similarly, the cytosolic Ca^{2+} concentration is connected to the mitochondrial Ca^{2+} concentration via connections between the ER and mitochondria (reviewed in [80]) and Ca^{2+}-binding proteins in mitochondria, likely including EFhd1 (reviewed in [69]). The spontaneous pH$_{mito}$ elevations that coincide with drops in Δψm can occur in single mitochondria or discrete regions of the mitochondrial network and have been termed "mitoflashes" [81,82]. Mitoflashes are modulated by, for example, metabolic state, oxidative stress, developmental stage, aging and Ca^{2+} [82]. It is currently a matter of debate as to whether mitoflashes do or do not represent superoxide bursts [83] but there appears to be some consensus that they are related to pH$_{mito}$ [81,82]. It thus has been proposed that the term "MitopHlash" [81,84] should be used and we will not address superoxide here. In general, during a mitoflash, spontaneous drops in Δψm are coupled to mitochondrial matrix alkalinization, thereby preserving an intact Δp and enabling ongoing ATP production [36]. Experiments performed in 293 cells have shown that EFhd1 is a mitochondrial Ca^{2+} sensor of Ca^{2+}-dependent mitoflashes induced by ionomycin [85]. This function of EFhd1 depends on its EF-hands, as shown in experiments with point mutations (i.e., E116A and E152A) at critical residues [86] in each of its two EF-hands. Furthermore, EFhd1 does not induce alterations in mitochondrial Ca^{2+} handling [85]. These data raise the possibility that EFhd1 is a Ca^{2+} sensor that monitors Ca^{2+} flux from the ER to mitochondria [69] and that EFhd1 may be coupled to complex I, III or IV of the respiratory chain or H$^+$ coupled Ca^{2+} transporters located in mitochondria. In fact, pH$_{mito}$ and mitochondrial Ca^{2+} are coupled [87]. There is room for speculation as to what the function of mitoflashes in early B cells might be. It is conceivable that external signals such as CXCL12 in pro-B cells [7] or successful expression of pre-BCR could activate mitoflashes by inducing signalling cascades, resulting in increased intracellular Ca^{2+} [23–26]. Assuming that mitoflashes represent mitochondrial bioenergetic phenomena that sustain the proton motive force and ATP production, the fast changes in pH$_{mito}$ and Δψm may maintain quick metabolic adaption, which is vital for the activation and differentiation processes that occur in pro- and pre-B cells. EFhd1 expression in pro- and likely very early pre-B cells may contribute to the Ca^{2+}-dependent control of mitoflashes, thereby, modulating CXCL12 and pre-BCR induced metabolic survival and expansion signals (Figure 3). In the case of the pre-BCR, it would, however, do so only transiently because surface expression of the pre-BCR (see above; [21]) leads to downregulation of EFhd1 via as yet unknown mechanisms. We propose that the downregulation of EFhd1 represents a sensor for optimal pre-BCR signalling strength. Pre-B cells downregulating EFhd1 faster because they express an appropriate μHC idiotype on the cell surface may switch to glycolysis faster and this may provide a competitive advantage. Hence, EFhd1 might integrate mitochondrial metabolism, glycolysis and μHC selection in BM. We envision an intimate interplay among pre-BCR-controlled genes involved in metabolism and Ca^{2+} sensing mechanisms, such as *efhd1*. This may be important for the fitness of early B cells.

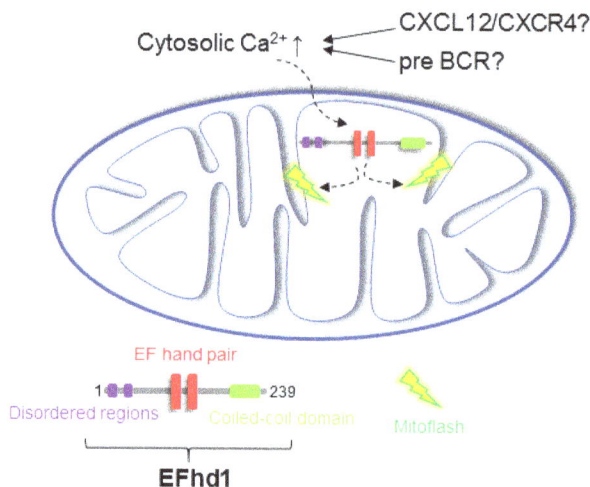

Figure 3. Putative mechanism for EFhd1-controlled mitoflashes in early B cells. Mitoflashes (also MitopHlashes) are bioenergetic responses to stochastic drops in the mitochondrial membrane potential ($\Delta\psi_m$). Their origin is unclear but research performed using pH-sensitive probes showed that flashes represent matrix alkalinization transients and are therefore linked to pH. It has been proposed that mitoflashes control mitochondrial metabolism and signalling in both healthy and disease states and can be triggered by increased mitochondrial Ca^{2+} concentrations. EFhd1 has been shown to mediate mitoflash activity in response to increases in mitochondrial Ca^{2+} concentrations via its two EF hands. The existence and, moreover, the consequences of this event in pro-B cells, which express EFhd1, is currently unclear. EFhd1 might translate a CXCL12 or pre-BCR induced increase in Ca^{2+} into a mitoflash, thereby coupling Ca^{2+} to mitochondrial pH regulation and proton motive force.

7. Conclusions and Perspectives

The metabolic regulation of pro- and pre-B cell development has important consequences for the expansion of normal and malignant pre-B cell clones. Under healthy conditions, it affects the normal BCR repertoire and contributes directly to adaptive immunity. To understand how the growth of normal or transformed pro- and pre-B cells is regulated, we have reviewed recent data on the regulation of glycolysis and oxidative phosphorylation in early B cells. It has become clear that pre-BCR ultimately induces metabolic quiescence because it leads to a reduction in glycolysis in late pre-B cells. The current data also suggest that large pre-B cells utilize glycolysis to sustain ATP but previous extracellular flux analyses of primary pro- and pre-B cells have missed large pre-B cells for technical reasons. Large pre-B cells may in the future be enriched from Irf4/8 double or BLNK/SLP-65 knock-out mice to clarify this issue. It remains unclear whether large pre-B cells (a) perform more glycolysis than is found in pro-B cells; (b) depend solely on glycolysis for their growth or (c) require pyruvate generated by glycolysis for ATP production via OxPhos. Hence, it remains completely unknown whether pro-B cells and pre-B cells depend in vivo on the mitochondrial respiratory chain. It will also be important to study metabolism in early B cells under hypoxic conditions, which occur naturally in BM. It is clear that HIF1α plays a role in early B cell development by promoting glycolysis in B cell progenitors. Here, we introduce the idea that recently described bioenergetic events in mitochondria ("mitoflashes") that have been shown to maintain the proton motive force required to generate mitochondrial ATP may also occur in pro- and early pre-B cells. A Ca^{2+}-binding protein that localizes on the IMM, Swiprosin-2/EFhd1 (EFhd1), might be involved in this phenomenon because it is expressed in pro-B cells and controls Ca^{2+}-dependent mitoflashes. We have shown that EFhd1 becomes downregulated by cell surface expression of pre-BCR and that EFhd1 limits glycolysis in pro-B cells. We propose that EFhd1 might integrate Ca^{2+} signals with gene regulation and metabolic activity in pro- and early B cells and EFhd1 may thereby serve as a sensor for optimal pre-BCR signalling strength. The interplay between pre-BCR signalling and metabolism was clearly revealed in mice lacking mTORC1 activity in B cells. mTORC1 is required to maintain the stability of the IgM heavy chain and it therefore controls pre-BCR signalling at this level. The mechanism behind this function remains unknown but it may involve the generation of anaplerotic intermediate factors produced during glycolysis. In summary, we suggest that more extensively examining the regulation of metabolism in pro- and pre-B cells would increase our understanding of growth control and the generation of a healthy BCR repertoire.

Funding: This research was funded by the Deutsche Forschungsgemeinschaft (DFG, TRR130).
Acknowledgments: We thank Julia Jellusova for comments during the writing process and the reviewers for helpful suggestions. We acknowledge support by Deutsche Forschungsgemeinschaft and Friedrich-Alexander-Universität Erlangen-Nürnberg (FAU) within the funding programme Open Access Publishing.

Conflicts of Interest: The authors declare no conflict of interest.

Abbreviations

ALL	Acute lymphoblastic leukaemia
AMP	Adenosine monophosphate
ADP	Adenosine diphosphate
ATP	Adenosine triphosphate
6-NBDG	6-(*N*-(7-nitrobenz-2-oxa-1, 3-diazol-4-yl)amino)-2-deoxyglucose
BLP	B cell-biased lymphoid progenitor
BCR	B cell receptor
BM	Bone marrow
CAR cell	CXCL12-abundant reticular cell
CLP	Common lymphoid progenitor
ECAR	Extracellular acidification rate
EFhd1	Swiprosin-2/EFhd1
EFhd1*tg*	EFhd1 transgenic
ETC	Electron transport chain
Erk	Extracellular signal regulated kinase
FA	Fatty acid
Flt	Fms-like tyrosine kinase
Fnip1	Folliculin-interacting protein 1
HC	Heavy chain
HSC	Hematopoietic stem cell
Ig	Immunoglobulin
IMM	Inner mitochondrial membrane
JAK	Janus kinase
Ki	Knock-in
LC	Light chain
LPS	Lipopolysaccharide
$\Delta\psi$m	Mitochondrial membrane potential
OCR	Oxygen consumption rate
OxPhos	Oxidative phosphorylation
Pi	Inorganic phosphate
PI3K	Phosphatidyl-inositol-3-kinase
ΔpH	Proton gradient
PTEN	Phosphatase and Tensin homologue
ROS	Reactive oxygen species
SCF	Stem cell factor

SDF-1   Stromal cell derived factor
STAT   Signal transducer of activation and transcription
TCA   Tricarbon cycle
TF   Transcription factor
WT   Wildtype

References

1. Green, A.C.; Rudolph-Stringer, V.; Chantry, A.D.; Wu, J.Y.; Purton, L.E. Mesenchymal lineage cells and their importance in B lymphocyte niches. *Bone* **2017**. [CrossRef] [PubMed]
2. Boulais, P.E.; Frenette, P.S. Making sense of hematopoietic stem cell niches. *Blood* **2015**, *125*, 2621–2630. [CrossRef] [PubMed]
3. Tokoyoda, K.; Egawa, T.; Sugiyama, T.; Choi, B.; Nagasawa, T. Cellular niches controlling B lymphocyte behavior within bone marrow during development. *Immunity* **2004**, *20*, 707–718. [CrossRef] [PubMed]
4. Morrison, S.J.; Scadden, D.T. The bone marrow niche for haematopoietic stem cells. *Nature* **2014**, *505*, 327–334. [CrossRef] [PubMed]
5. Breton, C.; Tellier, J.; Narang, P.; Chasson, L.; Jorquera, A.; Coles, M.; Schiff, C. Galectin-1—Expressing stromal cells constitute a specific niche for pre-BII cell development in mouse bone marrow. *Blood* **2011**, *117*, 6552–6561. [CrossRef]
6. Spencer, J.A.; Ferraro, F.; Roussakis, E.; Klein, A.; Wu, J.; Runnels, J.M.; Zaher, W.; Mortensen, L.J.; Alt, C.; Turcotte, R.; et al. Direct measurement of local oxygen concentration in the bone marrow of live animals. *Nature* **2014**, *508*, 269–273. [CrossRef] [PubMed]
7. D'Apuzzo, M.; Rolink, A.; Loetscher, M.; Hoxie, J.A.; Clark-Lewis, I.; Melchers, F.; Baggiolini, M.; Moser, B. The chemokine SDF-1, stromal cell-derived factor 1, attracts early stage B cell precursors via the chemokine receptor CXCR4. *Eur. J. Immunol.* **1997**, *27*, 1788–1793. [CrossRef] [PubMed]
8. Park, S.-Y.; Wolfram, P.; Canty, K.; Harley, B.; Nombela-Arrieta, C.; Pivarnik, G.; Manis, J.; Beggs, H.E.; Silberstein, L.E. Focal Adhesion Kinase Regulates the Localization and Retention of Pro-B Cells in Bone Marrow Microenvironments. *J. Immunol.* **2013**, *190*, 1094–1102. [CrossRef] [PubMed]
9. Egawa, T.; Kawabata, K.; Kawamoto, H.; Amada, K.; Okamoto, R.; Fujii, N.; Kishimoto, T.; Katsura, Y.; Nagasawa, T. The earliest stages of B cell development require a chemokine stromal cell-derived factor/pre-B cell growth-stimulating factor. *Immunity* **2001**, *15*, 323–334. [CrossRef]
10. Hagman, J.; Lukin, K. Transcription factors drive B cell development. *Curr. Opin. Immunol.* **2006**, *18*, 127–134. [CrossRef] [PubMed]
11. Hardy, R.R.; Carmack, C.E.; Shinton, S.A.; Kemp, J.D.; Hayakawa, K. Resolution and characterization of pro-B and pre-pro-B cell stages in normal mouse bone marrow. *J. Exp. Med.* **1991**, *173*, 1213–1225. [CrossRef] [PubMed]
12. Vettermann, C.; Herrmann, K.; Jäck, H.M. Powered by pairing: The surrogate light chain amplifies immunoglobulin heavy chain signaling and pre-selects the antibody repertoire. *Semin. Immunol.* **2006**, *18*, 44–55. [CrossRef] [PubMed]
13. Clark, M.R.; Mandal, M.; Ochiai, K.; Singh, H. Orchestrating B cell lymphopoiesis through interplay of IL-7 receptor and pre-B cell receptor signalling. *Nat. Rev. Immunol.* **2014**, *14*, 69–80. [CrossRef] [PubMed]
14. Hamel, K.M.; Mandal, M.; Karki, S.; Clark, M.R. Balancing proliferation with Igκ recombination during B-lymphopoiesis. *Front. Immunol.* **2014**, *5*, 139. [CrossRef] [PubMed]
15. Hedrick, S.M. The cunning little vixen: Foxo and the cycle of life and death. *Nat. Immunol.* **2009**, *10*, 1057–1063. [CrossRef] [PubMed]
16. Prieyl, J.A.; LeBien, T.W. Interleukin 7 independent development of human B cells. *Proc. Natl.Acad. Sci. USA* **1996**, *93*, 10348–10353. [CrossRef] [PubMed]
17. Milne, C.D.; Paige, C.J. IL-7: A key regulator of B lymphopoiesis. *Semin. Immunol.* **2006**, *18*, 20–30. [CrossRef] [PubMed]
18. Milne, C.D.; Fleming, H.E.; Paige, C.J. IL-7 does not prevent pro-B/pre-B cell maturation to the immature/sIgM(+) stage. *Eur. J. Immunol.* **2004**, *34*, 2647–2655. [CrossRef] [PubMed]
19. Mombaerts, P.; Johnson, R.S.; Herrup, K.; Tonegawa, S.; Papaioannouo, V.E. RAG-1-Deficient Mice Have No Mature B and T Lymphocytes. *Cell* **1992**, *68*, 869–877. [CrossRef]

20. Young, F.; Ardman, B.; Shinkai, Y.; Lansford, R.; Blackwell, T.K.; Mendelsohn, M.; Rolink, A.; Melchers, F.; Alt, F.W. Influence of immunoglobulin heavy- and light-chain expression on B-cell differentiation. *Genes Dev.* **1994**, *8*, 1043–1057. [CrossRef] [PubMed]

21. Stein, M.; Dütting, S.; Mougiakakos, D.; Bösl, M.; Fritsch, K.; Reimer, D.; Urbanczyk, S.; Steinmetz, T.; Schuh, W.; Bozec, A.; et al. A defined metabolic state in pre B cells governs B-cell development and is counterbalanced by Swiprosin-2/EFhd1. *Cell Death Differ.* **2017**, *24*, 1239–1252. [CrossRef] [PubMed]

22. Zeng, H.; Yu, M.; Tan, H.; Li, Y.; Su, W.; Shi, H.; Dhungana, Y.; Guy, C.; Neale, G.; Cloer, C.; et al. Discrete roles and bifurcation of PTEN signaling and mTORC1-mediated anabolic metabolism underlie IL-7-driven B lymphopoiesis. *Sci. Adv.* **2018**, *4*. [CrossRef] [PubMed]

23. Guo, B.; Kato, R.M.; Garcia-Lloret, M.; Wahl, M.I.; Rawlings, D.J. Engagement of the Human Pre-B Cell Receptor Generates a Lipid Raft-Dependent Calcium Signaling Complex. *Immunity* **2000**, *13*, 243–253. [CrossRef]

24. Su, Y.W.; Jumaa, H. LAT links the pre-BCR to calcium signaling. *Immunity* **2003**, *19*, 295–305. [CrossRef]

25. Taguchi, T.; Kiyokawa, N.; Takenouch, H.; Matsui, J.; Tang, W.R.; Nakajima, H.; Suzuki, K.; Shiozawa, Y.; Saito, M.; Katagiri, Y.U.; et al. Deficiency of BLNK hampers PLC-γ2 phosphorylation and Ca2+ influx induced by the pre-B-cell receptor in human pre-B cells. *Immunology* **2004**, *112*, 575–582. [CrossRef] [PubMed]

26. Feldhahn, N.; Klein, F.; Mooster, J.L.; Hadweh, P.; Sprangers, M.; Wartenberg, M.; Bekhite, M.M.; Hofmann, W.-K.; Herzog, S.; Jumaa, H.; et al. Mimicry of a constitutively active pre–B cell receptor in acute lymphoblastic leukemia cells. *J. Exp. Med.* **2005**, *201*, 1837–1852. [CrossRef] [PubMed]

27. Hess, J.; Werner, A.; Wirth, T.; Melchers, F.; Jack, H.M.; Winkler, T.H. Induction of pre-B cell proliferation after de novo synthesis of the pre-B cell receptor. *Proc. Natl. Acad. Sci. USA* **2001**, *98*, 1745–1750. [CrossRef] [PubMed]

28. Melchers, F. The pre-B-cell receptor: Selector of fitting immunoglobulin heavy chains for the B-cell repertoire. *Nat. Rev. Immunol.* **2005**, *5*, 578–584. [CrossRef] [PubMed]

29. Müschen, M. Autoimmunity checkpoints as therapeutic targets in B cell malignancies. *Nat. Rev. Cancer* **2018**, *18*, 103–116. [CrossRef] [PubMed]

30. Zouali, M. Transcriptional and metabolic pre-B cell receptor-mediated checkpoints: Implications for autoimmune diseases. *Mol. Immunol.* **2014**, *62*, 315–320. [CrossRef] [PubMed]

31. Vettermann, C.; Jäck, H.M. The pre-B cell receptor: Turning autoreactivity into self-defense. *Trends Immunol.* **2010**, *31*, 176–183. [CrossRef] [PubMed]

32. Herzog, S.; Reth, M.; Jumaa, H. Regulation of B-cell proliferation and differentiation by pre-B-cell receptor signalling. *Nat. Rev. Immunol.* **2009**, *9*, 195–205. [CrossRef] [PubMed]

33. Carsetti, R. Characterization of B-Cell Maturation in the Peripheral Immune System. In *B Cell Protocols*; Humana Press: Totowa, NJ, USA, 2004; pp. 25–36, E-ISBN: 1-59259-796-3.

34. Reth, M.; Nielsen, P. Signaling Circuits in Early B-Cell Development. *Adv. Immunol.* **2014**, *122*, 129–175. [CrossRef] [PubMed]

35. Donnelly, R.P.; Finlay, D.K. Glucose, glycolysis and lymphocyte responses. *Mol. Immunol.* **2015**, *68*, 513–519. [CrossRef] [PubMed]

36. Santo-Domingo, J.; Demaurex, N. Perspectives on: SGP symposium on mitochondrial physiology and medicine: The renaissance of mitochondrial pH. *J. Gen. Physiol.* **2012**, *139*, 415–423. [CrossRef] [PubMed]

37. Hardie, D.G. AMP-activated protein kinase: A key regulator of energy balance with many roles in human disease. *J. Intern. Med.* **2014**, *276*, 543–559. [CrossRef] [PubMed]

38. Andris, F.; Leo, O. AMPK in Lymphocyte Metabolism and Function. *Int. Rev. Immunol.* **2015**, *34*, 67–81. [CrossRef] [PubMed]

39. Iwata, T.N.; Ramírez-Komo, J.A.; Park, H.; Iritani, B.M. Control of B lymphocyte development and functions by the mTOR signaling pathways. *Cytokine Growth Factor Rev.* **2017**, *35*, 47–62. [CrossRef] [PubMed]

40. Kojima, H.; Kobayashi, A.; Sakurai, D.; Kanno, Y.; Hase, H.; Takahashi, R.; Totsuka, Y.; Semenza, G.L.; Sitkovsky, M.V.; Kobata, T. Differentiation stage-specific requirement in hypoxia-inducible factor-1alpha-regulated glycolytic pathway during murine B cell development in bone marrow. *J. Immunol.* **2010**, *184*, 154–163. [CrossRef] [PubMed]

41. Park, H.; Staehling, K.; Tsang, M.; Appleby, M.W.; Brunkow, M.E.; Margineantu, D.; Hockenbery, D.M.; Habib, T.; Liggitt, H.D.; Iritani, B.M. Disruption of Fnip1 reveals a metabolic checkpoint controlling B lymphocyte development. *Immunity* **2012**, *36*, 769–781. [CrossRef] [PubMed]

42. Xi, H.; Kurtoglu, M.; Lampidis, T.J. The wonders of 2-deoxy-D-glucose. *IUBMB Life* **2014**, *66*, 110–121. [CrossRef] [PubMed]

43. Yao, C.-H.; Liu, G.-Y.; Wang, R.; Moon, S.H.; Gross, R.W.; Patti, G.J. Identifying off-target effects of etomoxir reveals that carnitine palmitoyltransferase I is essential for cancer cell proliferation independent of β-oxidation. *PLOS Biol.* **2018**, *16*, e2003782. [CrossRef] [PubMed]

44. Shojaee, S.; Chan, L.N.; Buchner, M.; Cazzaniga, V.; Cosgun, K.N.; Geng, H.; Qiu, Y.H.; von Minden, M.D.; Erns, T.; Hochhaus, A.; et al. PTEN opposes negative selection and enables oncogenic transformation of pre-B cells. *Nat. Med.* **2016**, *4*, 379–387. [CrossRef] [PubMed]

45. Wofford, J.A.; Wieman, H.L.; Jacobs, S.R.; Zhao, Y.; Rathmell, J.C.; Jeffrey, C. IL-7 promotes Glut1 trafficking and glucose uptake via STAT5-mediated activation of Akt to support T cell survival IL-7 promotes Glut1 trafficking and glucose uptake via STAT5-mediated activation of Akt to support T cell survival. *Blood* **2007**, *111*, 2101–2112. [CrossRef] [PubMed]

46. Baracho, G.V.; Miletic, A.V.; Omori, S.A.; Cato, M.H.; Rickert, R.C. Emergence of the PI3-kinase pathway as a central modulator of normal and aberrant B cell differentiation. *Curr. Opin. Immunol.* **2011**, *23*, 178–183. [CrossRef] [PubMed]

47. Ochiai, K.; Maienschein-Cline, M.; Mandal, M.; Triggs, J.R.; Bertolino, E.; Sciammas, R.; Dinner, A.R.; Clark, M.R.; Singh, H. A self-reinforcing regulatory network triggered by limiting IL-7 activates pre-BCR signaling and differentiation. *Nat. Immunol.* **2012**, *13*, 300–307. [CrossRef] [PubMed]

48. Caro-Maldonado, A.; Wang, R.; Nichols, A.G.; Kuraoka, M.; Milasta, S.; Sun, L.D.; Gavin, A.L.; Abel, E.D.; Kelsoe, G.; Green, D.R.; et al. Metabolic Reprogramming Is Required for Antibody Production That Is Suppressed in Anergic but Exaggerated in Chronically BAFF-Exposed B Cells. *J. Immunol.* **2014**, *192*, 3626–3636. [CrossRef] [PubMed]

49. Doughty, C.A.; Bleiman, B.F.; Wagner, D.J.; Dufort, F.J.; Mataraza, J.M.; Roberts, M.F.; Chiles, T.C. Antigen receptor-mediated changes in glucose metabolism in B lymphocytes: Role of phosphatidylinositol 3-kinase signaling in the glycolytic control of growth. *Blood* **2006**, *107*, 4458–4465. [CrossRef] [PubMed]

50. Woodland, R.T.; Fox, C.J.; Schmidt, M.R.; Hammerman, P.S.; Opferman, J.T.; Korsmeyer, S.J.; Hilbert, D.M.; Thompson, C.B. Multiple signaling pathways promote B lymphocyte stimulator dependent B-cell growth and survival. *Blood* **2008**, *111*, 750–760. [CrossRef] [PubMed]

51. Clarke, A.J.; Riffelmacher, T.; Braas, D.; Cornall, R.J.; Simon, A.K. B1a B cells require autophagy for metabolic homeostasis and self-renewal. *J. Exp. Med.* **2018**, *215*, 399–413. [CrossRef] [PubMed]

52. Lam, W.Y.; Becker, A.M.; Kennerly, K.M.; Wong, R.; Curtis, J.D.; Llufrio, E.M.; McCommis, K.S.; Fahrmann, J.; Pizzato, H.A.; Nunley, R.M.; et al. Mitochondrial Pyruvate Import Promotes Long-Term Survival of Antibody-Secreting Plasma Cells. *Immunity* **2016**, *45*, 60–73. [CrossRef] [PubMed]

53. Lu, R.; Medina, K.L.; Lancki, D.W.; Singh, H. IRF-4, 8 orchestrate the pre-B-to-B transition in lymphocyte development. *Genes Dev.* **2003**, 1703–1708. [CrossRef] [PubMed]

54. Jumaa, H.; Wollscheid, B.; Mitterer, M.; Wienands, J.; Reth, M.; Nielsen, P.J. Abnormal development and function of B lymphocytes in mice deficient for the signaling adaptor protein SLP-65. *Immunity* **1999**, *11*, 547–554. [CrossRef]

55. Zhang, S.; Yang, C.; Yang, Z.; Zhang, D.; Ma, X.; Mills, G.; Liu, Z. Homeostasis of redox status derived from glucose metabolic pathway could be the key to understanding the Warburg effect. *Am. J. Cancer Res.* **2015**, *5*, 1265–1280. [PubMed]

56. Sitkovsky, M.; Lukashev, D. Regulation of immune cells by local-tissue oxygen tension: HIF1 alpha and adenosine receptors. *Nat. Rev. Immunol.* **2005**, *5*, 712–721. [CrossRef] [PubMed]

57. Kojima, H.; Gu, H.; Nomura, S.; Caldwell, C.C.; Kobata, T.; Carmeliet, P.; Semenza, G.L.; Sitkovsky, M.V. Abnormal B lymphocyte development and autoimmunity in hypoxia-inducible factor 1α-deficient chimeric mice. *Proc. Natl. Acad. Sci. USA* **2002**, *99*, 2170–2174. [CrossRef] [PubMed]

58. Meng, X.; Grötsch, B.; Luo, Y.; Knaup, K.X.; Wiesener, M.S.; Chen, X.X.; Jantsch, J.; Fillatreau, S.; Schett, G.; Bozec, A. Hypoxia-inducible factor-1α is a critical transcription factor for IL-10-producing B cells in autoimmune disease. *Nat. Commun.* **2018**, *9*, 251. [CrossRef] [PubMed]

59. Kraus, H.; Kaiser, S.; Aumann, K.; Bonelt, P.; Salzer, U.; Vestweber, D.; Erlacher, M.; Kunze, M.; Burger, M.; Pieper, K.; et al. A Feeder-Free Differentiation System Identifies Autonomously Proliferating B Cell Precursors in Human Bone Marrow. *J. Immunol.* **2014**, *192*, 1044–1054. [CrossRef] [PubMed]

60. Schuh, W.; Meister, S.; Herrmann, K.; Bradl, H.; Jäck, H.M. Transcriptome analysis in primary B lymphoid precursors following induction of the pre-B cell receptor. *Mol. Immunol.* **2008**, *45*, 362–375. [CrossRef] [PubMed]

61. Chan, L.N.; Müschen, M. B-cell identity as a metabolic barrier against malignant transformation. *Exp. Hematol.* **2017**, *53*, 1–6. [CrossRef] [PubMed]

62. Yasuda, T.; Sanjo, H.; Pagès, G.; Kawano, Y.; Karasuyama, H.; Pouysségur, J.; Ogata, M.; Kurosaki, T. Erk Kinases Link Pre-B Cell Receptor Signaling to Transcriptional Events Required for Early B Cell Expansion. *Immunity* **2008**, *28*, 499–508. [CrossRef] [PubMed]

63. Rowh, M.A.W.; Bassing, C.H. Foxos around make B cells tolerable. *Nat. Immunol.* **2008**, *9*, 586–588. [CrossRef] [PubMed]

64. Klotz, L.O.; Sánchez-Ramos, C.; Prieto-Arroyo, I.; Urbánek, P.; Steinbrenner, H.; Monsalve, M. Redox regulation of FoxO transcription factors. *Redox Biol.* **2015**, *6*, 51–72. [CrossRef] [PubMed]

65. Siggs, O.M.; Stockenhuber, A.; Deobagkar-Lele, M.; Bull, K.R.; Crockford, T.L.; Kingston, B.L.; Crawford, G.; Anzilotti, C.; Steeples, V.; Ghaffari, S.; et al. Mutation of *Fnip1* is associated with B-cell deficiency, cardiomyopathy, and elevated AMPK activity. *Proc. Natl. Acad. Sci. USA* **2016**, *113*, E3706–E3715. [CrossRef] [PubMed]

66. Hobeika, E.; Thiemann, S.; Storch, B.; Jumaa, H.; Nielsen, P.J.; Pelanda, R.; Reth, M. Testing gene function early in the B cell lineage in mb1-cre mice. *Proc. Natl. Acad. Sci. USA* **2006**, *103*, 13789–13794. [CrossRef] [PubMed]

67. Iwata, T.N.; Ramírez, J.A.; Tsang, M.; Park, H.; Margineantu, D.H.; Hockenbery, D.M.; Iritani, B.M. Conditional Disruption of Raptor Reveals an Essential Role for mTORC1 in B Cell Development, Survival, and Metabolism. *J. Immunol.* **2016**, *197*, 2250–2260. [CrossRef] [PubMed]

68. Dütting, S.; Brachs, S.; Mielenz, D. Fraternal twins: Swiprosin-1/EFhd2 and Swiprosin-2/EFhd1, two homologous EF-hand containing calcium binding adaptor proteins with distinct functions. *Cell Commun. Signal.* **2011**, *9*, 2. [CrossRef] [PubMed]

69. Hajnóczky, G.; Booth, D.; Csordás, G.; Debattisti, V.; Golenár, T.; Naghdi, S.; Niknejad, N.; Paillard, M.; Seifert, E.L.; Weaver, D. Reliance of ER-mitochondrial calcium signaling on mitochondrial EF-hand Ca2+ binding proteins: Miros, MICUs, LETM1 and solute carriers. *Curr. Opin. Cell Biol.* **2014**, *29*, 133–141. [CrossRef] [PubMed]

70. Mootha, V.K.; Bunkenborg, J.; Olsen, J.V.; Hjerrild, M.; Wisniewski, J.R.; Stahl, E.; Bolouri, M.S.; Ray, H.N.; Sihag, S.; Kamal, M.; et al. Integrated analysis of protein composition, tissue diversity, and gene regulation in mouse mitochondria. *Cell* **2003**, *115*, 629–640. [CrossRef]

71. Tominaga, M.; Kurihara, H.; Honda, S.; Amakawa, G.; Sakai, T.; Tomooka, Y. Molecular characterization of mitocalcin, a novel mitochondrial Ca2+-binding protein with EF-hand and coiled-coil domains. *J. Neurochem.* **2006**, *96*, 292–304. [CrossRef] [PubMed]

72. Bossen, C.; Murre, C.S.; Chang, A.N.; Mansson, R.; Rodewald, H.-R.; Murre, C. The chromatin remodeler Brg1 activates enhancer repertoires to establish B cell identity and modulate cell growth. *Nat. Immunol.* **2015**, *16*, 775–784. [CrossRef] [PubMed]

73. Chow, K.T.; Timblin, G.A.; McWhirter, S.M.; Schlissel, M.S. MK5 activates Rag transcription via Foxo1 in developing B cells. *J. Exp. Med.* **2013**, *210*, 1621–1634. [CrossRef] [PubMed]

74. Henzi, T.; Schwaller, B. Antagonistic Regulation of Parvalbumin Expression and Mitochondrial Calcium Handling Capacity in Renal Epithelial Cells. *PLoS ONE* **2015**, *10*, e0142005. [CrossRef] [PubMed]

75. Takane, K.; Midorikawa, Y.; Yagi, K.; Sakai, A.; Aburatani, H.; Takayama, T.; Kaneda, A. Aberrant promoter methylation of PPP1R3C and EFHD1 in plasma of colorectal cancer patients. *Cancer Med.* **2014**, *3*, 1235–1245. [CrossRef] [PubMed]

76. Fleming, H.E.; Paige, C.J. Cooperation between IL-7 and the pre-B cell receptor: A key to B cell selection. *Semin. Immunol.* **2002**, *14*, 423–430. [CrossRef] [PubMed]

77. Flemming, A.; Brummer, T.; Reth, M.; Jumaa, H. The adaptor protein SLP-65 acts as a tumor suppressor that limits pre-B cell expansion. *Nat. Immunol.* **2003**, *4*, 38–43. [CrossRef] [PubMed]

78. Amin, R.H.; Schlissel, M.S. Foxo1 directly regulates the trancription of recombination activating genes during B cell development. *Nat. Immunol.* **2008**, *9*, 613–622. [CrossRef] [PubMed]

79. Hurt, E.M.; Thomas, S.B.; Peng, B.; Farrar, W.L. Molecular consequences of SOD2 expression in epigenetically silenced pancreatic carcinoma cell lines. *Br. J. Cancer* **2007**, *97*, 1116–1123. [CrossRef] [PubMed]

80. Bagur, R.; Hajnóczky, G. Intracellular Ca2+ Sensing: Its Role in Calcium Homeostasis and Signaling. *Mol. Cell* **2017**, *66*, 780–788. [CrossRef] [PubMed]

81. Rosselin, M.; Santo-Domingo, J.; Bermont, F.; Giacomello, M.; Demaurex, N. L-OPA1 regulates mitoflash biogenesis independently from membrane fusion. *EMBO Rep.* **2017**, *18*, 451–463. [CrossRef] [PubMed]

82. Hou, T.; Wang, X.; Ma, Q.; Cheng, H. Mitochondrial flashes: New insights into mitochondrial ROS signaling and beyond. *J. Physiol.* **2014**, *592*, 3703–3713. [CrossRef] [PubMed]

83. Schwarzländer, M.; Wagner, S.; Ermakova, Y.G.; Belousov, V.V.; Radi, R.; Beckman, J.S.; Buettner, G.R.; Demaurex, N.; Duchen, M.R.; Forman, H.J.; et al. The 'mitoflash' probe cpYFP does not respond to superoxide. *Nature* **2014**, *514*, E12–E14. [CrossRef] [PubMed]

84. Schwarzländer, M.; Murphy, M.P.; Duchen, M.R.; Logan, D.C.; Fricker, M.D.; Halestrap, A.P.; Müller, F.L.; Rizzuto, R.; Dick, T.P.; Meyer, A.J.; et al. Mitochondrial "flashes": A radical concept repHined. *Trends Cell Biol.* **2012**, *22*, 503–508. [CrossRef] [PubMed]

85. Hou, T.; Jian, C.; Xu, J.; Huang, A.Y.; Xi, J.; Hu, K.; Wei, L.; Cheng, H.; Wang, X. Identification of EFHD1 as a novel Ca2+ sensor for mitoflash activation. *Cell Calcium* **2016**, *59*, 262–270. [CrossRef] [PubMed]

86. Hagen, S.; Brachs, S.; Kroczek, C.; Fürnrohr, B.G.; Lang, C.; Mielenz, D. The B cell receptor-induced calcium flux involves a calcium mediated positive feedback loop. *Cell Calcium* **2012**, *51*, 411–417. [CrossRef] [PubMed]

87. Poburko, D.; Santo-Domingo, J.; Demaurex, N. Dynamic regulation of the mitochondrial proton gradient during cytosolic calcium elevations. *J. Biol. Chem.* **2011**, *286*, 11672–11684. [CrossRef] [PubMed]

MDPI

St. Alban-Anlage 66

4052 Basel

Switzerland

Tel. +41 61 683 77 34

Fax +41 61 302 89 18

www.mdpi.com

International Journal of Molecular Sciences Editorial Office

E-mail: ijms@mdpi.com

www.mdpi.com/journal/ijms

www.ingramcontent.com/pod-product-compliance
Lightning Source LLC
Chambersburg PA
CBHW041217220326
41597CB00033BA/5997